Stress-Induced Mutagenesis

David Mittelman
Editor

Stress-Induced Mutagenesis

Springer

Editor
David Mittelman
Virginia Bioinformatics Institute
Blacksburg, VA, USA

Department of Biological Sciences
Virginia Tech
Blacksburg, VA, USA

Department of Basic Science
Virginia Tech Carilion School of Medicine
and Research Institute
Blacksburg, VA, USA

ISBN 978-1-4614-6279-8 ISBN 978-1-4614-6280-4 (eBook)
DOI 10.1007/978-1-4614-6280-4
Springer New York Heidelberg Dordrecht London

Library of Congress Control Number: 2012956438

Printed on acid-free paper

Springer is part of Springer Science+Business Media (www.springer.com)

For Sarah, Joseph, Joshua, and Sophia

Preface

In 1942, Julian Huxley referred to the cross-disciplinary, unified evolutionary theory as the *modern evolutionary synthesis*. In the early part of the twentieth century, Fisher, Haldane, and Wright—and later Mayr, Dobzhansky, and others—produced a revised model of Darwinian evolution that rationalizes Mendelian genetics in the context of natural selection. The presence of pre-existing heritable variation is key for selection to be effective and is a contingency of adaptability. At the same time, mutation is proposed to accumulate at a constant rate, regardless of selective pressures and environmental cues.

The discoveries of epigenetic inheritance and stress-induced mutation have challenged the claim of independence between mutation and selection processes. The ability of an organism to undergo genetic or epigenetic change in response to environmental stresses suggests an ability to alter the *rate* of mutation, which could take effect globally or at specific parts of the genome, and temporarily or permanently. Mutation rates can be altered globally and permanently by the presence of mutator alleles, temporarily due to transient events such as environmental stress, or locally at "hotspot" locations in the genome. Such mechanisms are clearly valuable from an evolutionary perspective, as it is advantageous for mutation to be restricted in both time and space, if most selectable mutations produce deleterious outcomes.

This volume compiles key evidence for stress-induced genetic and epigenetic mutation, integrating cross-disciplinary observations from a number of species and biological systems, including human. The observations have vast implications for evolutionary biology but also for human medicine. For example, genomic instability is now recognized as a hallmark of most cancers. Tumor resistance and recurrence are modeled within the context of clonal expansions. The comprehensive understanding of stress-induced mutagenesis and the processes underlying evolvability, studied across many biological systems, will enable gains in the treatment and management of cancer, as well as other human disorders that result from damaged or unstable genomes.

Stress-induced mutagenesis has been most widely studied in bacterial systems. In the first chapter, Ivan Matic describes the role (and regulation) of mutator alleles and stress-induced mutagenesis pathways in the evolution of bacterial populations. The chapter by Susan Rosenberg and colleagues dissects the pathways to stress-induced mutagenesis, focusing specifically on the localization of mutagenic repair to double-strand breaks. This chapter provides some exciting new evidence that argues mutagenesis is not an inevitable consequence of DNA repair. This has been a long-standing point of debate. In his classic critic, "Adaptation and Natural Selection," George Williams argued that mutation rate is a "mechanical inevitability," the byproduct of physical limitations in the fidelity of DNA repair processes and not the product of natural selection. Rosenberg and colleagues now show that stress-induced mutagenic repair is activated by repair components that are not required for the proper resolution of a DNA break. They also discuss the localization of mutation, during stress, to DNA breaks, which minimizes the impact of deleterious mutations to the genome. In the third chapter, Eduardo Robleto and Ronald Yasbin describe transcription-coupled mutagenesis pathways that illustrate another way in which mutagenesis can be triggered temporally by environmental cues, and then localized to focused portions of the genome. In the fourth chapter, Milton Saier and colleagues present intriguing evidence that stress can activate transposons, which play roles in gene regulation and disease.

From the seminal work of Lindquist and Rutherford, the stress-activated Hsp90 chaperone is now known to participate in the canalization of traits, something first described by Conrad Waddington more than a half-century ago. Hsp90 normally functions to buffer client proteins against the effects of genetic variation. Severe environmental stress can overwhelm the chaperone's buffering capacity, causing previously cryptic genetic variation to be expressed. In Chap. 5, Douglas Ruden and colleagues share very exciting evidence that, in flies, Hsp90 can induce novel epigenetic changes in addition to exposing existing variation. Shunsuke Ishii and colleagues continue the discussion of epigenetics in a very compelling chapter that maps a molecular pathway by which the ATF-2 family of transcription factors facilitates the inheritance of stress-induced epigenetic changes.

As mentioned above, focusing mutagenesis in time and space minimizes the impact of deleterious mutations across the genome. Tandem repeats are an important source of functional variation that also fit these constraints. Tandem repeat mutation rate is modulated by global and local factors and triggered by temporal events such as stress. Many tandem repeats affect morphological, behavioral, and life-history traits through subtle and quantitative effects on gene function. Most interestingly, the incremental functional impact of repeat mutation even further decreases the frequency of catastrophically deleterious effects. In Chap. 7, John Wilson and colleagues outline key lines of evidence in human cells that reveal how repeat mutation is modulated by local and global factors, as well as stress, transcription, and DNA methylation.

In Chap. 8, Peter Glazer and colleagues outline the mechanistic details by which the genomes of human cells become unstable as a result of exposure to hypoxia, or oxygen deprivation. This is a particularly important stressor in the context of cancer, as developing tumors experience hypoxic stress prior to the recruitment of dedicated blood supplies through angiogenesis. The finding that hypoxia and other stressors can destabilize cancer genomes is very significant, as it is likely that tumors draw upon this variation to adapt to their microenvironments and to resist drug treatment. In the following chapter, Jac Nickoloff and colleagues describe a mechanism for the fascinating and equally frightening observation that in human cells, delayed transgenerational genomic instability can be induced by low-dose radiation treatment. The possibility that radiotherapy can be a trigger for future cancer development raises important questions about the safety and appropriateness of such a therapy. In the related and following chapter, Carmel Mothersill and colleagues explore stress-induced bystander effects and highlight the relevance of this phenomenon for cancer and adaptive evolution. Denise Montell and colleagues present in Chap. 11 some rather surprising studies that document the reversal of the apoptotic process. They present evidence that dying cells with damaged genomes can revert to living and proliferating cells, in a process they cleverly term *anastasis*. One implication of this phenomenon is that it could be a possible mechanism for tumor cells to survive and even resist treatment. In Chap. 12, Yuri Dubrova documents provocative evidence, in rodents, for transgenerational instability induced by radiation or chemical toxins. The findings are particularly compelling as the genomic instability increases in subsequent generations. The emerging evidence for this phenomenon in humans is frightening, particularly in the context of the recent Fukushima Daiichi nuclear disaster in Japan.

The final chapter by Subhajyoti De summarizes the emerging revolution in high-throughput sequencing. The affordability and availability of high-throughput sequencing has created an unprecedented surge in the use of genomic data in basic, translational, and clinical research. The ability to rapidly sequence and analyze entire genomes or populations of genomes is transforming the study of mutagenesis and genome evolution. For the first time, rather than utilizing engineered assays or genomic markers, the complete and unbiased spectrum of stress-induced changes can be directly measured genome-wide. The implications for cancer are vast as well, since tumor sequencing now enables the identification of trigger mutations and as well as passenger mutations that could serve as targets for tumor susceptibilities.

I am so grateful to the chapter authors for their enthusiasm and for helping me assemble this volume. As a graduate student, "DNA Repair and Mutagenesis" was my bible. The textbook, helmed by Errol Friedberg, is one of the most complete resources for understanding the intricacies of DNA repair and genome stability. In preparing this volume I strived to produce a worthy and complementary resource documenting the evidence for stress-induced genetic and epigenetic mutation across all biological systems; and the implications of these processes to evolutionary theory and cancer genetics. The authors of these chapters are leaders in their respective disciplines, and I am incredibly thrilled and grateful that so many of them were able to collaboratively assemble what I think is one of the most comprehensive cross-disciplinary resources for this exciting and relevant field.

Specific thanks must go to Susan Rosenberg for encouraging me to tackle the daunting task of assembling this volume and to John Wilson for advice and support. Both are from Baylor College of Medicine and I have to acknowledge the Department of Molecular and Human Genetics at BCM as an amazing environment in which to train and incubate ideas about science. I would like to thank my lab at my current institution, Virginia Tech, for their feedback and suggestions—and I would like to single out R. Matthew Ward, who carefully read through all the chapters with me, provided lots of valuable input and coauthored the last chapter. I also would like to thank Maureen Lawrence-Kuether, my administrative specialist, and Melanie Tucker and Meredith Clinton, both from Springer, for technical support and assistance during the writing process. Lastly, I would like to thank my wife, Kristen, for her patience, love, and support.

Blacksburg, VA, USA David Mittelman

Contents

Contributors

Christopher P. Allen Department of Environmental and Radiological Health Sciences, Colorado State University, Fort Collins, CO, USA

Research Center for Charged Particle Therapy, and International Open Laboratory, National Institute of Radiological Sciences, Chiba, Japan

Nimrat Chatterjee Verna and Marrs McLean Department of Biochemistry and Molecular Biology, Baylor College of Medicine, Houston, TX, USA

Subhajyoti De Department of Medicine, University of Colorado School of Medicine and University of Colorado Cancer Center, Aurora, CO, USA

Yuri E. Dubrova Department of Genetics, University of Leicester, Leicester, UK

Akira Fujimori Research Center for Charged Particle Therapy, and International Open Laboratory, National Institute of Radiological Sciences, Chiba, Japan

Mark D. Garfinkel Department of Biological Sciences, University of Alabama in Huntsville, Huntsville, AL, USA

Peter M. Glazer Department of Therapeutic Radiology, Yale University School of Medicine, New Haven, CT, USA

P.J. Hastings Department of Molecular and Human Genetics, Baylor College of Medicine, Houston, TX, USA

Shunsuke Ishii Laboratory of Molecular Genetics, RIKEN Tsukuba Institute, Tsukuba, Ibaraki, Japan

Xiangyi Lu Institute of Environmental Health Sciences, Wayne State University, Detroit, MI, USA

Toshio Maekawa Laboratory of Molecular Genetics, RIKEN Tsukuba Institute, Tsukuba, Ibaraki, Japan

Holly A. Martin School of Life Sciences, University of Nevada, Las Vegas, NV, USA

Ivan Matic INSERM U1001, Université Paris Descartes, Sorbonne Paris Cité Faculté de Médecine Paris Descartes, France

Denise J. Montell Department of Biological Chemistry, Center for Cell Dynamics, Johns Hopkins University School of Medicine, Baltimore, MD, USA

Carmel Mothersill Medical Physics and Applied Radiation Sciences, McMaster University, Hamilton, ON, Canada

Jac A. Nickoloff Department of Environmental and Radiological Health Sciences, Colorado State University, Fort Collins, CO, USA

Ryuichi Okayasu Research Center for Charged Particle Therapy, and International Open Laboratory, National Institute of Radiological Sciences, Chiba, Japan

Mario Pedraza-Reyes Division of Natural and Exact Sciences, Department of Biology, University of Guanajuato, Guanajuato, Mexico

Eduardo Robleto School of Life Sciences, University of Nevada, Las Vegas, NV, USA

Douglas M. Ruden Institute of Environmental Health Sciences, Wayne State University, Detroit, MI, USA

Department of Obstetrics and Gynecology, C. S. Mott Center for Human Growth and Development, Wayne State University, Detroit, MI, USA

Milton H. Saier Jr. Division of Biological Sciences, Department of Molecular Biology, University of California, San Diego, La Jolla, CA, USA

Beatriz A. Santillan Department of Molecular and Human Genetics, Baylor College of Medicine, Houston, TX, USA

Susan E. Scanlon Department of Therapeutic Radiology, Yale University School of Medicine, New Haven, CT, USA

Ki-Hyeon Seong Laboratory of Molecular Genetics, RIKEN Tsukuba Institute, Tsukuba, Ibaraki, Japan

Colin Seymour McMaster University, Hamilton, ON, Canada

Chandan Shee Departments of Molecular and Human Genetics, Biochemistry and Molecular Biology, Molecular Virology and Microbiology, Dan L Duncan Cancer Center, Baylor College of Medicine, Houston, TX, USA

Maksim A. Shlykov Division of Biological Sciences, Department of Molecular Biology, University of California, San Diego, La Jolla, CA, USA

Vincent E. Sollars Department of Biochemistry and Microbiology, Joan C. Edwards School of Medicine, Marshall University, One John Marshall Drive, Huntington, WV, USA

Susan M. Rosenberg Departments of Molecular and Human Genetics, Biochemistry and Molecular Biology, Molecular Virology and Microbiology, Dan L Duncan Cancer Center, Baylor College of Medicine, Houston, TX, USA

Ho Lam Tang Department of Biological Chemistry, Center for Cell Dynamics, Johns Hopkins University School of Medicine, Baltimore, MD, USA

Ho Man Tang Department of Biological Chemistry, Center for Cell Dynamics, Johns Hopkins University School of Medicine, Baltimore, MD, USA

Carmen Vallin School of Life Sciences, University of Nevada, Las Vegas, NV, USA

Jing Wang Division of Biological Sciences, Department of Molecular Biology, University of California, San Diego, La Jolla, CA, USA

Luan Wang Institute of Environmental Health Sciences, Wayne State University, Detroit, MI, USA

Department of Obstetrics and Gynecology, C. S. Mott Center for Human Growth and Development, Wayne State University, Detroit, MI, USA

R. Matthew Ward Virginia Bioinformatics Institute, Virginia Tech, Blacksburg, VA, USA

John H. Wilson Verna and Marrs McLean Department of Biochemistry and Molecular Biology, Department of Molecular and Human Genetics, Baylor College of Medicine, Houston, TX, USA

Ronald Yasbin College of Arts and Sciences, University of Missouri, St. Louis, MO, USA

Zhongge Zhang Division of Biological Sciences, Department of Molecular Biology, University of California, San Diego, La Jolla, CA, USA

Chapter 1
Stress-Induced Mutagenesis in Bacteria

Ivan Matic

Abstract Under stress, high mutation rates can be advantageous because they increase the probability of generation of the adaptive mutations. Mutation rates can be modulated by changing the proportion of constitutive mutator versus non-mutator bacteria at the population level, or by inducing stress responses, which increase mutation rates transiently in individual cells. Constitutive mutator alleles are selected because they hitchhike with the adaptive mutations they generate. There are two nonexclusive hypotheses concerning the nature of selective pressure acting on the molecular mechanisms controlling stress-induced mutagenesis: stress induced mutagenesis could be an unavoidable by-product of mechanisms involved in survival under stress, or stress-induced mutator phenotypes could be selected for in the same way as constitutive mutator alleles; that is, via hitchhiking with the adaptive mutations they generate. However, regardless of the nature of selective pressure acting on stress-induced mutagenesis, it is very likely that the resulting increased genetic variability plays an important role in the bacterial evolution.

Stress is a disturbance of the normal functioning of a biological system, provoked by environmental factors whose amplitude and persistence are such that they cause a reduction in growth rate and increased mortality (Bijlsma and Loeschcke 1997). Some organisms react to stress by inducing behavioral or physiological responses, while others increase production of genetically diverse offspring. For a long time, increased production of genetic diversity was not considered adaptive, because, given the huge population size of most bacterial species, it was thought the adaptation was rarely limited by the available genetic diversity. However, strong bottlenecks or stressful environmental conditions can severely reduce bacterial population size and, consequently, genetic diversity. Under such circumstances, it could be advantageous to have

I. Matic (✉)
INSERM U1001, Université Paris Descartes, Sorbonne Paris Cité, Faculté de Médecine Paris Descartes, 156 Rue de Vaugirard, 75730 Paris Cedex 15, France
e-mail: ivan.matic@inserm.fr

D. Mittelman (ed.), *Stress-Induced Mutagenesis*, DOI 10.1007/978-1-4614-6280-4_1,

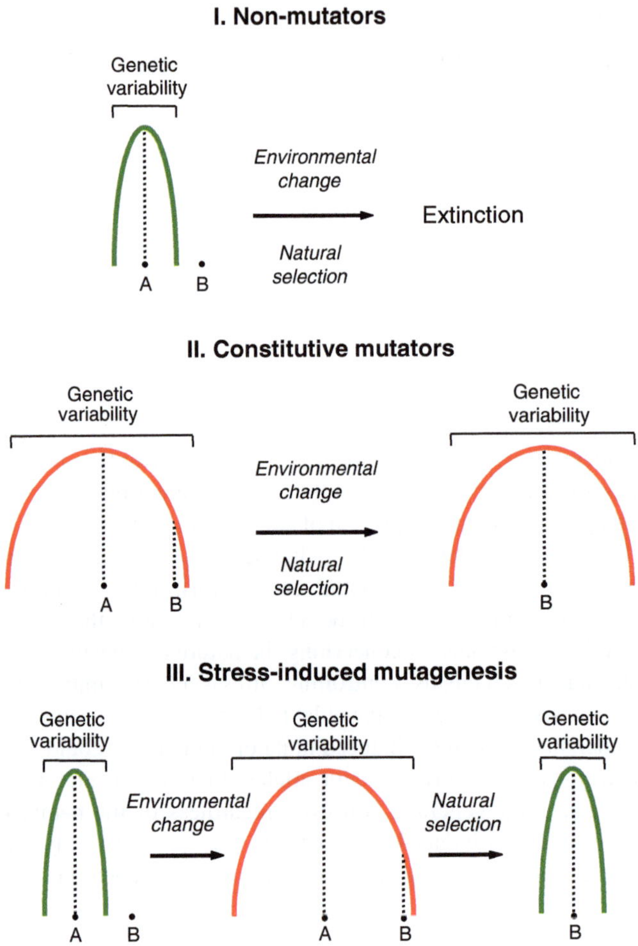

Fig. 1.1 Interplay between genetic variability and population evolvability. Allele A has maximal fitness under starting non-stressful environmental condition and consequently dominates population. Allele B is required for the adaptation to new environmental condition. **(I)** Because genetic variability is low, allele B is not present in the starting nonmutator population. When such population face environmental change, it goes to extinction. **(II)** Mutator population generates higher genetic variability. Consequently, allele B is present in the starting population. Upon environmental change, the individuals carrying allele B are favored by the natural selection. While this strategy increases probability for the rapid adaptation to the environmental change, on the longer run, the cost associated with high mutation rates is expected to cause the loss of fitness of the mutator cells. **(III)** The ability to regulate mutation rates in response to the environmental cues might be advantageous from the evolutionary point of view because the limitation of the increase in the mutation rates to stress phases should reduce the overall cost of a high rate of mutation and, in the same time, increase probability of generation of the adaptive alleles

high mutation rates (Fig. 1.1). Experimental (LeClerc et al. 1998; Mao et al. 1997) and theoretical (Boe et al. 2000) studies indicate that the observed frequency of strains with high constitutive mutation rates (mutators) in bacterial natural populations is much higher than would be expected from mutation/selection equilibrium alone.

This suggests that there are situations in nature where mutator phenotypes confer a selective advantage. Constitutive mutator phenotypes generally result from mutations in genes coding for DNA repair enzymes or proteins that assure the accuracy of DNA replication. These mutant genes are called mutator alleles. Mutagenesis can also increase during stress as a result of direct alteration of the DNA molecule, and/or from a genetic program that is induced under stress. Two stress responses are known to increase mutagenesis when induced: the SOS system induced by genotoxic stresses (Friedberg et al. 2006); and the RpoS-regulated general stress response triggered by many different stresses (Battesti et al. 2011). In this chapter, the role of molecular, ecological, and evolutionary factors involved in shaping mutation rates in bacterial populations, using the *Escherichia coli* species as an example, is discussed.

Selection of Constitutive Mutators

Newly arisen mutations can have very different impacts on the fitness of the organism, ranging from deleterious to neutral to beneficial. However, they appear at very different rates. For example, for *E. coli* K-12, the rate of deleterious mutations per genome per replication is at least $2–8 \times 10^{-4}$ (Boe et al. 2000; Kibota and Lynch 1996), while that of beneficial mutations is about 2×10^{-9} (Imhof and Schlotterer 2001). It is generally believed that the rate of spontaneous mutations results from a balance between the effects of deleterious mutations and the metabolic costs of reducing mutation rates (Drake et al. 1998). Indeed, the replication error rate in *E. coli* is low, i.e., about 10^{-10} per base per replication, and it seems that it cannot be reduced further (Drake 1993). Any variant that has increased mutation rates is expected to have reduced fitness due to increased production of deleterious mutations. However, when adaptation is limited by the available genetic variability, natural selection favors mutator cells. Mutator alleles are carried at high frequency through hitchhiking with the beneficial mutations they generate when the fitness gain provided by beneficial mutations counterbalances the fitness loss of the increased generation of deleterious mutations (Taddei et al. 1997a). The linkage between beneficial mutations and mutator alleles is particularly strong in bacteria because the rate of gene exchange in these asexual organisms is, in general, very low. Mutators are particularly favored when several beneficial mutations are required for the adaptation (Mao et al. 1997). For example, if the probability of generating each beneficial mutation is 10^2-fold higher in mutator populations than in non-mutator populations, then the probability that two beneficial mutations will be generated in mutator population is 10^4-higher than in non-mutator population.

The selection of mutator alleles also depends on many other parameters. For example, the increase in frequency of mutators depends on the total population size (Tenaillon et al. 1999), on mutator strength (i.e., the mutator mutation rate relative to the non-mutator mutation rate (Taddei et al. 1997a)), and on the rate of gene exchange (Tenaillon et al. 1999). It also depends on the stability of the environment. For example, mutator alleles are particularly advantageous upon a shift in environmental conditions (Tanaka et al. 2003; Travis and Travis 2002). The dynamics of mutator selection depend also on environmental spatial heterogeneity, which can allow or prevent the competition between the cells carrying different adaptive mutations. Therefore, theoretical modeling predicts that mutators will be particularly favored in temporally and spatially heterogeneous environmental conditions (Travis and Travis 2004).

Conditions favoring strong constitutive mutators must be frequent in nature because mutators have been found in populations of *E. coli* (Matic et al. 1997), *Salmonella enterica* (LeClerc et al. 1996), *Neisseria meningitidis* (Richardson et al. 2002), *Haemophilus influenzae* (Watson et al. 2004), *Staphylococcus aureus* (Prunier et al. 2003), *Helicobacter pylori* (Bjorkholm et al. 2001), *Streptococcus pneumoniae* (del Campo et al. 2005), and *Pseudomonas aeruginosa* (Oliver et al. 2000), with frequencies ranging from 0.1% to above 60%.

Mismatch-Repair Deficient Mutators

The vast majority of strong constitutive mutators found in the laboratory (*E. coli* (Sniegowski et al. 1997), *S. enterica* serovar Typhimurium (LeClerc et al. 1998), *Pseudomonas fluorescens* (Pal et al. 2007) and in nature (*E. coli* (Matic et al. 1997; Sniegowski et al. 1997), *S. enterica* (LeClerc et al. 1996), *N. meningitides* (Richardson et al. 2002), *P. aeruginosa* (Oliver et al. 2002)) have a defective mismatch-repair system due to the inactivation of *mutS* or *mutL* genes. Molecular characterization of *E. coli* and *P. aeruginosa* natural *mutS* and *mutL* mutants has revealed that these genes are inactivated by a variety of mechanisms: frameshifts, insertions, premature stop codons, and deletions (Oliver et al. 2002; Li et al. 2003). The mismatch repair system controls the fidelity of DNA replication by eliminating biosynthetic errors (Friedberg et al. 2006), and by participating in DNA lesion processing during transcription-coupled repair (Mellon and Champe 1996). In addition, the mismatch repair system is involved in the maintenance of chromosomal structural integrity and in the control of horizontal gene transfer by preventing recombination between nonidentical DNA sequences (Matic et al. 1995). The mismatch-repair system involves several proteins, of which two—MutS and MutL—have been highly conserved during evolution. The MutS protein recognizes seven of eight possible base pair mismatches. Only C-C mismatches, which represent the least frequent replication error, are not recognized. In addition, the MutS protein binds up to four unpaired bases, allowing for repair of frameshift errors. The efficiency with which different mismatches are repaired is determined by the affinities of MutS protein for various mismatches. MutL plays the role of "molecular matchmaker" between MutS-mismatch complexes and other proteins involved in the repair process. The inactivation of *mutS* or *mutL* genes results in a strong mutator phenotype, with a 10^2-fold increased rate of transition mutations (G:C->A:T and A:T->G:C), and 10^3-fold increased rate of frameshift mutations. In addition, *mutS* or *mutL* knockout mutants have a strong hyper-recombination effect, resulting in a 10^1–10^3-fold increase in the rate of chromosomal rearrangements.

Any bacterial population is expected to harbor a subpopulation of mismatch repair mutants due to spontaneous mutations in the mismatch repair genes. The frequency of mismatch repair-deficient mutators was estimated to be less than 3×10^{-5} in cultures of *E. coli* K-12 that were not subjected to selective pressure (Mao et al. 1997; Boe et al. 2000). For *S. enterica* serovar Typhimurium, the frequency of mutators in the unselected population is even lower, $1–4 \times 10^{-6}$ (LeClerc et al. 1998). It was experimentally and theoretically demonstrated that the mismatch repair mutators do not have a selective advantage because of the absence of the metabolic load imposed by the production and activity of these DNA repair enzymes. If there were a selective advantage due to decreased metabolic load, then this advantage should be independent of the initial ratio of mutator to non-mutator cells. However, this is not the case. The mutator outgrows the non-mutator strain only when the ratio of mutator/non-mutator population size is above a certain threshold. Such a threshold was observed for *mutS* (Giraud et al. 2001; Labat et al. 2005) and *mutT* (Chao and Cox 1983) mutators. This threshold is determined by the ratio of the frequency of mutants carrying beneficial alleles in mutator populations to the frequency in non-mutator population. In each population, the frequency of these mutants depends on the mutation rate and the population size (Le Chat et al. 2006). Therefore, mismatch repair deficient mutators are selected because they produce more adaptive mutations. This selective advantage occurs despite the fact that mismatch repair defective mutators start with a small selective disadvantage (about 1%) relative to non-mutators (Boe et al. 2000; Trobner and Piechocki 1984). The selective advantage of mutators over non-mutator strains is not restricted to *mutS* mutants, but is also observed for *mutL* mismatch repair deficient mutants (Tröbner and Piechocki 1981).

Low spontaneous mutation rates are maintained by the activity of many molecular mechanisms that protect and repair DNA, as well as by the mechanisms that assure high-fidelity DNA replication. Inactivation of over 20 different *E. coli* genes can confer mutator phenotypes of different strengths (for a review, see Horst et al. 1999). So, why do the vast majority of strong mutators found in nature and in laboratory systems have defective mismatch-repair systems? One explanation for this phenomenon is that inactivation of the other genes involved in important aspects of DNA or RNA metabolism might have too high a cost for advantageous mutations to compensate. For example, competition experiments in chemostats have shown that *E. coli mutT* mutators can also be selected for by the beneficial mutations they generate (Chao and Cox 1983), but are never found in *E. coli* natural populations. This may be explained by the fact that inactivation of *mutT* gene, which codes for the protein that eliminates 8-oxo-G from the nucleotide pool (Friedberg et al. 2006), increases replication, but also increases transcriptional errors (Taddei et al. 1997b) and sensitivity to oxidative stress (Guelfo et al. 2010), which might considerably reduce the fitness of the mutant cell.

Another specific advantage of mismatch repair deficient strains over other mutator alleles, which might also explain their abundance in nature, is their hyper-recombination phenotype (Friedberg et al. 2006). Recombination can also increase adaptability by increasing genetic variability. Consequently, genotypes with increased recombination rates might be selected due to favorable genotypes generated by the association of beneficial mutations from different individuals. Indeed, the frequency of mismatch repair-deficient mutants has been reported to increase rapidly in *E. coli* populations by hitchhiking with the recombination events they generate (Funchain et al. 2001).

Counter-Selection of the Constitutive Mutators

Experimental and theoretical studies showed that the frequency of mutator strains in a population could rapidly increase to almost 100%. However, the majority of natural isolates are not mutators. A major factor that diminishes the fitness of constitutive mutators is the continuous production of deleterious mutations once adaptation is achieved (Giraud et al. 2001). A second factor is that neutral, beneficial, and deleterious mutations can impact fitness differently in different environments. Consequently, no single genotype is optimally adapted to all environments. For example, an adaptive mutation in one environment can be deleterious in another, a phenomenon called "antagonistic pleiotropy" (Cooper an. Lenski 2000). Therefore, migration from one environment to another might contribute to the reduction of mutator fitness in natural populations, as observed in in vivo laboratory experiments for *E. coli mutS* (Giraud et al. 2001) and *S. enterica* serovar Typhimurium *mutS* (Nilsson et al. 2004) mutators. Continuous passage through strong bottlenecks results in the accumulation of deleterious mutations due to genetic drift, a phenomenon, called Muller's ratchet, which is particularly deleterious to strong mutator populations. For example, when wild-type and *mutS* defective cells were passaged through single-cell bottlenecks, only 3% of the wild-type lineages had phenotypically detectable mutations after 40 cycles. Contrarily, 4% of *mutS* lineages had died out, 55% had auxotrophic requirements, 70% had defects in at least one sugar or catabolic pathway, 33% had a defect in cell motility, and 26% were either temperature-sensitive or cold-sensitive lethal (Funchain et al. 2000).

Therefore, in the long run, the fitness cost associated with high mutation rates is expected to cause the elimination of the mutator genome with a consequential loss of adaptive mutations from bacterial populations. However, some adaptive mutations generated in mutator backgrounds can be saved either by horizontal transfer to a non-mutator background, or by a reduction in the mutation rate of the adapted mutator strain before the load of deleterious mutations becomes too high. This reduction of mutation rate might be achieved by the reversion of the mutator mutation, or by the acquisition of suppressor mutations as observed in the populations of *mutT* mutators (Tröbner and Piechocki 1984). The probability of acquiring these two types of mutations is higher in mutator backgrounds. The hyper-recombination phenotype of mismatch repair-deficient strains might also facilitate the reacquisition of the functional mismatch-repair genes from non-mutator bacteria via horizontal gene transfer. Such events seem to happen frequently during *E. coli* evolution (Denamur et al. 2000).

Some bacterial species, like *H. influenzae* and *N. meningitides* (Moxon et al. 2006), possess mechanisms allowing them to permanently maintain high mutation rates at some loci, while at the same time avoiding the fitness costs associated with high genome-wide mutation rates. The hypermutability of these loci results from the mutational properties of repetitive DNA sequences located within the gene, or within its controlling elements. These genes code for evasins, LPS biosynthesis enzymes, adhesins, iron acquisition proteins, and restriction-modification systems. Repetitive DNA sequences experience high rates of insertion and deletion mutations through replication slippage, which results in alternating loss of function and reversion mutations. Such mutagenesis can increase bacteria fitness by allowing evasion of the host's immune system. However, the type of variation produced by localized mutator activity might not always be sufficient for adaptation, as suggested by the presence of the strong mismatch repair deficient mutators in the natural populations of *N. meningitides* and *H. influenzae* (Richardson et al. 2002; Watson et al. 2004).

Stress-Induced Mutagenesis

Because the cost of constitutive mutator alleles comes largely from deleterious mutations generated outside the adaptation phase, limitation of increased mutation rates to phases of adaptation could be particularly advantageous by reducing the overall cost of a high mutation rate (Fig. 1.1) (Bjedov et al. 2003; Tenaillon et al. 2004). The possibility of adapting the mutation rate to environmental conditions could be interesting from an evolutionary point of view. Another advantage of limiting the increase in the mutation rate to stressful periods is the fact that environmental stresses can alleviate the average deleterious effect of mutations (Kishony and Leibler 2003). Computer simulations have shown that stress-induced mutator alleles could be selected for almost as efficiently as constitutive mutators (Bjedov et al. 2003). The strength of the selected stress-induced mutator alleles is positively correlated with the strength of selection. Mutations produced under stress could represent a large proportion of overall mutations and may have evolutionary consequences. For example, a population with a tenfold or 100-fold (relative to the non mutator mutation rates) stress-inducible mutator phenotype will adapt up to 15% or 38% faster, respectively, than a non-mutator population. An increase in the adaptation rate due to stress-inducible mutagenesis may also limit the selection of the constitutive mutator alleles. It was previously described that the fixation of an allele improving the rate of adaptation decreases the selection for other alleles improving the rate of adaptation (Tanaka et al. 2003). Simulations have shown that fixation frequency of a constitutive mutator decreases as a function of the strength of the stress-inducible mutagenesis. The relative decrease in the fixation of a constitutive mutator allele was almost perfectly correlated with the relative improvement of the adaptation rate due to stress-inducible mutagenesis. Therefore, computer simulations suggest that stress-induced mutagenesis could be the result of selection because of the beneficial mutations that such a process can generate. However, analysis of the molecular mechanisms involved in the control of the stress-induced mutagenesis suggests that mutagenesis might also increase as a byproduct of the survival strategy.

Molecular Mechanisms Associated with Stress-Induced Mutagenesis

In *E. coli*, different stresses increase the generation of mutations via different mechanisms. (1) Various chemical and physical agents can generate mutagenic miscoding DNA structures that cause DNA replication errors. For example, reactive oxygen species generate 8-oxo-guanine, while methylating agents generate O^6-methyl-guanine (Friedberg et al. 2006). (2) Some environmental agents directly affect DNA, but also inhibit anti-mutator DNA repair enzymes, thus increasing mutation rates. For example, the nitric oxide produced by macrophages damages DNA and inhibits Fpg DNA glycosylase, O^6-methyl-guanine-DNA methyltransferase, and DNA ligase (Graziewicz et al. 1996; Laval and Wink 1994; Wink and Laval 1994). (3) DNA lesions, such as the pyrimidine dimers produced by the UV irradiation, which block replicative DNA polymerase, induce the SOS system, resulting in increased mutagenesis (Friedberg et al. 2006). (4) Different stresses—such as starvation, high osmolarity, low temperature and low pH—induce the RpoS-regulon (Battesti et al. 2011). The induction of this regulon increases the capacity of cells to resist different stresses and survive, but also results in increased mutagenesis (Bjedov et al. 2003; McKenzie et al. 2001). (5) Stresses have also been shown to induce the mobility of transposons and insertion sequences, which can lead to gene activation or inactivation (Levy et al. 1993).

Cases (3), (4), and (5) imply genetic control of the mutation rate. Concerning the selective pressure acting on such mechanisms, the case of insertion sequences and transposons is peculiar. It has been shown that transposons could be selected for as mutator genes by hitchhiking with the mutations they produce (Chao et al. 1983). Nevertheless, it is hard to tell whether transposon mobility is the result of a selection acting to enhance the chance of survival of the bacterial strain carrying them, or if it is the result of an inherently selfish nature of transposons. Transposon mobility can increase the opportunity for transmission to other bacteria by increasing their copy number in the chromosome, or by jumping on the conjugative plasmids and bacteriophages. Bacteria are likely to die under stress, but transposons might be transmitted before or after the cell death.

SOS Response

All living organisms possess inducible genetic networks capable of responding to, and coping with, genotoxic stresses. The paradigm for such a network is the *E. coli* SOS system (Friedberg et al. 2006), which is induced in response to stresses that damage DNA and/or interfere with the replication catalyzed by the replicative DNA polymerase (Sassanfar and Roberts 1990). All these stresses increase the intracellular concentration of single-stranded DNA (ssDNA), the SOS inducing signal (Sassanfar and Roberts 1990). ssDNA is the substrate for the RecA protein, which binds it to form a RecA-nucleofilament (RecA*). Depending on the nature of the DNA substrate, RecA loading requires either RecFOR or RecBCD complexes. The RecFOR complex facilitates the formation of RecA* on ssDNA gaps covered with SSB (Morimatsu and Kowalczykowski 2003), while the substrate for the RecBCD complex is a blunt, or nearly blunt, double-stranded DNA end, from which it produces ssDNA, on to which it loads RecA (Kowalczykowski et al. 1994). RecA* acts as a co-protease, promoting the self-cleavage of the SOS repressor LexA, thus inducing the SOS response (Little et al. 1980). At least 40 genes belong to the SOS regulon (Courcelle et al. 2001; Fernandez De Henestrosa et al. 2000). The timing of expression of different SOS genes is controlled by the affinity of the LexA repressor for the SOS boxes of those genes. The level and length of the induction depends on the amount and persistence of the single-stranded DNA in the cell. Once DNA lesions are repaired and replication restored, SOS functions are again repressed.

Most SOS functions are implicated in dealing with the DNA lesions. These functions can be loosely grouped into two categories: the elimination of DNA lesions, aims to restore the original genetic information, and the tolerance such lesions, which allows continuation of the genome replication without eliminating the lesions. Damage tolerance is a measure of last resort to rescue cells from DNA damage because persistent lesions block the replicative polymerase, which is potentially lethal event. Without it cells would become highly sensitive to death by either external or endogenously generated DNA-damaging agents. DNA lesions can be tolerated via different pathways, of which the two best studied are homologous recombination, and replicative lesion bypass. Replicative lesion bypass requires specialized DNA polymerases (Rattray and Strathern 2003), most of which belong to the Y-family of DNA polymerases that are found in prokaryotes, eukaryotes, and archaea (Ohmori et al. 2001). *E. coli* possesses two Y family DNA polymerases, regulated by the SOS system, capable of performing translesion synthesis (TLS): PolIV and PolV, which are encoded by *dinB* and *umuDC* genes, respectively. Y-family DNA polymerases lack $3' \rightarrow 5'$ exonuclease activity, but have a more open catalytic site compared to the replicative polymerases, and have low processivity (Yang 2003). These features enable the Y-family DNA polymerase to successfully bypass lesions, but also compromise the accuracy of replication of an undamaged template. Lesion bypass can be either error-free or error-prone, depending on whether the correct or incorrect nucleotide, respectively, is incorporated opposite the damage (Friedberg et al. 2002).

The most ubiquitous branch of the Y-family of DNA polymerases, the DinB branch, is typified by *Escherichia coli* PolIV, human Polκ, and the archaeal Dbh/Dpo4 enzymes (Ohmori et al. 2001). Such remarkable conservation throughout evolution strongly suggests that the Y-family DNA polymerases from the DinB branch are extremely important for cell survival and fitness. In addition to SOS, the transcription of the *dinB* gene is controlled by RpoS, a sigma subunit of RNA polymerase, which regulates a general stress response (Layton and Foster 2003). PolIV is also regulated by the heat shock chaperone GroE (Layton and Foster 2005). Therefore, PolIV is a component of several cellular stress responses. When the SOS regulon is induced, the number of PolIV molecules rapidly increases to 250 and

2,500 per cell (Nohmi 2006). The overexpression of the *dinB* gene substantially increases spontaneous mutagenesis (Kim et al. 1997), probably by competing with PolIII for binding to the β-clamp (Lenne-Samuel et al. 2002). PolIV is a low fidelity enzyme with a misincorporation frequency in the range of 10^{-3} to 10^{-5}. In stressed cells, PolIV was shown to considerably contribute to mutagenesis. For example, PolIV is responsible for the untargeted mutagenesis of non-irradiated lambda phage in UV irradiated cells (Brotcorne-Lannoye and Maenhaut-Michel 1986), and for the increased generation of mutations under carbon source starvation in the stationary phase (McKenzie et al. 2001; Tompkins et al. 2003; Foster 2000).

PolV, is regulated both transcriptionally and posttranslationally, and in addition needs other proteins in order to perform translesion synthesis (Sutton et al. 2000). PolV is a protein complex composed of three subunits: UmuC, and two truncated UmuD proteins. The transcriptional induction of the *umuDC* operon and the synthesis of the UmuC and UmuD proteins do not result in the production of an active lesion-bypass DNA polymerase. The UmuD protein must first undergo RecA* assisted self-cleavage (mechanistically similar to that undergone by LexA) which removes the N-terminal 24 residues and yields a UmuD' derivative. UmuD and UmuD' each form homodimers and, in addition, interact with each other to form a UmuD–UmuD' heterodimer that is more stable than either of the homodimers. All three classes of dimers interact with UmuC and greatly influence its action. Despite the fact that UmuC has catalytic DNA polymerase activity, the protein strictly requires UmuD' homodimer to function as a polymerase on damaged DNA. In addition to RecA's roles in SOS response induction and UmuD cleavage, RecA also plays a direct role in the PolV mediated TLS. PolV concentration increases from zero to about 60 molecules per cell only 1 h after SOS induction (Nohmi 2006). Thus, *E. coli* possess different mechanisms of control in order to ensure that active PolV is only present in cells that have suffered DNA damage that cannot be repaired otherwise. Like PolIV, PolV is a low fidelity enzyme, with a misincorporation frequency in the range of 10^{-3} to 10^{-4}. In addition, PolV is predominantly error-prone when promoting TLS. Consequently, inactivation of the *umuDC* operon eliminates mutagenesis induced by many genotoxic agents (Friedberg et al. 2006).

Y family DNA polymerases are an excellent illustration of how molecular constraints on survival functions can lead to mutagenesis. They can bypass noncoding lesions that modify the structure of the DNA and block replicative polymerases, thus allowing survival. But because this bypass is performed with low fidelity, it introduces mutations. Hence, the maintenance of genetic integrity is sacrificed for survival. Why did such polymerases not evolve to be error-free—i.e., to add the proper nucleotide opposite the cognate DNA lesions? There are two possible, non-exclusive explanations. First, because each TLS polymerase recognizes several types of lesions, the reduced fidelity could be the optimal solution for the trade-off between the ability to bypass different lesions and the fidelity of the bypass. Another possible explanation is that the cost of the resulting deleterious mutations is lower than the selective cost associated with the activity of error-free DNA repair systems. Hence, there is no strong selective pressure to reduce the error-rate.

Some other SOS-associated phenomena that are not involved in DNA repair can also increase genetic variability in stressed bacterial populations; for example, the increased transposition frequency of Tn*5* and Tn*10*, and the induction of temperate bacteriophages such as lambda, 434, 21, P22, f80, and coliphage 186 (Roberts and Devoret 1983). Induction of bacteriophages results in cell lysis, but bacteriophages can transfer host genes to new cells. Some conjugative plasmids carry genes encoding PolV orthologs that are even more active than those encoded by the host chromosomes, and therefore can confer increased cellular resistance to genotoxic agents, but also increased mutagenesis (Sedgwick and Goodwin 1985). Numerous SOS functions can be implicated in the genetic exchange. During interspecies conjugation, the DNA sequence divergence between genomes of different species slows down the RecA-mediated recombination steps, resulting in the induction of the SOS response by the RecA*. The induction of the SOS response enhances interspecies recombination via an overproduction of the proteins involved in homologous replication (Matic et al. 1995). Thus, interspecies conjugation acts as an intracellular stress inducer in the recipient cells. Paradoxically, DNA sequence divergence—a major component of the interspecies genetic barrier (Matic et al. 1996)—helps cells to partially overcome this obstacle by triggering the SOS response (Matic et al. 2000). The SOS response-dependent restriction alleviation can increase the frequency of transduction and conjugation. Furthermore, it has been found that double-strand exonuclease (ExoV) is inhibited in SOS induced cells, which might confer a hyper-recombinogenic phenotype (Kannan and Dharmalingam 1990; Rinken and Wackernagel 1992).

RpoS-Regulated General Stress Response

RpoS is one of seven *E. coli* RNA polymerase sigma factors, which compete for the association with the core polymerase subunit (Hengge-Aronis 2002; Eisenstark et al. 1996). The outcome of the competition is influenced by the varying number of each sigma factor and by different molecules that can affect the binding of sigma factors to the RNA polymerase. Each sigma factor coordinates the transcription of a set of genes, thus allowing fine control of adaptation to different physiological condition. The production of RpoS is regulated at every step of gene expression: transcription, translation, protein stability, and activity (Battesti et al. 2011). Transcription of the *rpoS* gene is controlled by the cAMP receptor protein and through the signaling of ppGpp and polyphosphate. The *rpoS* mRNA is translated at low levels because the long 5′ untranslated region of the *rpoS* transcript folds into a stem-loop that occludes the ribosome binding site. The stability of this *rpoS* mRNA secondary structure is modulated by a cascade of interacting factors, including Hfq, HU, H-NS (histone-like nucleoid structuring protein), LeuO (transcription regulator), and small noncoding RNAs: *dsrA* RNA, *rprA* RNA, and *oxyS* RNA. In growing cells, RpoS levels are maintained at a low level due to degradation by the ClpXP protease in a reaction that is promoted by RssB (proteolytic targeting factor) and inhibited by the chaperone DnaK. RssB activity is modulated by three proteins—IraP, IraM, and IraD—produced under specific stress conditions that interact with RssB and prevent RpoS degradation. Various stress conditions differentially affect the RpoS concentration control mechanisms (Hengge-Aronis 2002; Peterson et al. 2005). Thus, a reduced growth rate results in increased *rpoS* transcription whereas high cell density, high osmolarity, low temperature, phosphorus starvation, and low pH stimulate the translation of already present *rpoS* mRNA. Low pH, carbon source starvation, and high temperature modulate RpoS proteolysis.

When present at high concentration, RpoS outcompetes the vegetative sigma factor, RpoD, and regulates transcription of hundreds of genes with unrelated physiological functions (Patten et al. 2004; Weber et al. 2005). Several factors have been shown to determine the outcome of this competition: Lrp (leucine-responsive regulatory protein) affects the selectivity of these two sigma factors for many promoters; Rsd (regulator of Sigma D), an anti-RpoD factor, controls the level of functional RpoD holoenzyme (Jishage and Ishihama 1998); and *rsd* gene expression is inversely correlated with growth rate. The intracellular concentrations of glutamate and polyphosphate, as well as decreased DNA superhelicity, have also been shown to enhance the activity of RpoS holoenzyme, and to repress that of RpoD in stationary phase *E. coli* cells. Because the RpoS regulon is not induced only in stationary phase, but rather responds to many different stress conditions, it is considered a general stress response (Hengge-Aronis 2000).

The induction of the RpoS regulon, which concerns about 10% of the *E. coli* genes, results in morphological and metabolic modifications, and provides resistance to a variety of stresses (e.g., resistance to UV, heath shock, oxidative stress, and extreme osmolarity). Intriguingly, while the key priority of this regulon is to assure survival, conservation of original genetic information is not. For example, RpoS stimulates transposition of the *Pseudomonas putida* transposon Tn*465* during stationary phase (Ilves et al. 2001). The overproduction of the RpoS-regulated *hha* gene increases the frequency of transposition of insertion elements within the *E. coli* chromosome as well (Mikulskis and Cornelis 1994). Overproduction of the *rpoS* gene results in increased mutagenesis in growing cells (Yang et al. 2004). Furthermore, the RpoS regulon has been implicated in stationary phase mutagenesis in the *E. coli* and in *P. putida* (Saumaa et al. 2002).

Two molecular mechanisms have been described as being are responsible for the RpoS regulon-dependent increase of stationary phase mutagenesis: induction of the *dinB* gene, and downregulation of the mismatch repair system. Such noncanonical regulation, i.e., LexA independent, of the PolIV TLS DNA polymerase may help the cells to survive certain DNA damages without new protein synthesis. This could be a case with cytotoxic alkylating DNA lesions (Bjedov et al. 2007), which can accumulate in DNA because of RpoS-dependent downregulation of the *alkA* gene. This gene codes for the DNA glycosylase that removes replication-blocking 3-methyladenine and 3-methylguanine (Landini and Busby 1999). In stationary phase cells, the transcription of *mutS* and *mutH* genes and the concentration of MutS and MutH proteins decreases to very low levels via an RpoS- and Hfq-dependent mechanism, compared to the growing cells (Feng et al. 1996; Tsui et al. 1997;

Harris et al. 1997). Consequently, the activity of the mismatch repair system is reduced during stationary phase. The overproduction of the MutS protein—but not the MutL protein—in wild-type cells significantly decreases stationary phase mutagenesis (Bjedov et al. 2003). Similarly, overexpression of the MutS repair protein significantly decreased the rate of G:C T:A transversion mutation in stationary-phase wild-type, *mutY* and *mutM* strains (Zhao and Winkler 2000).

What would be the benefit of inactivating the mismatch repair system, a major contributor to DNA replication and recombination fidelity? Once again, there are two possibilities; either the downregulation of the mismatch repair system activity contributes to survival, or it is selected for because it facilitates adaptation via the beneficial mutations it generates, as is the case for the mismatch repair mutants. Currently, there is no definite answer, but the most likely hypothesis is that the contribution to survival of the downregulation of the mismatch repair activity may be as simple as the energy saved by not expressing these repair functions when nutrients are limited. The absence of the mismatch repair system surveillance results in increased mutagenesis, but the fitness cost of the resulting deleterious mutations is probably much less important than that incurred by energy "exhaustion" due to the production of "useless" proteins under starvation conditions.

Environmental Tuning of Mutation Rates Among *E. coli* Natural Isolates

The availability of essential nutrients—including carbon, nitrogen, and phosphorus—and frequency of different stresses both between different ecological niches. Therefore, it is not surprising that different *E. coli* ecotypes, which have different lifestyles, have different nutritional capabilities and respond differently to the same stresses. In many cases, such variability was shown to result from allelic variation of *rpoS* gene or from polymorphism of its regulatory elements (Ferenci 2003). For example, strains with high constitutive intracellular levels of RpoS protein metabolize fewer substrates and poorly compete for low concentrations of nutrients, and have increased stress resistance (King et al. 2004). Constitutively low intracellular concentration of RpoS has exactly the opposite effects on nutritional competence and stress resistance. These phenotypes are, to a large degree, the consequence of a competition between different sigma factors within the cell; e.g., the absence of RpoS allows a higher level of transcription of RpoD-dependent genes (Nystrom 2004). Therefore, nutritional limitations and environmental stress conditions impose conflicting choices that result in selection of the loss or modification of RpoS function in different environmental niches as a function of nutrient availability, and of the frequency and nature of stresses.

E. coli strains that have increased stress resistance due to high intracellular levels of the RpoS protein have also increased mutation rates (Yang et al. 2004; King et al. 2004). This can be explained by the RpoS dependent regulation of DNA repair genes discussed in this chapter. As we observed that there is a positive correlation between intracellular concentrations of RpoS protein found in natural *E. coli* isolates and spontaneous mutation frequency, it can be concluded that different amounts of RpoS protein differently affect DNA repair activity. It was also observed that the level of stationary phase-dependent mismatch repair system downregulation varies between *E. coli* natural isolates (Li et al. 2003). Hence, the variable rates of natural isolate genetic variability generation can be viewed as a byproduct of selection for improved fitness (Tenaillon et al. 2004). This can explain why the ecological niche from which the strain was isolated appears to be the major determinant of a strain's mutator phenotype. For example, host diet and bacterial lifestyle—i.e., commensal and pathogenic—was shown to correlate with the variability of stationary phase mutation frequencies among *E. coli* natural isolates (Bjedov et al. 2003). The absence of correlation between bacterial phylogeny and the mutator phenotype suggests that the evolution of stress-inducible mutation rates is rapid on an evolutionary time scale.

Conclusion

Organisms live in constantly changing environments in which, the nature, severity and frequency of environmental stresses are very variable. For this reason, organisms possess multiple strategies for coping with the environmental fluctuations. One such strategy is modulation of mutation rates as a function of the degree of adaptation to the environment. Mutation rates are low when environment is stable and high when environment is unstable and stressful. High mutation rates can be advantageous because they increase the probability of generating adaptive mutations (Fig. 1.1). Mutation rates can be modulated by changing the proportion of constitutive mutator versus non-mutator bacteria at the population level, and by induction of stress responses, which increase mutation rates transiently in individual cells. It was clearly demonstrated that constitutive mutator alleles are selected for because they hitchhike with the adaptive mutations they generate. There are two nonexclusive hypotheses concerning the nature of the selective pressure acting on the molecular mechanisms controlling stress-induced mutagenesis. Stress-induced mutagenesis could be an unavoidable by-product of mechanisms involved in survival under stress, either due to constraints at the molecular level, or due to a trade-off between survival functions and the costs of repair and protection functions. It could be also that stress-induced mutator phenotypes are selected for in the same way as constitutive mutator alleles, via hitchhiking with adaptive mutations they generate. However, regardless of the nature of selective pressure acting on stress-induced mutagenesis, it is very likely that the resulting increased genetic variability plays an important role in the bacterial evolution (Fig. 1.1).

Acknowledgments This work was supported by FP7-HEALTH-F3-2010-241476 and ANR-09-BLAN-0251 grants.

References

Battesti A, Majdalani N, Gottesman S (2011) The RpoS-mediated general stress response in Escherichia coli. Annu Rev Microbiol 65:189–213

Bijlsma R, Loeschcke V (eds) (1997) Environmental stress, adaptation, and evolution, vol EXS 83. Birkhäuser, Basel

Bjedov I, Tenaillon O, Gerard B, Souza V, Denamur E, Radman M, Taddei F, Matic I (2003) Stress-induced mutagenesis in bacteria. Science 300:1404–1409

Bjedov I, Dasgupta CN, Slade D, Le Blastier S, Selva M, Matic I (2007) Involvement of *Escherichia coli* DNA polymerase IV in tolerance of cytotoxic alkylating DNA lesions in vivo. Genetics 176:1431–1440

Bjorkholm B, Sjolund M, Falk PG, Berg OG, Engstrand L, Andersson DI (2001) Mutation frequency and biological cost of antibiotic resistance in Helicobacter pylori. Proc Natl Acad Sci USA 98:14607–14612

Boe L, Danielsen M, Knudsen S, Petersen JB, Maymann J, Jensen PR (2000) The frequency of mutators in populations of *Escherichia coli*. Mutat Res 448:47–55

Brotcorne-Lannoye A, Maenhaut-Michel G (1986) Role of RecA protein in untargeted UV mutagenesis of bacteriophage lambda: evidence for the requirement for the *dinB* gene. Proc Natl Acad Sci USA 83:3904–3908

Chao L, Cox EC (1983) Competition between high and low mutating strains of *Escherichia coli*. Evolution 37:125–134

Chao L, Vargas C, Spear BB, Cox EC (1983) Transposable elements as mutator genes in evolution. Nature 303:633–635

Cooper VS, Lenski RE (2000) The population genetics of ecological specialization in evolving *Escherichia coli* populations. Nature 407:736–739

Courcelle J, Khodursky A, Peter B, Brown PO, Hanawalt PC (2001) Comparative gene expression profiles following UV exposure in wild-type and SOS-deficient *Escherichia coli*. Genetics 158:41–64

del Campo R, Morosini MI, de la Pedrosa EG, Fenoll A, Munoz-Almagro C, Maiz L, Baquero F, Canton R (2005) Population structure, antimicrobial resistance, and mutation frequencies of Streptococcus pneumoniae isolates from cystic fibrosis patients. J Clin Microbiol 43:2207–2214

Denamur E, Lecointre G, Darlu P, Tenaillon O, Acquaviva C, Sayada C, Sunjevaric I, Rothstein R, Elion J, Taddei F et al (2000) Evolutionary implications of the frequent horizontal transfer of mismatch repair genes. Cell 103:711–721

Drake JW (1993) General antimutators are improbable. J Mol Biol 229:8–13

Drake JW, Charlesworth B, Charlesworth D, Crow JF (1998) Rates of spontaneous mutation. Genetics 148:1667–1686

Eisenstark A, Calcutt MJ, Becker-Hapak M, Ivanova A (1996) Role of *Escherichia coli rpoS* and associated genes in defense against oxidative damage. Free Radic Biol Med 21:975–993

Feng G, Tsui HC, Winkler ME (1996) Depletion of the cellular amounts of the MutS and MutH methyl-directed mismatch repair proteins in stationary-phase *Escherichia coli* K-12 cells. J Bacteriol 178:2388–2396

Ferenci T (2003) What is driving the acquisition of mutS and rpoS polymorphisms in Escherichia coli? Trends Microbiol 11:457–461

Fernandez De Henestrosa AR, Ogi T, Aoyagi S, Chafin D, Hayes JJ, Ohmori H, Woodgate R (2000) Identification of additional genes belonging to the LexA regulon in Escherichia coli. Mol Microbiol 35:1560–1572

Foster PL (2000) Adaptive mutation in *Escherichia* coli. Cold Spring Harb Symp Quant Biol 65:21–29

Friedberg EC, Wagner R, Radman M (2002) Specialized DNA polymerases, cellular survival, and the genesis of mutations. Science 296:1627–1630

Friedberg EC, Walker GC, Siede W, Wood RD, Schultz RA, Ellenberger T (2006) DNA repair and mutagenesis. ASM Press, Washington, DC

Funchain P, Yeung A, Stewart JL, Lin R, Slupska MM, Miller JH (2000) The consequences of growth of a mutator strain of *Escherichia coli* as measured by loss of function among multiple gene targets and loss of fitness. Genetics 154:959–970

Funchain P, Yeung A, Stewart J, Clendenin WM, Miller JH (2001) Amplification of mutator cells in a population as a result of horizontal transfer. J Bacteriol 183:3737–3741

Giraud A, Matic I, Tenaillon O, Clara A, Radman M, Fons M, Taddei F (2001) Costs and benefits of high mutation rates: adaptive evolution of bacteria in the mouse gut. Science 291:2606–2608

Graziewicz M, Wink DA, Laval F (1996) Nitric oxide inhibits DNA ligase activity: potential mechanisms for NO-mediated DNA damage. Carcinogenesis 17:2501–2505

Guelfo JR, Rodriguez-Rojas A, Matic I, Blazquez J (2010) A MATE-family efflux pump rescues the Escherichia coli 8-oxoguanine-repair-deficient mutator phenotype and protects against H(2)O(2) killing. PLoS Genet 6:e1000931

Harris RS, Feng G, Ross KJ, Sidhu R, Thulin C, Longerich S, Szigety SK, Winkler ME, Rosenberg SM (1997) Mismatch repair protein MutL becomes limiting during stationary-phase mutation. Genes Dev 11:2426–2437

Hengge-Aronis R (2000) The general stress response in *Escherichia coli*. In: Storz G, Hengge-Aronis R (eds) Bacterial stress responses. ASM Press, Washington DC, pp 161–178

Hengge-Aronis R (2002) Signal transduction and regulatory mechanisms involved in control of the sigma(S) (RpoS) subunit of RNA polymerase. Microbiol Mol Biol Rev 66:373–395

Horst JP, Wu TH, Marinus MG (1999) *Escherichia coli* mutator genes. Trends Microbiol 7:29–36

Ilves H, Horak R, Kivisaar M (2001) Involvement of sigma(S) in starvation-induced transposition of Pseudomonas putida transposon Tn4652. J Bacteriol 183:5445–5448

Imhof M, Schlotterer C (2001) Fitness effects of advantageous mutations in evolving *Escherichia coli* populations. Proc Natl Acad Sci USA 98:1113–1117

Jishage M, Ishihama A (1998) A stationary phase protein in *Escherichia coli* with binding activity to the major sigma subunit of RNA polymerase. Proc Natl Acad Sci USA 95:4953–4958

Kannan P, Dharmalingam K (1990) Induction of the inhibitor of the RecBCD enzyme in *Escherichia coli* is a *lexA*-independent SOS response. Curr Microbiol 21:7–15

Kibota TT, Lynch M (1996) Estimate of the genomic mutation rate deleterious to overall fitness in *E. coli*. Nature 381:694–696

Kim SR, Maenhaut-Michel G, Yamada M, Yamamoto Y, Matsui K, Sofuni T, Nohmi T, Ohmori H (1997) Multiple pathways for SOS-induced mutagenesis in *Escherichia coli*: an overexpression of *dinB*/*dinP* results in strongly enhancing mutagenesis in the absence of any exogenous treatment to damage DNA. Proc Natl Acad Sci USA 94:13792–13797

King T, Ishihama A, Kori A, Ferenci T (2004) A regulatory trade-off as a source of strain variation in the species *Escherichia coli*. J Bacteriol 186:5614–5620

Kishony R, Leibler S (2003) Environmental stresses can alleviate the average deleterious effect of mutations. J Biol 2:14

Kowalczykowski SC, Dixon DA, Eggleston AK, Lauder SD, Rehrauer WM (1994) Biochemistry of homologous recombination in Escherichia coli. Microbiol Rev 58:401–465

Labat F, Pradillon O, Garry L, Peuchmaur M, Fantin B, Denamur E (2005) Mutator phenotype confers advantage in *Escherichia coli* chronic urinary tract infection pathogenesis. FEMS Immunol Med Microbiol 44:317–321

Landini P, Busby SJ (1999) Expression of the *Escherichia coli ada* regulon in stationary phase: evidence for *rpoS*-dependent negative regulation of *alkA* transcription. J Bacteriol 181:6836–6839

Laval F, Wink DA (1994) Inhibition by nitric oxide of the repair protein, O6-methylguanine-DNA-methyltransferase. Carcinogenesis 15:443–447

Layton JC, Foster PL (2003) Error-prone DNA polymerase IV is controlled by the stress-response sigma factor, RpoS, in Escherichia coli. Mol Microbiol 50:549–561

Layton JC, Foster PL (2005) Error-prone DNA polymerase IV is regulated by the heat shock chaperone GroE in *Escherichia coli*. J Bacteriol 187:449–457
Le Chat L, Fons M, Taddei F (2006) *Escherichia coli* mutators: selection criteria and migration effect. Microbiology 152:67–73
LeClerc JE, Li B, Payne WL, Cebula TA (1996) High mutation frequencies among *Escherichia coli* and *Salmonella* pathogens. Science 274:1208–1211
LeClerc JE, Payne WL, Kupchella E, Cebula TA (1998) Detection of mutator subpopulations in *Salmonella typhimurium* LT2 by reversion of his alleles. Mutat Res 400:89–97
Lenne-Samuel N, Wagner J, Etienne H, Fuchs RP (2002) The processivity factor beta controls DNA polymerase IV traffic during spontaneous mutagenesis and translesion synthesis in vivo. EMBO Rep 3:45–49
Levy MS, Balbinder E, Nagel R (1993) Effect of mutations in SOS genes on UV-induced precise excision of Tn10 in Escherichia coli. Mutat Res 293:241–247
Li B, Tsui HC, LeClerc JE, Dey M, Winkler ME, Cebula TA (2003) Molecular analysis of mutS expression and mutation in natural isolates of pathogenic Escherichia coli. Microbiology 149:1323–1331
Little JW, Edmiston SH, Pacelli Z, Mount DW (1980) Cleavage of the *Escherichia coli lexA* protein by the *recA* protease. Proc Natl Acad Sci USA 77:3225–3229
Mao EF, Lane L, Lee J, Miller JH (1997) Proliferation of mutators in a cell population. J Bacteriol 179:417–422
Matic I, Rayssiguier C, Radman M (1995) Interspecies gene exchange in bacteria: the role of SOS and mismatch repair systems in evolution of species. Cell 80:507–515
Matic I, Taddei F, Radman M (1996) Genetic barriers among bacteria. Trends Microbiol 4:69–72
Matic I, Radman M, Taddei F, Picard B, Doit C, Bingen E, Denamur E, Elion J (1997) Highly variable mutation rates in commensal and pathogenic *Escherichia coli*. Science 277:1833–1834
Matic I, Taddei F, Radman M (2000) No genetic barriers between Salmonella enterica serovar typhimurium and Escherichia coli in SOS-induced mismatch repair-deficient cells. J Bacteriol 182:5922–5924
McKenzie GJ, Lee PL, Lombardo MJ, Hastings PJ, Rosenberg SM (2001) SOS mutator DNA polymerase IV functions in adaptive mutation and not adaptive amplification. Mol Cell 7:571–579
Mellon I, Champe GN (1996) Products of DNA mismatch repair genes mutS and mutL are required for transcription-coupled nucleotide-excision repair of the lactose operon in Escherichia coli. Proc Natl Acad Sci USA 93:1292–1297
Mikulskis AV, Cornelis GR (1994) A new class of proteins regulating gene expression in enterobacteria. Mol Microbiol 11:77–86
Morimatsu K, Kowalczykowski SC (2003) RecFOR proteins load RecA protein onto gapped DNA to accelerate DNA strand exchange: a universal step of recombinational repair. Mol Cell 11:1337–1347
Moxon R, Bayliss C, Hood D (2006) Bacterial contingency loci: the role of simple sequence DNA repeats in bacterial adaptation. Annu Rev Genet 40:307–333
Nilsson AI, Kugelberg E, Berg OG, Andersson DI (2004) Experimental adaptation of Salmonella typhimurium to mice. Genetics 168:1119–1130
Nohmi T (2006) Environmental stress and lesion-bypass DNA polymerases. Annu Rev Microbiol 60:231–253
Nystrom T (2004) Growth versus maintenance: a trade-off dictated by RNA polymerase availability and sigma factor competition? Mol Microbiol 54:855–862
Ohmori H, Friedberg EC, Fuchs RP, Goodman MF, Hanaoka F, Hinkle D, Kunkel TA, Lawrence CW, Livneh Z, Nohmi T et al (2001) The Y-family of DNA polymerases. Mol Cell 8:7–8
Oliver A, Canton R, Campo P, Baquero F, Blazquez J (2000) High frequency of hypermutable *Pseudomonas aeruginosa* in cystic fibrosis lung infection. Science 288:1251–1254
Oliver A, Baquero F, Blazquez J (2002) The mismatch repair system (mutS, mutL and uvrD genes) in Pseudomonas aeruginosa: molecular characterization of naturally occurring mutants. Mol Microbiol 43:1641–1650

Pal C, Macia MD, Oliver A, Schachar I, Buckling A (2007) Coevolution with viruses drives the evolution of bacterial mutation rates. Nature 450:1079–1081

Patten CL, Kirchhof MG, Schertzberg MR, Morton RA, Schellhorn HE (2004) Microarray analysis of RpoS-mediated gene expression in *Escherichia coli* K-12. Mol Genet Genomics 272:580–591

Peterson CN, Mandel MJ, Silhavy TJ (2005) *Escherichia coli* starvation diets: essential nutrients weigh in distinctly. J Bacteriol 187:7549–7553

Prunier AL, Malbruny B, Laurans M, Brouard J, Duhamel JF, Leclercq R (2003) High rate of macrolide resistance in Staphylococcus aureus strains from patients with cystic fibrosis reveals high proportions of hypermutable strains. J Infect Dis 187:1709–1716

Rattray AJ, Strathern JN (2003) Error-prone DNA polymerases: when making a mistake is the only way to get ahead. Annu Rev Genet 37:31–66

Richardson AR, Yu Z, Popovic T, Stojiljkovic I (2002) Mutator clones of *Neisseria meningitidis* in epidemic serogroup A disease. Proc Natl Acad Sci USA 99:6103–6107

Rinken R, Wackernagel W (1992) Inhibition of the RecBCD-dependent activation of Chi recombinational hot spots in SOS-induced cells of *Escherichia coli*. J Bacteriol 174:1172–1178

Roberts JW, Devoret R (1983) Lysogenic induction. In: Hendrix RW, Roberts JW, Stahl FW, Weisberg RA (eds) Lambda II. Cold Spring Harbor Laboratory, Cold Spring Harbor, pp 123–144

Sassanfar M, Roberts JW (1990) Nature of the SOS-inducing signal in Escherichia coli. The involvement of DNA replication. J Mol Biol 212:79–96

Saumaa S, Tover A, Kasak L, Kivisaar M (2002) Different spectra of stationary-phase mutations in early-arising versus late-arising mutants of *Pseudomonas putida*: involvement of the DNA repair enzyme MutY and the stationary-phase sigma factor RpoS. J Bacteriol 184:6957–6965

Sedgwick SG, Goodwin PA (1985) Differences in mutagenic and recombinational DNA repair in enterobacteria. Proc Natl Acad Sci USA 82:4172–4176

Sniegowski PD, Gerrish PJ, Lenski RE (1997) Evolution of high mutation rates in experimental populations of *E. coli*. Nature 387:703–705

Sutton MD, Smith BT, Godoy VG, Walker GC (2000) The SOS response: recent insights into *umuDC*-dependent mutagenesis and DNA damage tolerance. Annu Rev Genet 34:479–497

Taddei F, Radman M, Maynard-Smith J, Toupance B, Gouyon PH, Godelle B (1997a) Role of mutator alleles in adaptive evolution. Nature 387:700–702

Taddei F, Hayakawa H, Bouton M, Cirinesi A, Matic I, Sekiguchi M, Radman M (1997b) Counteraction by MutT protein of transcriptional errors caused by oxidative damage. Science 278:128–130

Tanaka MM, Bergstrom CT, Levin BR (2003) The evolution of mutator genes in bacterial populations: the roles of environmental change and timing. Genetics 164:843–854

Tenaillon O, Toupance B, Le Nagard H, Taddei F, Godelle B (1999) Mutators, population size, adaptive landscape and the adaptation of asexual populations of bacteria. Genetics 152:485–493

Tenaillon O, Denamur E, Matic I (2004) Evolutionary significance of stress-induced mutagenesis in bacteria. Trends Microbiol 12:264–270

Tompkins JD, Nelson JL, Hazel JC, Leugers SL, Stumpf JD, Foster PL (2003) Error-prone polymerase, DNA polymerase IV, is responsible for transient hypermutation during adaptive mutation in *Escherichia coli*. J Bacteriol 185:3469–3472

Travis JM, Travis ER (2002) Mutator dynamics in fluctuating environments. Proceedings 269:591–597

Travis ER, Travis JM (2004) Mutators in space: the dynamics of high-mutability clones in a two-patch model. Genetics 167:513–522

Tröbner W, Piechocki R (1981) Competition growth between *Escherichia coli mutL* and *mut*⁺ in continuously growing cultures. Z Allg Mikrobiol 21:347–349

Trobner W, Piechocki R (1984) Selection against hypermutability in Escherichia coli during long term evolution. Mol Gen Genet 198:177–178

Tröbner W, Piechocki R (1984) Selection against hypermutability in *Escherichia coli* during long term evolution. Mol Gen Genet 198:177–178

Tsui HC, Feng G, Winkler ME (1997) Negative regulation of *mutS* and *mutH* repair gene expression by the Hfq and RpoS global regulators of *Escherichia coli* K-12. J Bacteriol 179: 7476–7487
Watson ME Jr, Burns JL, Smith AL (2004) Hypermutable Haemophilus influenzae with mutations in mutS are found in cystic fibrosis sputum. Microbiology 150:2947–2958
Weber H, Polen T, Heuveling J, Wendisch VF, Hengge R (2005) Genome-wide analysis of the general stress response network in *Escherichia coli*: sigmaS-dependent genes, promoters, and sigma factor selectivity. J Bacteriol 187:1591–1603
Wink DA, Laval J (1994) The Fpg protein, a DNA repair enzyme, is inhibited by the biomediator nitric oxide in vitro and in vivo. Carcinogenesis 15:2125–2129
Yang W (2003) Damage repair DNA polymerases Y. Curr Opin Struct Biol 13:23–30
Yang H, Wolff E, Kim M, Diep A, Miller JH (2004) Identification of mutator genes and mutational pathways in Escherichia coli using a multicopy cloning approach. Mol Microbiol 53:283–295
Zhao J, Winkler ME (2000) Reduction of GC®TA transversion mutation by overexpression of MutS in Escherichia coli K-12. J Bacteriol 182:5025–5028

Chapter 2
Mutagenesis Associated with Repair of DNA Double-Strand Breaks Under Stress

Chandan Shee, P.J. Hastings, and Susan M. Rosenberg

Abstract Evolutionary theory predicted that mutations occur randomly both in time and in genomic space. This expectation has been revised by the discoveries of stress-induced mutation mechanisms, which activate mutagenesis pathways under the control of stress responses. Stress-induced mutation mechanisms produce mutations preferentially when cells or organisms are maladapted to their environment, i.e., when they are stressed, potentially accelerating evolution. We review stress-induced mutagenesis associated with repair of double-strand breaks in *Escherichia coli*. In this mechanism, the process of DNA break repair by homologous recombination is high-fidelity in unstressed cells, but is switched to a mutagenic mode using the error-prone DNA polymerase DinB, and other error-prone DNA polymerases, under the control of the RpoS general stress response. The switch to mutagenic repair occurs during starvation or if RpoS is upregulated artificially in unstressed cells, and presumably during the many different stresses that activate the RpoS response. Recent work shows that this mechanism accounts for most spontaneous base-substitution and frameshift mutagenesis during starvation in *E. coli*, acts not only in plasmid DNA but also in the chromosomes of plasmid-free cells, illustrates the generality of this mechanism in many organisms and circumstances, and resolves some other old tensions in the field. Stress-induced mutation mechanisms studied in the laboratory are likely to provide supe-

C. Shee • S.M. Rosenberg (✉)
Departments of Molecular and Human Genetics,
Biochemistry and Molecular Biology, Molecular Virology and Microbiology,
Dan L Duncan Cancer Center, Baylor College of Medicine, Houston TX, USA
e-mail: smr@bcm.edu

P.J. Hastings
Department of Molecular and Human Genetics,
Baylor College of Medicine, Houston, TX, USA

D. Mittelman (ed.), *Stress-Induced Mutagenesis*, DOI 10.1007/978-1-4614-6280-4_2,

rior models for mutagenesis underlying pathogen-host adaptation, antibiotic resistance, and cancer progression and resistance mechanisms, all problems of evolution under stress driven by mutations.

Stress-Induced Mutation Is Changing Ideas About Evolution

Before the acceptance of Darwin's theory of natural selection, the impact of environmental stress on induction of natural variation was a popular idea (Mayr 1982). However, lack of strong experimental evidence of environmental influence on induction of genetic variation diminished the idea's traction and it was replaced by the neo-Darwinian modern synthesis of the 1930s. In the neo-Darwinian view (Mayr 1982), mutations were imagined to be random, and to occur randomly in time and in genomic space, with natural selection being the sole driver of evolution. The rise of the neutralists added genetic drift (random survival and proliferation of individuals with mutations) as an additional driver of evolution (Kimura 1991), but did not change the general view of the randomness of mutations.

However, recent discoveries of special mechanisms of mutation induced in bacteria, yeast, plant, and human cells under growth-limiting stress is changing this view (Galhardo et al. 2007; Hastings 2007; Heidenreich 2007; Saint-Ruf et al. 2007; Cirz and Romesberg 2007; Bindra et al. 2007; Fonville et al. 2011; Yao and Kovalchuk 2011; Rosenberg et al. 2012). These mechanisms increase genetic diversity and, potentially, the ability to evolve, specifically when cells are maladapted to their environment, i.e., when they are stressed. The kinds of mutations generated include base substitutions, small deletions and insertions, gross chromosomal rearrangements including copy-number variations (CNVs, amplifications, and large deletions), and transpositions. Although the mechanisms that produce stress-induced mutations vary, the overriding common theme in these pathways is their activation by cellular stress responses, resulting in increased, and sometimes different kinds of, mutations under stress. The kinds of stressors that can provoke stress-induced mutation mechanisms include starvation, hypoxia, oxidative stress, antibiotics, and, presumably, many others that activate the stress responses that promote stress-induced mutations. Understanding of molecular mechanisms of stress-induced mutagenesis is changing how we think about evolution. Stress-induced mutation mechanisms may provide superior models for genetic changes that drive pathogen–host adaptation, antibiotic resistance, aging, cancer progression, and therapy-resistance mechanisms, and, possibly much of evolution, generally. In this chapter, we focus on recent advances in understanding a molecular mechanism of stress-induced mutation, dissected initially in starving *Escherichia coli* cells: double-strand-break-dependent stress-induced mutation. Other stress-induced mutation mechanisms are reviewed by (Galhardo et al. 2007; Hastings 2007; Heidenreich 2007; Saint-Ruf et al. 2007; Cirz and Romesberg 2007; Bindra et al. 2007; Fonville et al. 2011; Yao and Kovalchuk 2011).

Cellular and Environmental Stresses and Stress Responses

Microbes are challenged continuously by environmental stressors that result from a constantly changing environment. At the cellular level, stress can disturb development, affecting structure, function, stability, growth, and survival (Tiligada 2006). Environmental factors that act as potential stressors include starvation, radiation, hypoxia, reactive oxygen species, hyper- or hypothermia, hyper- or hypo-osmotic conditions, factors underlying metabolic deficiencies and other metabolic conditions, heavy metals, toxic agents, and drugs (Tiligada et al. 2002). In nature, microbes constantly face challenges from their environment, from nutrient starvation to variations in temperature, pH, and osmolarity. Antimicrobial agents also act as stressors for microorganisms, because they interfere with important pathways and threaten survival. To overcome these various stresses, all cells have evolved specific adaptive stress responses. Many of these stress-response pathways are quite specific, involving sets of regulated genes that help cells survive specific stress conditions. In addition, all or most cells employ global stress responses that are activated by multiple stressors and provide resistance to many stresses when activated.

The RpoS General Stress Response

In *E. coli*, the general or starvation stress response controlled by the RpoS (σ^S) transcriptional activator is such a general stress response (Weber et al. 2005; Battesti et al. 2011). RpoS (σ^S) is a sigma factor: an interchangeable subunit of the bacterial RNA polymerase. When present, σ^S directs transcription to a large set of stress-responsive promoters and away from "housekeeping" promoters recognized by the RNA polymerase carrying the housekeeping sigma factor. The RpoS response is activated, and σ^S synthesized and stabilized in cells, in response to starvation, stationary phase, cold shock, osmotic shock, and oxidative and pH stress. RpoS directly or indirectly upregulates and downregulates the transcription of approximately 500 genes—about 10% of *E. coli* genes (Weber et al. 2005; Battesti et al. 2011). This program is a general stress response, in that cells challenged with one stressor then display greater resistance to multiple stressors when the RpoS response is activated. In this chapter, the RpoS response is important because, in addition to its immediate stress-tolerating properties, we see that it throws the critical switch that activates stress-induced mutation during repair of DNA breaks.

DNA Damage and Its Repair Are Important to Mutagenesis

DNA damage and repair are important to mutagenesis because repair pathways can create mutations. We explore a mutation mechanism in which the mutagenicity of DNA break repair is upregulated by a stress response, causing stress-induced mutations.

DNA is frequently exposed to DNA-damaging agents such as ultraviolet light, mutagenic chemicals, reactive oxygen species generated by cellular metabolism, ionizing radiation, and radio-mimetic drugs (Friedberg et al. 2005; Hoeijmakers 2001). Frequent types of DNA damage include single- and double-strand breaks (SSBs and DSBs), base lesions, sugar modifications, apurinic/apyrimidinic sites (AP sites), and DNA-DNA and DNA-protein cross-links (Friedberg et al. 2005; Hoeijmakers 2001). If left unrepaired, DSBs can be lethal (Meulle et al. 2008). Both single-strand gaps and DSBs induce the SOS DNA-damage response. The SOS response upregulates transcription of about 40 *E. coli* genes that function in DNA repair, DNA-damage tolerance, mutagenesis, and cell-cycle checkpoint control (Friedberg et al. 2005; Courcelle et al. 2001; Kenyon and Walker 1980). *E. coli* possesses five DNA polymerases. Along with DNA polymerases II and V, the error-prone DNA polymerase Pol IV, encoded by *dinB,* is among the genes upregulated by the SOS response in *E. coli* (Kenyon and Walker 1980; Kim et al. 1997; Wagner et al. 1999). DNA polymerase Pol II is a fairly high-fidelity polymerase (Banach-Orlowska et al. 2005), whereas Pol IV and Pol V are Y-family error-prone DNA polymerases (Ohmori et al. 2001). All three SOS DNA polymerases allow replication forks that are blocked at sites of damaged bases to insert a few bases and move on, a process called translesion synthesis (TLS) (Nohmi 2006). In addition, we see that, although they are not required for efficient repair of DNA breaks, Pol IV, Pol II, and Pol V are permitted to participate in break repair specifically during stress, and make mutations when they do.

Homologous recombination (HR) is the predominant mechanism for DSB repair in *E. coli*, and in some eukaryotes, including baker's yeast (Haber 1999). Mammalian cells use two main mechanisms for the repair of DSBs: HR and non-homologous repair mechanisms, including non-homologous end joining (NHEJ) (Hefferin and Tomkinson 2005; Weterings and van Gent 2004; Wyman and Kanaar 2004) and micro-homology-mediated break-induced replication (MMBIR) (Hastings et al. 2009a, b).

Discovery of Proteins Required for DNA Break-Dependent Stress-Induced Mutation in *E. coli*

Much of the molecular mechanism of double-strand break-dependent stress-induced mutation has been revealed from studies of the *E. coli* Lac assay for detecting mutation of a plasmid-borne gene, and then generalized using other assays for chromosomal mutations in plasmid-free cells.

The Lac Assay

In the Lac assay, *E. coli* carrying a *lac* +1 bp frameshift allele in an F′ conjugative plasmid are grown to stationary phase in liquid medium with a non-lactose carbon source (Cairns and Foster 1991). During this growth, spontaneous "generation-dependent" mutations, like those of Luria and Delbruck (Luria and Delbrück 1943), occur. The cells are then spread on solid medium with lactose as the sole carbon source, on which only Lac+ reversion mutants can form colonies. Generation-dependent Lac+ revertant colonies appear after 2 days of incubation (Cairns and Foster 1991). Over subsequent days starving on lactose medium, additional Lac+ colonies arise continuously (Cairns and Foster 1991), and these have been shown to be stress-induced mutants: cells with mutations formed dependently on the activation of three stress responses: the RpoS, SOS and RpoE responses (below).

Most Lac+ revertant colonies visible by day five carry compensatory frameshift mutations ("point mutations") (Rosenberg et al. 1994; Foster and Trimarchi 1994). These are mostly −1 bp deletions in simple repeat sequences, a sequence spectrum more homogeneous than that of generation-dependent reversions (Rosenberg et al. 1994; Foster and Trimarchi 1994). Arising later, and ultimately becoming 40% of Lac+ colonies by day eight, are *lac*-amplified clones (Hastings et al. 2000). These carry 20–100 copies of 7–40 kb tandem repeats containing the leaky *lac* allele in the F′ plasmid, and produce sufficient β-galactosidase activity to allow growth. Both point mutations and amplifications are formed after exposure to the stress condition (Hastings et al. 2000; McKenzie et al. 1998), and both form dependently on activation of the RpoS response, indicating that they are stress-induced (Lombardo et al. 2004), and suggesting that the particular stress was starvation. In new work, we show the specific starvation signaling pathway that activates RpoS during stress-induced mutagenesis (Al Mamun et al. 2012).

The molecular mechanism of stress-induced point mutagenesis is an RpoS-controlled switch from high-fidelity to error-prone repair of DNA double-strand breaks or ends (DSBs/DSEs) (Ponder et al. 2005; Shee et al. 2011a, 2012). This was deduced first by identification of the proteins required for stress-induced point mutagenesis in the Lac assay, and, second, using specific molecular/biochemical demonstrations in living cells to show that mutagenesis is part of DSB/DSE repair.

Proteins

Stress-induced point mutation in the Lac assay requires the proteins used in repair of DSBs/DSEs by homologous recombination (Harris et al. 1994; Foster et al. 1996; Harris et al. 1996), DNA polymerases DinB/Pol IV (McKenzie et al. 2001) and Pol II (Frisch et al. 2010), and the activators of three stress responses: the RpoS transcriptional activator of the general/starvation stress response (Lombardo et al. 2004; Layton and Foster 2003), the SOS DNA-damage response (Cairns and Foster 1991; McKenzie et al. 2000), and the RpoE (σ^E) envelope protein stress responses (Gibson et al. 2010). In addition, DSBs/DSEs are required (Ponder et al. 2005; Harris et al. 1994).

The RpoE envelope protein stress response somehow contributes to the formation of spontaneous DSBs/DSEs (Gibson et al. 2010), at least at some genomic locations. Because DSBs are required for point mutation (Harris et al. 1994) and amplification (Slack et al. 2006), RpoE is also required for both (Gibson et al. 2010). That is, RpoE is required for both point mutation and amplification; the requirement is for formation of mutations, not for growth of the colony after a mutation is formed, and RpoE is no longer required if a DSB is provided in the same molecule as *lac* using a restriction enzyme expressed in the cell to make a site-specific DSB in that molecule (Gibson et al. 2010).

The SOS response is required solely for upregulation of DinB/Pol IV (Galhardo et al. 2009). Pol IV is present at about 250 molecules per cell and is upregulated about tenfold when the SOS response is induced. This tenfold upregulation of is necessary (Galhardo et al. 2009), but not sufficient for mutation during DSB repair (Ponder et al. 2005; Shee et al. 2011a); the RpoS response must also be activated for Pol IV-dependent mutations to arise (Ponder et al. 2005; Shee et al. 2011a). We will see below that the molecular mechanism of point mutagenesis is that during RpoS-inducing stress, Pol IV, and other error-prone DNA pols are allowed to participate in DNA break repair and make mutations.

The mechanism of DSB/DSE-dependent stress-induced point mutation was suggested to be peculiar to F′ conjugative plasmids (e.g., Foster and Trimarchi 1995; Radicella et al. 1995; Galitski and Roth 1995; Roth et al. 2006). However, recently we have demonstrated that DSB-dependent stress-induced mutation occurs in starved plasmid-free cells (Shee et al. 2011a) (discussed below). Thus, the same DSB repair proteins, error-prone DNA polymerases SOS, and RpoS stress responses are required for DSB-dependent stress-induced mutation of chromosomal genes in plasmid-free cells (Shee et al. 2011a, 2012; Al Mamun et al. 2012) (discussed below).

DSBs Are Repaired by Homologous Recombination in *E. coli*

DSB-repair in *E. coli* occurs via RecA/RecBCD-mediated homologous recombination. The RecBCD enzyme loads onto DNA at double-strand ends and degrades both strands as exonuclease (Exo) V. RecBCD ceases double-strand degradation and creates single-strand DNA most often at Chi sites (5′ GCTGGTGG 3′) (Dillingham and Kowalczykowski 2008). Single-stranded ends created at Chi are then coated with RecA strand-exchange protein in preparation for recombination. The 3′ end of a RecA nucleoprotein filament invades a homologous DNA sequence (usually in a sister chromosome) to produce a heteroduplex recombination intermediate (Camerini-Otero and Hsieh 1993; Kowalczykowski and Eggleston 1994). Any DNA lost or degraded from the broken molecule is copied from the intact sister chromosome. After repair synthesis, the inter-molecular recombination intermediate is resolved to yield two intact DNA molecules, often by the Holiday-junction resolution proteins RuvABC (West 2003). RecA, RecBC, and RuvABC are all required for stress-induced point mutation (Harris et al. 1994; Foster et al. 1996; Harris et al. 1996). The high-fidelity major replicative DNA polymerase, DNA Pol III, is required for replicative repair of DSBs in unstressed *E. coli* cells (Motamedi et al. 1999), and, in unstressed cells, repair synthesis is non-mutagenic (Ponder et al. 2005; Shee et al. 2011a).

In addition to promoting recombination, the RecA-coated single-stranded DNA filament also activates the SOS DNA-damage response in about 25% of successful DSB repair events (Pennington and Rosenberg 2007). The RecA/ssDNA filament promotes auto-proteolysis of the LexA transcriptional repressor, which upregulates the transcription of about 40 SOS genes that function in DNA repair, DNA-damage tolerance, mutagenesis, and cell-cycle checkpoint control (Courcelle et al. 2001; Sutton et al. 2000).

RpoS Throws a Switch to Mutagenic Repair of Double-Strand Breaks

We showed that stress-induced point mutations are formed in acts of DNA break repair by creating *E. coli* cells that produce site-specific DNA breaks. We cloned the I-*Sce*I double-strand endonuclease under a bacterial regulatable promoter, and placed it into the *E. coli* chromosome (Gumbiner-Russo et al. 2001). Using this tool, we engineered Lac-assay cells to have a single DSB in their F′ near *lac* (Ponder et al. 2005), which could be repaired by homologous recombination with an uncleaved sister DNA molecule. We found that DSBs in the same molecule as *lac* produced 6,000-fold more Lac$^+$ point mutation; DSBs made in a different plasmid in the cell produced only threefold more Lac reversions. However, if we engineered the DNA near *lac* to carry a sequence identical to one end of the cleaved other plasmid, then DSBs in that other molecule also greatly stimulated Lac reversion during starvation (Ponder et al. 2005). These data show that the mutations occur during acts of DSB repair by homologous recombination. These I-*Sce*I-instigated mutations require DSB-repair proteins, Pol IV, and the SOS and RpoS responses to occur (Ponder et al. 2005), just as standard DSB-dependent stress-induced mutations do. They do not require RpoE, which indicated that the role of RpoE was in spontaneous DSB/DSE formation (Gibson et al. 2010).

Importantly, the repair of I-*Sce*I-induced DSBs is mutagenic only if the RpoS response is induced; that is, if cells sense another (non-DSB) stress. When the I-*Sce*I cuts were made and repaired in log-phase unstressed cells growing in liquid, mutations were not stimulated (this time using reversion of a *tet* frameshift allele as the mutation reporter) (Ponder et al. 2005; Shee et al. 2011a). If the cells were either allowed to go stationary, or if RpoS was upregulated artificially in the log-phase unstressed cells, then Pol IV-dependent mutagenesis occurred during repair of the I-*Sce*I-induced DSBs (Ponder et al. 2005; Shee et al. 2011a). Thus, the RpoS response throws a switch that causes the otherwise high-fidelity (non-mutagenic) process of DSB repair by homologous recombination to become mutagenic, using Pol IV. This means that even though Pol IV is present at ~2,500–5,000 molecules per cell after a DSB is made and induces the SOS response, it appears not to participate in DSB repair or cause DSB-dependent mutations unless the RpoS response is also activated (Ponder et al. 2005; Shee et al. 2011a).

Thus, put simply, the mutation mechanism requires three simultaneous events: (1) a DSB and its repair by homologous recombination; (2) induction of the SOS response, which DSBs induce, and which upregulates Pol IV; and (3) a second stress that activates the RpoS response, which licenses use of Pol IV and other error-prone DNA pols in DSB repair, causing mutations.

RpoS-Controlled Switch to Mutagenic Break Repair in Chromosomes of Plasmid-Free Cells

These findings hold true in the F′-based Lac assay (Ponder et al. 2005), and also when chromosomal mutations in plasmid-free cells were assayed (Shee et al. 2011a). We showed that chromosomal reversion of a *tet* +1 bp frameshift allele was stimulated 50- to 100-fold by nearby DSBs made by I-*Sce*I endonuclease. These mutations occur by the same DSB-dependent mutation pathway as in the F′, requiring DSB-repair proteins RecA, RecBCD and RuvABC, Pol IV error-prone DNA polymerase, the SOS response (which upregulates Pol IV), and the RpoS stress response (Shee et al. 2011a). This mutagenesis required either a prolonged stationary phase, or an artificial upregulation of RpoS in log-phase growing cells, again demonstrating the RpoS-controlled switch to mutagenic break repair (Shee et al. 2011a). These data put to rest previous concerns that DSB-dependent stress-induced mutation might be peculiar to plasmids (e.g., Foster and Trimarchi 1995; Radicella et al. 1995; Galitski and Roth 1995; Roth et al. 2006).

DNA Polymerases Used

Under stress, the RpoS response causes DSB repair to switch to a mutagenic mode using error-prone DNA polymerases, principally Pol IV (*dinB*), but also Pol II (*polB*) and Pol V (*umuDC*) (Ponder et al. 2005; Shee et al. 2011a). In the *E. coli* Lac assay, ~85% of stress-induced point mutation requires Pol IV (McKenzie et al. 2001) and the remaining ~15% requires Pol II (Frisch et al. 2010), either when I-*Sce*I-promoted (Ponder et al. 2005; Frisch et al. 2010), or when spontaneous (Frisch et al. 2010) stress-induced mutations are assayed. All DSB-dependent chromosomal *tet* reversion requires Pol IV (Shee et al. 2011a). Pol V is partially required for the Pol IV-dependent chromosomal *tet* frameshift reversions (Shee et al. 2011a), and for DSB-repair-protein-, SOS-, and RpoS-dependent forward mutations in a chromosomal *ampD* gene during starvation (Petrosino et al. 2009). Interestingly, DSB repair in unstressed cells requires high-fidelity DNA Pol III (Motamedi et al. 1999) and is non-mutagenic (Ponder et al. 2005; Shee et al. 2011a). We have suggested that the mechanism of the RpoS-controlled switch to mutagenic DSB repair might be downregulation of Pol III, which might then let the other error-prone DNA polymerases compete more effectively for a spot at the DSB repair replisome (Frisch et al. 2010; Rosenberg et al. 2012). Though, the other DNA polymerases compete with Pol IV during DSB repair under stress (Hastings et al. 2010), the hypothesis that RpoS promotes mutation by downregulation Pol III remains to be tested.

DSB-Dependent Stress-Induced Mutagenesis Produces Spontaneous Mutations

The RpoS-controlled switch to mutagenic DSB repair was demonstrated with artificially created DSBs made by I-*Sce*I (Ponder et al. 2005; Shee et al. 2011a). Importantly, we showed that half of spontaneous frameshift reversions and base-substitution mutations in the chromosomes of starved plasmid-free *E. coli* occur by the DSB-dependent stress-induced mutation pathway when no I-*Sce*I is present: the process requires DSB-repair proteins, RpoS, SOS, and Pol IV (Shee et al. 2011a). Thus, this mechanism is important to evolution. Apparently, without I-*Sce*I, the mutations occur during acts of repair of spontaneous DSBs/DSEs.

Mutagenesis Is Not an Unavoidable Consequence of DNA Break Repair

The idea that the upregulation of mutagenesis by a stress response would accelerate evolution, and that this might be selected for its evolution-enhancing ability, was suggested first for the SOS DNA-damage response by Radman (Radman 1975) then Echols (McPartland et al. 1980). However, because the SOS response upregulates DNA repair and damage-survival functions, from its original proposal until the present, others have argued that mutagenesis is an unavoidable consequence of repairing DNA. They argue that high-fidelity (non-mutagenic) DNA repair cannot evolve (e.g., Sniegowski et al. 2000; Erill et al. 2007; Andersson et al. 2010; Lynch 2010). This argument cannot be resolved in the context of the SOS response, because SOS *is* required for survival of DNA damage. However, our work in DNA break-dependent stress-induced mutation shows that survival of the break does not require mutagenesis; neither RpoS nor Pol IV, which throw the switch to mutagenic repair, is required for DSB survival, even under RpoS-inducing stress conditions (Ponder et al. 2005; Shee et al. 2011a). Thus, use of the error-prone DNA polymerase during repair synthesis is not an unavoidable consequence of repair. It is a regulated response that boosts mutagenesis, potentially accelerating evolution, and it remains possible that it might have been selected by that property.

Other Old Problems Resolved

Recent results with the chromosomal Tet assay for DSB-dependent stress-induced mutation resolve other old problems.

DSB-dependent stress-induced mutation was studied initially in *E. coli* cells carrying a specific F′ conjugative plasmid that carries an extra copy of *dinB* encoding Pol IV, prompting concerns that the mutation mechanism might be peculiar to the assay system (Foster and Trimarchi 1995; Radicella et al. 1995; Galitski and Roth 1995; Roth et al. 2006; Slechta et al. 2002a), conjugative plasmids, or the specific F′ used (Radicella et al. 1995; Slechta et al. 2003). Our recent work shows that DSB-dependent stress-induced mutagenesis occurs in the chromosome of starved, plasmid-free *E. coli*, and even accounts for half of spontaneous frameshift and base-substitution mutations there, putting this concern to rest (Shee et al. 2011a).

A related concern was that perhaps the stress of starvation selects, rather than induces, the mutations—a possibility in the Lac assay in which Lac^+ reversions were selected during starvation on lactose plates. When Lac^+ mutants are selected, spontaneous gene amplifications of the *lac* region could be selected, causing multiple *lac* copies, and increasing mutation rate per cell by simply increasing copy number (Roth et al. 2006; Slechta et al. 2003; Roth 2010). A preexisting *lac* gene duplication might undergo amplification, allowing slow growth by production of beta-galactosidase from the weakly functional *lac* gene, and increased *lac* mutation per cell. This concern was addressed by use of the Tet assay in the F′ (Ponder et al. 2005) or chromosome (Shee et al. 2011a), in which cells are starved without selection for function of the defective *tet* gene, then assayed for Tet^R mutations after rescue from starvation. In the Tet assay, mutagenesis occurred by the same DSB-repair-protein-, SOS-, RpoS-, and Pol IV-dependent mechanism (Ponder et al. 2005; Shee et al. 2011a), and the amount of mutagenesis related to the length of time the cells were starved (Shee et al. 2011a). Thus, DSB-dependent stress-induced mutation occurs independently of selection for the function of mutated gene, so that selected amplification of the target gene can be ruled out as a component of the mechanism.

Duplications Are Attractive

Although the specific amplification-selection-mutation model discussed above is not supported, spontaneous gene duplications may nevertheless be important. Duplications are an attractive solution to the problem of which segment of DNA is used for repair of a chromosomal DSB during homologous recombination in starving, haploid *E. coli* (Shee et al. 2011b; Rosenberg et al. 2012). It could be that spontaneously duplicated segments are the source of homology for repair by recombination in chromosomes of starving cells.

Hypermutable Cell Subpopulation

We and others found that during DSB-dependent stress-induced mutation, the cells with Lac^+ mutations have more mutations in other genes in their genome than Lac^- starved cells taken from the same petri plates (Torkelson et al. 1997; Rosche and Foster 1999; Godoy et al. 2000; Slechta et al. 2002b). This indicates that a cell subpopulation has an increased mutation rate in unselected chromosomal genes: that there is hypermutable cell subpopulation (HMS). Further, those Lac^+ mutants with the additional, unselected "secondary" mutations did not have heritably higher mutation rates once the Lac^+ colonies were picked. The HMS appears to be a transiently differentiated subpopulation, not rare hypermutator mutants. Secondary mutations are increased proportionately in hyper-recombining DSB repair mutants *recD* and *recG,* which also increase Lac^+ mutation, implying that they form by a similar DSB-repair-protein-dependent mechanism (Bull et al. 2000).

These results ended a previous Lamarck versus Darwin debate for stress-induced mutation (Bridges 1997), but are now the focus of a different problem. Those who favor models of constant, gradual evolutionary change dislike the idea of stress-induced increase in mutation rate that potentially accelerates evolution (Chicurel 2001). They, and some others (Rosche and Foster 1999), suggest that although the HMS exists, it is not important: that most Lac^+ adaptive mutants arise from cells not in the HMS—specifically that 90% do not, and only 10% of Lac^+ mutants do arise from the HMS (Slechta et al. 2003; Slechta et al. 2002b; Roth and Andersson 2004; Hendrickson et al. 2002). In contrast, the following evidence supports the hypothesis that the HMS generates most stress-induced mutants. First, the sequences of Lac reversions from cells demonstrably from the HMS (with a secondary mutation) are identical to the majority of Lac^+ stress-induced mutations, indicating that both arise from a similar mechanism (Gonzalez et al. 2008). Second, when the dominant DSB-dependent mutation mechanism is increased by providing I-*Sce*I cuts, the associated secondary mutations increase proportionately (Gonzalez et al. 2008). This shows that the HMS cannot be uncoupled from the major stress-induced mutation pathway. In the future, possible isolation of the HMS during mutagenesis may be possible, and could provide direct evidence concerning this point.

The discovery of a Pol IV-dependent, but DSB-independent, stress-induced mutation mechanism activated by the ComK-controlled competence stress response in *Bacillus subtilis* (Sung and Yasbin 2002) suggests that a different bacterium might also create a transiently differentiated HMS. The competence response is famously induced in only a subpopulation of starving cells, which it makes competent for uptake of DNA from the environment, allowing natural transformation (Dubnau and Losick 2006). Thus this stress-induced mutation mechanism seems likely to occur in a transiently differentiated hypermutating subpopulation.

DSBs and Mutation Hotspots

Mutation hotspots, clusters, and showers are apparent in cases ranging from phage to human (Drake 2007a; Drake 2007b; Caporale 2006; Wang et al. 2007), but the mechanisms that form hotspots are unknown. We and others have suggested that a consequence of linking mutagenesis to DSB repair might be that mutations could be formed in localized hotspots, near the sites of repair (Galhardo et al. 2007; Ponder et al. 2005; Ninio 1996; Yang et al. 2008). Mutational hotspotting is important because it could promote evolution, including evolution of tumors and pathogens: first, by potentially targeting regions in which variability might provide a growth advantage, as occurs with somatic hypermutation of immunoglobulin genes (Di Noia and Neuberger 2007) and pathogen "contingency" genes (Moxon et al. 1994), and, as is seen in the cancer-driving Philadelphia chromosome (Albano et al. 2010). Additionally, hotspot formation could allow high-level mutagenesis that can promote concerted evolution (multiple mutations) within genes or gene clusters, without accumulation of deleterious mutations throughout the genome (Ponder et al. 2005; Ninio 1996; Yang et al. 2008). Recent data showing mutation clusters in human cancers have been interpreted in support of the hypothesis that DSBs cause hotspots (Nik-Zainal et al. 2012; Roberts et al. 2012). However, we, and others (Nik-Zainal et al. 2012), note that this is not demonstrated, and other sources of the hotspots are possible. DSB-dependent mutation was discovered in *E. coli* (Rosenberg et al. 1994; Harris et al. 1994), then demonstrated in yeast (Yang et al. 2008; Strathern et al. 1995; Deem et al. 2011; Hicks et al. 2010), in which it is not known to be stress-inducible, and might be constitutive. Until recently, no experiments in either organism had shown whether or not the strong stimulation of mutagenesis near DSB sites also occurs distantly from the DSB in the same molecule. Hotspots would be produced only if the mutagenesis were localized. In the sole yeast study to address this point, mutations occurred equally well next to and 36 kb away from a DSB, and greater distances were not assayed (Deem et al. 2011). Further, these yeast experiments were done in conditions under which long-distance replication was demanded, which might make them unusual. In *E. coli*, we have recently demonstrated that DSBs provoke both strong local mutation hotspots like in cancers (Nik-Zainal et al. 2012; Roberts et al. 2012) and weak long-distance hotspots (Shee et al. 2012). This work provides a plausible molecular mechanism for hotspotting, an important force in genome evolution.

Cancer Cells Display a Switch to Mutagenic Break Repair Under Stress

Cancer is an evolutionary process, driven by mutations that fuel oncogenesis, tumor progression, and development of resistance to chemotherapies. Human cancer cells display a stress-induced mutation mechanism somewhat analogous to DSB-dependent stress-induced mutagenesis in *E. coli*. In *E. coli*, the RpoS stress response throws a switch from high fidelity to mutagenic repair of DSBs, causing mutations preferentially under stress (Ponder et al. 2005; Shee et al. 2011a). In cancer cells, hypoxic stress activates hypoxic stress responses, which have been shown to downregulate the *BRCA1* and *RAD51* genes required for DSB repair by homologous recombination (HR) (Bindra et al. 2005; Bindra et al. 2004). In humans, non-homologous end-joining (NHEJ) pathways take over when HR is not operative. NHEJ can cause genomic rearrangements. Thus, downregulation of *BRCA1* and *RAD51* is expected to cause a switch from genome-stabilizing repair by HR in unstressed cells to genome-rearranging end-joining under hypoxic stress (Bindra et al. 2005; Bindra et al. 2004). Though not the same mechanism as in *E. coli*, it is analogous in using a switch from high-fidelity to mutagenic repair of DNA breaks under stress.

Antibiotic Resistance

Mutation is a major route to resistance to antibiotics, an important and urgent clinical problem (Davies and Davies 2010). Recent work has shown that antibiotics themselves are stressors (Davies and Davies 2010; Kohanski et al. 2007; Miller et al. 2004) that induce mutagenesis, causing mutations that confer resistance to the same antibiotic (Cirz et al. 2005), and to different antibiotics, or in other genes (Perez-Capilla et al. 2005; Kohanski et al. 2010). In two instances, this mutagenesis appears to occur by mechanisms similar to DSB-dependent stress induced mutation. Mutagenesis induced by the fluoroquinolone antibiotic ciprofloxacin (cipro) causes cipro resistance via a mutation pathway that requires DSB-repair proteins, the SOS response, and all three SOS DNA polymerases (including Pol IV) (Cirz et al. 2005). Whether RpoS is also required is not known, but this implicates DSB-dependent mutagenesis. In pathogenic Salmonella, bile-resistance mutations are induced by exposure to bile (an antibacterial agent produced in our bodies), and this mutation pathway also requires DSB-repair proteins, Pol IV, the SOS response, and, additionally, RpoS (Prieto et al. 2006; J Casadesus Pers. commun.). Bile-induced mutagenesis bears all the hallmarks of DSB-dependent stress-induced mutagenesis, and in this case, the stressor appears to be bile, a membrane-disrupting agent. Understanding the molecular mechanisms by which antibiotics provoke mutation is critical to combating resistance. We have suggested that the development of new drugs that would block the induced mutagenesis would block the evolution of resistance (e.g., Rosenberg et al. 2012; Al Mamun et al. 2012).

Conclusion

Stress-induced mutation mechanisms appear to be both ubiquitous and important. Of 787 natural isolates of *E. coli* collected world-wide, about 80% showed induction of mutation under the generic laboratory stress of aging in a colony

(Bjedov et al. 2003). This ability appears to have been selected in that it correlates with ecological niche, as expected for traits under recent selection, implying that stress-induced mutation mechanisms are important to bacterial evolution. Regarding the more general question of the relevance of stress-induced mutation mechanisms as a whole, it seems clear that mutability induced by and allowing growth under stress is an important departure from the classical Luria/Delbrück paradigm (Luria and Delbrück 1943), and that many systems previously modeled on Luria/Delbrück principles may now need to be rethought, in the light of inducible genetic change mechanisms that might provide more appropriate models. Our understanding of evolution needs to move into the molecular age. Understanding stress-induced mutation mechanisms will propel treatment and prevention options for evolution-based diseases such as cancer and infectious disease, and is critical to a realistic picture of the molecular mechanisms of evolution.

Acknowledgment Supported by National Institutes of Health grant R01-GM53158.

References

Al Mamun AA, Lombardo MJ, Shee C, Lisewski AM, Gonzalez C, Lin D, Nehring R, Saint-Ruff C, Gibson GL, Frisch RL et al (2012) Identify and function of a large gene network underlying mutagenic repair of DNA breaks. Science 338:1344–1348

Albano F, Anelli L, Zagaria A, Coccaro N, Casieri P, Rossi AR, Vicari L, Liso V, Rocchi M, Specchia G (2010) Non random distribution of genomic features in breakpoint regions involved in chronic myeloid leukemia cases with variant t(9;22) or additional chromosomal rearrangements. Mol Cancer 9:120

Andersson DI, Koskiniemi S, Hughes D (2010) Biological roles of translesion synthesis DNA polymerases in eubacteria. Mol Microbiol 77:540–548

Banach-Orlowska M, Fijalkowska IJ, Schaaper RM, Jonczyk P (2005) DNA polymerase II as a fidelity factor in chromosomal DNA synthesis in *Escherichia coli*. Mol Microbiol 58:61–70

Battesti A, Majdalani N, Gottesman S (2011) The RpoS-mediated general stress response in *Escherichia coli*. Annu Rev Microbiol 65:189–213

Bindra RS, Schaffer PJ, Meng A, Woo J, Maseide K, Roth ME, Lizardi P, Hedley DW, Bristow RG, Glazer PM (2004) Down-regulation of Rad51 and decreased homologous recombination in hypoxic cancer cells. Mol Cell Biol 24:8504–8518

Bindra RS, Gibson SL, Meng A, Westermark U, Jasin M, Pierce AJ, Bristow RG, Classon MK, Glazer PM (2005) Hypoxia-induced down-regulation of BRCA1 expression by E2Fs. Cancer Res 65:11597–11604

Bindra RS, Crosby ME, Glazer PM (2007) Regulation of DNA repair in hypoxic cancer cells. Cancer Metastasis Rev 26:249–260

Bjedov I, Tenaillon O, Gerard B, Souza V, Denamur E, Radman M, Taddei F, Matic I (2003) Stress-induced mutagenesis in bacteria. Science 300:1404–1409

Bridges BA (1997) Microbial genetics. Hypermutation under stress. Nature 387:557–558

Bull HJ, McKenzie GJ, Hastings PJ, Rosenberg SM (2000) Evidence that stationary-phase hypermutation in the *Escherichia coli* chromosome is promoted by recombination. Genetics 154:1427–1437

Cairns J, Foster PL (1991) Adaptive reversion of a frameshift mutation in *Escherichia coli*. Genetics 128:695–701

Camerini-Otero RD, Hsieh P (1993) Parallel DNA triplexes, homologous recombination, and other homology-dependent DNA interactions. Cell 73:217–223
Caporale LH (2006) The implicit genome. Oxford University Press, USA
Casadesus. J Pers Commun.
Chicurel M (2001) Genetics. Can organisms speed their own evolution? Science 292:1824–1827
Cirz RT, Romesberg FE (2007) Controlling mutation: intervening in evolution as a therapeutic strategy. Crit Rev Biochem Mol Biol 42:341–354
Cirz RT, Chin JK, Andes DR, de Crecy-Lagard V, Craig WA, Romesberg FE (2005) Inhibition of mutation and combating the evolution of antibiotic resistance. PLoS Biol 3:e176
Courcelle J, Khodursky A, Peter B, Brown PO, Hanawalt PC (2001) Comparative gene expression profiles following UV exposure in wild-type and SOS-deficient *Escherichia coli*. Genetics 158:41–64
Davies J, Davies D (2010) Origins and evolution of antibiotic resistance. Microbiol Mol Biol Rev 74:417–433
Deem A, Keszthelyi A, Blackgrove T, Vayl A, Coffey B, Mathur R, Chabes A, Malkova A (2011) Break-induced replication is highly inaccurate. PLoS Biol 9:e1000594
Di Noia JM, Neuberger MS (2007) Molecular mechanisms of antibody somatic hypermutation. Annu Rev Biochem 76:1–22
Dillingham MS, Kowalczykowski SC (2008) RecBCD enzyme and the repair of double-stranded DNA breaks. Microbiol Mol Biol Rev 72:642–671
Drake JW (2007a) Mutations in clusters and showers. Proc Natl Acad Sci USA 104:8203–8204
Drake JW (2007b) Too many mutants with multiple mutations. Crit Rev Biochem Mol Biol 42:247–258
Dubnau D, Losick R (2006) Bistability in bacteria. Mol Microbiol 61:564–572
Erill I, Campoy S, Barbe J (2007) Aeons of distress: an evolutionary perspective on the bacterial SOS response. FEMS Microbiol Rev 31:637–656
Fonville NC, Ward RM, Mittelman D (2011) Stress-induced modulators of repeat instability and genome evolution. J Mol Microbiol Biotechnol 21:36–44
Foster PL, Trimarchi JM (1994) Adaptive reversion of a frameshift mutation in *Escherichia coli* by simple base deletions in homopolymeric runs. Science 265:407–409
Foster PL, Trimarchi JM (1995) Adaptive reversion of an episomal frameshift mutation in *Escherichia coli* requires conjugal functions but not actual conjugation. Proc Natl Acad Sci USA 92:5487–5490
Foster PL, Trimarchi JM, Maurer RA (1996) Two enzymes, both of which process recombination intermediates, have opposite effects on adaptive mutation in *Escherichia coli*. Genetics 142:25–37
Friedberg EC, Walker GC, Siede W, Wood RD, Schultz RA, Ellenberger T (2005) DNA repair and mutagenesis. ASM Press, Washington, DC
Frisch RL, Su Y, Thornton PC, Gibson JL, Rosenberg SM, Hastings PJ (2010) Separate DNA Pol II- and Pol IV-dependent pathways of stress-induced mutation during double-strand-break repair in *Escherichia coli* are controlled by RpoS. J Bacteriol 192:4694–4700
Galhardo RS, Hastings PJ, Rosenberg SM (2007) Mutation as a stress response and the regulation of evolvability. Crit Rev Biochem Mol Biol 42:399–435
Galhardo RS, Do R, Yamada M, Friedberg E, Hastings P, Nohmi T, Rosenberg S (2009) DinB up-regulation is the sole role of the SOS response in stress-induced mutagenesis in *Escherichia coli*. Genetics 182:55–68
Galitski T, Roth JR (1995) Evidence that F plasmid transfer replication underlies apparent adaptive mutation. Science 268:421–423
Gibson JL, Lombardo MJ, Thornton PC, Hu KH, Galhardo RS, Beadle B, Habib A, Magner DB, Frost LS, Herman C et al (2010) The sigma(E) stress response is required for stress-induced mutation and amplification in *Escherichia coli*. Mol Microbiol 77:415–430
Godoy VG, Gizatullin FS, Fox MS (2000) Some features of the mutability of bacteria during non-lethal selection. Genetics 154:49–59

Gonzalez CP, Hadany L, Ponder RG, Price M, Hastings PJ, Rosenberg SM (2008) Mutability and importance of a hypermutable cell subpopulation that produces stress-induced mutants in *Escherichia coli*. PLoS Genet 4:e10000208

Gumbiner-Russo LM, Lombardo M-J, Ponder RG, Rosenberg SM (2001) The TGV *t*ransgenic *v*ectors for single copy gene expression in the *Escherichia coli* chromosome. Gene 273:97–104

Haber JE (1999) DNA repair. Gatekeepers of recombination. Nature 398:665–667

Harris RS, Longerich S, Rosenberg SM (1994) Recombination in adaptive mutation. Science 264:258–260

Harris RS, Ross KJ, Rosenberg SM (1996) Opposing roles of the holliday junction processing systems of *Escherichia coli* in recombination-dependent adaptive mutation. Genetics 142: 681–691

Hastings PJ (2007) Adaptive amplification. Crit Rev Biochem Mol Biol 42:271–273

Hastings PJ, Bull HJ, Klump JR, Rosenberg SM (2000) Adaptive amplification: an inducible chromosomal instability mechanism. Cell 103:723–731

Hastings PJ, Ira G, Lupski JR (2009a) A microhomology-mediated break-induced replication model for the origin of human copy number variation. PLoS Genet 5:e1000327

Hastings PJ, Lupski JR, Rosenberg SM, Ira G (2009b) Mechanisms of change in gene copy number. Nat Rev Genet 10:551–564

Hastings PJ, Hersh MN, Thornton PC, Fonville NC, Slack A, Frisch RL, Ray MP, Harris RS, Leal SM, Rosenberg SM (2010) Competition of *Escherichia coli* DNA polymerases I, II and III with DNA Pol IV in stressed cells. PLoS One 5:e10862

Hefferin ML, Tomkinson AE (2005) Mechanism of DNA double-strand break repair by non-homologous end joining. DNA Repair (Amst) 4:639–648

Heidenreich E (2007) Adaptive mutation in *Saccharomyces cerevisiae*. Crit Rev Biochem Mol Biol 42:285–311

Hendrickson H, Slechta ES, Bergthorsson U, Andersson DI, Roth JR (2002) Amplification-mutagenesis: evidence that "directed" adaptive mutation and general hypermutability result from growth with a selected gene amplification. Proc Natl Acad Sci USA 99:2164–2169

Hicks WM, Kim M, Haber JE (2010) Increased mutagenesis and unique mutation signature associated with mitotic gene conversion. Science 329:82–85

Hoeijmakers JH (2001) Genome maintenance mechanisms for preventing cancer. Nature 411:366–374

Kenyon CJ, Walker GC (1980) DNA-damaging agents stimulate gene expression at specific loci in *Escherichia coli*. Proc Natl Acad Sci USA 77:2819–2823

Kim SR, Maenhaut-Michel G, Yamada M, Yamamoto Y, Matsui K, Sofuni T, Nohmi T, Ohmori H (1997) Multiple pathways for SOS-induced mutagenesis in *Escherichia coli*: an overexpression of *dinB*/*dinP* results in strongly enhancing mutagenesis in the absence of any exogenous treatment to damage DNA. Proc Natl Acad Sci USA 94:13792–13797

Kimura M (1991) The neutral theory of molecular evolution: a review of recent evidence. Jpn J Genet 66:367–386

Kohanski MA, Dwyer DJ, Hayete B, Lawrence CA, Collins JJ (2007) A common mechanism of cellular death induced by bactericidal antibiotics. Cell 130:797–810

Kohanski MA, DePristo MA, Collins JJ (2010) Sublethal antibiotic treatment leads to multidrug resistance via radical-induced mutagenesis. Mol Cell 37:311–320

Kowalczykowski SC, Eggleston AK (1994) Homologous pairing and DNA strand-exchange proteins. Annu Rev Biochem 63:991–1043

Layton JC, Foster PL (2003) Error-prone DNA polymerase IV is controlled by the stress-response sigma factor, RpoS, in *Escherichia coli*. Mol Microbiol 50:549–561

Lombardo MJ, Aponyi I, Rosenberg SM (2004) General stress response regulator RpoS in adaptive mutation and amplification in *Escherichia coli*. Genetics 166:669–680

Luria SE, Delbrück M (1943) Mutations of bacteria from virus sensitivity to virus resistance. Genetics 28:491–511

Lynch M (2010) Evolution of the mutation rate. Trends Genet 26:345–352

Mayr E (1982) The Growth of Biological Thought: Diversity, Evolution, and Inheritance. Harvard University Press, Cambridge, MA, pp 738–744

McKenzie GJ, Lombardo MJ, Rosenberg SM (1998) Recombination-dependent mutation in *Escherichia coli* occurs in stationary phase. Genetics 149:1163–1165

McKenzie GJ, Harris RS, Lee PL, Rosenberg SM (2000) The SOS response regulates adaptive mutation. Proc Natl Acad Sci USA 97:6646–6651

McKenzie GJ, Lee PL, Lombardo MJ, Hastings PJ, Rosenberg SM (2001) SOS mutator DNA polymerase IV functions in adaptive mutation and not adaptive amplification. Mol Cell 7:571–579

McPartland A, Green L, Echols H (1980) Control of recA gene RNA in E. coli: regulatory and signal genes. Cell 20:731–737

Meulle A, Salles B, Daviaud D, Valet P, Muller C (2008) Positive regulation of DNA double strand break repair activity during differentiation of long life span cells: the example of adipogenesis. PLoS One 3:e3345

Miller C, Thomsen LE, Gaggero C, Mosseri R, Ingmer H, Cohen SN (2004) SOS response induction by beta-lactams and bacterial defense against antibiotic lethality. Science 305:1629–1631

Motamedi MR, Szigety SK, Rosenberg SM (1999) Double-strand-break repair recombination in *Escherichia coli*: physical evidence for a DNA replication mechanism in vivo. Genes Dev 13:2889–2903

Moxon ER, Rainey PB, Nowak MA, Lenski RE (1994) Adaptive evolution of highly mutable loci in pathogenic bacteria. Curr Biol 4:24–33

Nik-Zainal S, Alexandrov LB, Wedge DC, Van Loo P, Greenman CD, Raine K, Jones D, Hinton J, Marshall J, Stebbings LA et al (2012) Mutational processes molding the genomes of 21 breast cancers. Cell 149:979–993

Ninio J (1996) Gene conversion as a focusing mechanism for correlated mutations: a hypothesis. Mol Gen Genet 251:503–508

Nohmi T (2006) Environmental stress and lesion-bypass DNA polymerases. Annu Rev Microbiol 60:231–253

Ohmori H, Friedberg EC, Fuchs RP, Goodman MF, Hanaoka F, Hinkle D, Kunkel TA, Lawrence CW, Livneh Z, Nohmi T et al (2001) The Y-family of DNA polymerases. Mol Cell 8:7–8

Pennington JM, Rosenberg SM (2007) Spontaneous DNA breakage in single living *Escherichia coli* cells. Nat Genet 39:797–802

Perez-Capilla T, Baquero MR, Gomez-Gomez JM, Ionel A, Martin S, Blazquez J (2005) SOS-independent induction of *dinB* transcription by beta-lactam-mediated inhibition of cell wall synthesis in *Escherichia coli*. J Bacteriol 187:1515–1518

Petrosino JF, Galhardo RS, Morales LD, Rosenberg SM (2009) Stress-induced beta-lactam antibiotic resistance mutation and sequences of stationary-phase mutations in the *Escherichia coli* chromosome. J Bacteriol 191:5881–5889

Ponder RG, Fonville NC, Rosenberg SM (2005) A switch from high-fidelity to error-prone DNA double-strand break repair underlies stress-induced mutation. Mol Cell 19:791–804

Prieto AI, Ramos-Morales F, Casadesus J (2006) Repair of DNA damage induced by bile salts in *Salmonella enterica*. Genetics 174:575–584

Radicella JP, Park PU, Fox MS (1995) Adaptive mutation in *Escherichia coli*: a role for conjugation. Science 268:418–420

Radman M (1975) SOS repair hypothesis: phenomenology of an inducible DNA repair which is accompanied by mutagenesis. Basic Life Sci 5A:355–367

Roberts SA, Sterling J, Thompson C, Harris S, Mav D, Shah R, Klimczak LJ, Kryukov GV, Malc E, Mieczkowski PA et al (2012) Clustered mutations in yeast and in human cancers can arise from damaged long single-strand DNA regions. Mol Cell 46:424–435

Rosche WA, Foster PL (1999) The role of transient hypermutators in adaptive mutation in *Escherichia coli*. Proc Natl Acad Sci USA 96:6862–6867

Rosenberg SM, Longerich S, Gee P, Harris RS (1994) Adaptive mutation by deletions in small mononucleotide repeats. Science 265:405–407

Rosenberg SM, Shee C, Frisch RL, Hastings PJ (2012) Stress-induced mutation via DNA breaks in *E. coli:* a molecular mechanism with implications for evolution and medicine. Bioessays 34:885–892

Roth JR (2010) The joys and terrors of fast adaptation: new findings elucidate antibiotic resistance and natural selection. Mol Microbiol 79:279–282

Roth JR, Andersson DI (2004) Adaptive mutation: how growth under selection stimulates Lac(+) reversion by increasing target copy number. J Bacteriol 186:4855–4860

Roth JR, Kugelberg E, Reams AB, Kofoid E, Andersson DI (2006) Origin of mutations under selection: the adaptive mutation controversy. Annu Rev Microbiol 60:477–501

Saint-Ruf C, Pesut J, Sopta M, Matic I (2007) Causes and consequences of DNA repair activity modulation during stationary phase in *Escherichia coli.* Crit Rev Biochem Mol Biol 42: 259–270

Shee C, Gibson JL, Darrow MC, Gonzalez C, Rosenberg SM (2011a) Impact of a stress-inducible switch to mutagenic repair of DNA breaks on mutation in *Escherichia coli.* Proc Natl Acad Sci USA 108:13659–13664

Shee C, Gibson JL, Rosenberg SM (2012) Two mechanisms produce mutation hotspots at DNA breaks in *Escherichia coli.* Cell Reports 2:714–721

Shee C, Ponder R, Gibson JL, Rosenberg SM (2011b) What limits the efficiency of double-strand break-dependent stress-induced mutation in *Escherichia coli*? J Mol Microbiol Biotechnol 21:8–19

Slack A, Thornton PC, Magner DB, Rosenberg SM, Hastings PJ (2006) On the mechanism of gene amplification induced under stress in *Escherichia coli.* PLoS Genet 2:e48

Slechta ES, Harold J, Andersson DI, Roth JR (2002a) The effect of genomic position on reversion of a *lac* frameshift mutation *(lacIZ33)* during non-lethal selection (adaptive mutation). Mol Microbiol 44:1017–1032

Slechta ES, Liu J, Andersson DI, Roth JR (2002b) Evidence that selected amplification of a bacterial *lac* frameshift allele stimulates Lac(+) reversion (adaptive mutation) with or without general hypermutability. Genetics 161:945–956

Slechta ES, Bunny KL, Kugelberg E, Kofoid E, Andersson DI, Roth JR (2003) Adaptive mutation: general mutagenesis is not a programmed response to stress but results from rare coamplification of *dinB* with *lac*. Proc Natl Acad Sci USA 100:12847–12852

Sniegowski PD, Gerrish PJ, Johnson T, Shaver A (2000) The evolution of mutation rates: separating causes from consequences. Bioessays 22:1057–1066

Strathern JN, Shafer BK, McGill CB (1995) DNA synthesis errors associated with double-strand-break repair. Genetics 140:965–972

Sung HM, Yasbin RE (2002) Adaptive, or stationary-phase, mutagenesis, a component of bacterial differentiation in *Bacillus subtilis.* J Bacteriol 184:5641–5653

Sutton MD, Smith BT, Godoy VG, Walker GC (2000) The SOS response: recent insights into *umuDC*-dependent mutagenesis and DNA damage tolerance. Annu Rev Genet 34:479–497

Tiligada E (2006) Chemotherapy: induction of stress responses. Endocr Relat Cancer 13(Suppl 1): S115–124

Tiligada E, Miligkos V, Delitheos A (2002) Cross-talk between cellular stress, cell cycle and anticancer agents: mechanistic aspects. Curr Med Chem Anticancer Agents 2:553–566

Torkelson J, Harris RS, Lombardo MJ, Nagendran J, Thulin C, Rosenberg SM (1997) Genome-wide hypermutation in a subpopulation of stationary-phase cells underlies recombination-dependent adaptive mutation. EMBO J 16:3303–3311

Wagner J, Gruz P, Kim SR, Yamada M, Matsui K, Fuchs RP, Nohmi T (1999) The dinB gene encodes a novel *E. coli* DNA polymerase, DNA pol IV, involved in mutagenesis. Mol. Cell 4:281–286

Wang J, Gonzalez KD, Scaringe WA, Tsai K, Liu N, Gu D, Li W, Hill KA, Sommer SS (2007) Evidence for mutation showers. Proc Natl Acad Sci USA 104:8403–8408

Weber H, Polen T, Heuveling J, Wendisch VF, Hengge R (2005) Genome-wide analysis of the general stress response network in *Escherichia coli*: sigmaS-dependent genes, promoters, and sigma factor selectivity. J Bacteriol 187:1591–1603

West SC (2003) Molecular views of recombination proteins and their control. Nat Rev Mol Cell Biol 4:435–445

Weterings E, van Gent DC (2004) The mechanism of non-homologous end-joining: a synopsis of synapsis. DNA Repair (Amst) 3:1425–1435

Wyman C, Kanaar R (2004) Homologous recombination: down to the wire. Curr Biol 14: R629–631

Yang Y, Sterling J, Storici F, Resnick MA, Gordenin DA (2008) Hypermutability of damaged single-strand DNA formed at double-strand breaks and uncapped telomeres in yeast *Saccharomyces cerevisiae*. PLoS Genet 4:e1000264

Yao Y, Kovalchuk I (2011) Abiotic stress leads to somatic and heritable changes in homologous recombination frequency, point mutation frequency and microsatellite stability in Arabidopsis plants. Mutat Res 707:61–66

Chapter 3
Transcription-Mediated Mutagenic Processes

Eduardo Robleto, Holly A. Martin, Carmen Vallin, Mario Pedraza-Reyes, and Ronald Yasbin

Abstract During growth, DNA is stable, easily repaired, replicated with high fidelity, and stably passed from generation to generation. Cellular division and growth require that DNA be replicated, transcribed, and translated. For several decades, the study of DNA metabolism has uncovered a vast number of mechanisms that generate mutations and promote evolution. While all DNA in a cell has to be replicated, the act of transcription is reduced to a subset of genes required for essential growth, or occurs in response to an environmental cue. Transcripts are then processed or translated to functional entities, which allow cell growth and adaptation to beneficial or detrimental conditions. While this view limits RNA to transient functions in the cell, our understanding of RNA metabolism in recent years has progressed from being a mere informational intermediate step in the central dogma to becoming a factor that mediates the production of genetic diversity, which modifies the genetic makeup of future generations and accelerates evolution. Here, we briefly review how gene transcriptional activation is coupled with mutagenic pathways; these processes are novel and improve our understanding of the evolutionary process.

E. Robleto (✉) • H.A. Martin • C. Vallin
School of Life Sciences, University of Nevada, Las Vegas, NV, USA
e-mail: eduardo.robleto@unlv.edu

M. Pedraza-Reyes
Division of Natural and Exact Sciences, Department of Biology, University of Guanajuato, Guanajuato, Mexico

R. Yasbin
College of Arts and Sciences, University of Missouri, St. Louis, MO, USA

D. Mittelman (ed.), *Stress-Induced Mutagenesis*, DOI 10.1007/978-1-4614-6280-4_3,

Introduction

The mechanisms underlying mutagenic events associated with transcriptional de-repression or transcriptional activation are reviewed here. (Transcription of a gene is repressed when an element—a repressor—prevents the RNA polymerase (RNAP) from synthesizing a full mRNA. A gene becomes de-repressed—i.e., transcribed—when repression is relieved. De-repression is usually mediated when the repressor element binds a corepressor or activator molecule.) We focus our attention to three different models that explain how the process of transcription may promote genetic diversity. The mechanisms discussed here are well conserved and have been implicated in generating beneficial and detrimental mutations in both bacteria and eukaryotes. Hence, the act of transcription should be considered not only as an intermediate step that provides immediate adaptation to a cell, but also as an element that can contribute to and change the genetic makeup of future generations.

One of the first experiments providing evidence that the process of transcription was facilitating the formation of mutations was conducted in *E. coli* in the early 1970s. This research detected mutations in the β-galactosidase gene (Brock 1971; Herman and Dworkin 1971). The experiments conducted by Herman and Dworkin measured reversion of a defective frameshift *lacZ* allele in the presence and absence of the transcriptional inducer IPTG, in combination with different levels of ICR-191, a frame-shift inducing mutagen. Those experiments showed a twofold increase in the rate of occurrence of Lac^+ reversion in the presence of IPTG when cultures were exposed to ICR-191. Since deficiencies in RecA did not affect the Lac^+ reversion frequency in the strain used, the influence of recombination was ruled out. Advances in the molecular mechanisms underlying the generation of mutations incited by the act of transcription point to three major pathways, which may operate separately or in combination (Fig. 3.1). First, the stress imposed by the act of transcription on DNA mediates the formation of single-stranded DNA, or secondary structures that are mutagenic. Second, conflicts with replication forks engender genome instability. Third, the events that follow the repair of transcribed genes produce mutations. In order to gain insight into these potential mechanisms, some of the experiments in selected model organisms that support transcription-mediated mutagenesis are discussed, as is each of the models in Fig. 3.1.

When appropriate, models presented in Fig. 3.1 are discussed as cellular strategies that allow cells to escape stressful conditions. Evolutionarily speaking, cells that evolve mechanisms to bias mutagenic events towards genetic regions that are under prolonged selective pressure have an advantage over cells that undergo rampant hypermutagenic states. General hypermutation is more likely to generate mutations at essential genes, and, consequently, lethal events (Galhardo et al. 2007; Robleto et al. 2007). This concept is illustrated in Fig. 3.2: under stressful conditions, cells display transcriptional activation of specific regulons that alleviate cellular stress. Prolonged stress then places sustained transcriptional activation on stress-related genes, which then mediates mutagenic events at stress-defined genetic regions. The elements of stress and gene activation limit increases in mutagenesis in time and

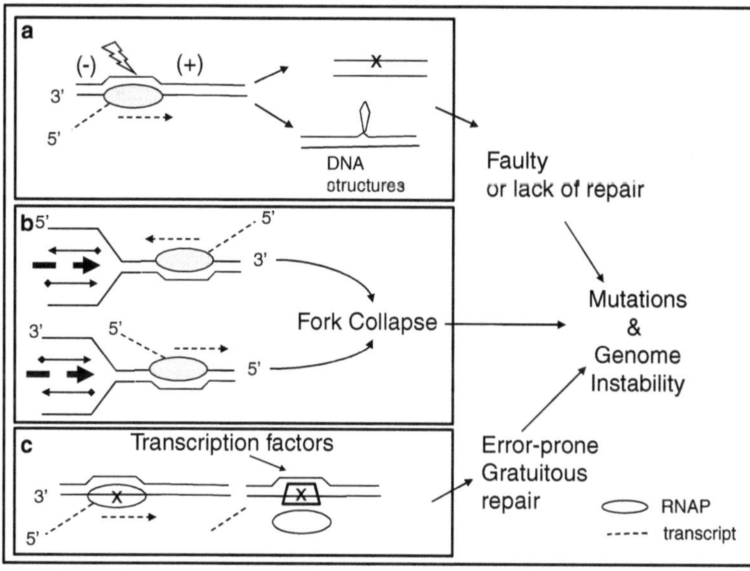

Fig. 3.1 Transcription-mediated mutagenic pathways. (**a**) The stress of transcription generates single-stranded or differential supercoiling (denoted by "+" and "–"), which increases exposure to solvent or mediates the formation of structures that make DNA more prone to lesions (lightning bolt) than double-stranded DNA. (**b**) Conflicts between transcription and replication cause mutagenic events or genome instability. (**c**) Error-prone or gratuitous repair at transcribed genes caused by stalling or lesions (x)

genomic space, which may increase the likelihood of acquiring a beneficial mutation compared to cellular states in which mutation occurs with no bias. A consequence of transcriptionally biased mutagenesis is the appearance of mutations at highly transcribed genes under neutral selection. There are several reports showing evidence that the generation of beneficial mutations is disproportionately higher than that observed for neutral or detrimental mutations, and some allude to the process of transcriptional activation as a cellular mechanism to bias beneficial events (Crozat et al. 2005; Hall and Joseph 2010; Kim et al. 2010).

Gene Induction and Mutagenesis

In this section, we present observations that link the act of transcription with mutagenic events. Experiments supporting the hypothesis that transcription promotes mutagenic events have been presented in diverse microorganisms and accrued for several years (discussed by Wright [2000, 2004]). These mutagenic processes span several species, from bacteria to mammalian cells. In *E. coli*, reversion to amino acid prototrophy (e.g., arginine, leucine) and carbon utilization (e.g., *lacZ*-β-galactosidase) has been shown to increase under conditions of gene activation

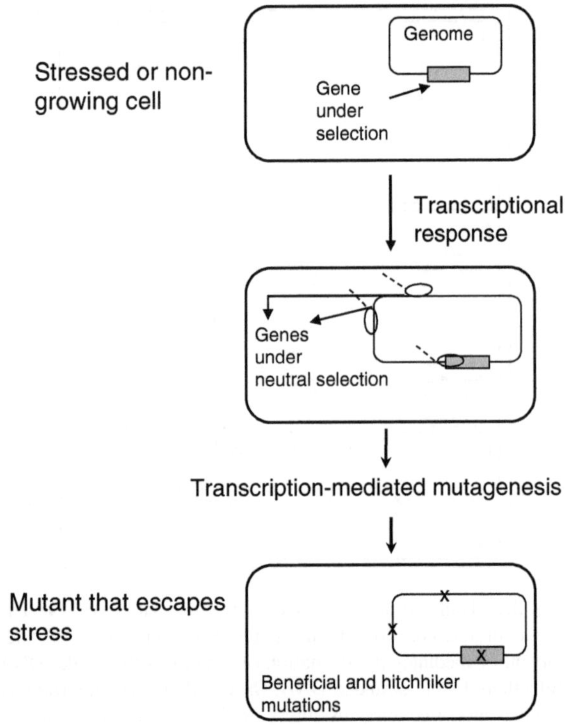

Fig. 3.2 Transcription-mediated mutagenesis as a bacterial strategy to escape stress. Cells respond to stress by activation of defined regulons, which contain genes that are under direct or neutral selection. Mutagenesis at highly transcribed genes accumulates in genes under direct selection, and in genes that are highly transcribed. Cells with beneficial mutations are selected and contain hitchhiker mutations

(Burkala et al. 2007; Reimers et al. 2004; Wright 1996; Wright and Minnick 1997). In *Salmonella enterica*, defective *lacZ* alleles containing mutations that could be reverted by different types of DNA base transitions and transversions were tested in both the presence and absence of gene induction. Those experiments showed that all the tested types of point mutations were increased under gene induction conditions (Hudson et al. 2003).

In the Gram-positive organism *Bacillus subtilis*, the first experiments that indicated an association between increased transcription and reversion to amino acid biosynthesis were conducted by Rudner et al. (Rudner et al. 1999) using the stringent response to alter transcription of the test genes. Other experiments also demonstrated that reversion to leucine biosynthesis was increased under conditions in which a point mutated *leu* allele was transcriptionally derepressed (Martin et al. 2011; Pybus et al. 2010). Interestingly, those experiments showed that the increases in mutagenic events only occurred in cells under starvation stress, and extended to reversion to histidine and methionine prototrophy. In yeast, high transcription levels of a defective lysine biosynthesis gene containing a frameshift increased the rate of spontaneous mutation more than tenfold (Datta and Jinks-Robertson 1995; Kim et al. 2007).

In mammalian cells, the role of transcriptional gene induction has been discussed as a process required for the production of immunogenic diversity, and as a process that contributes to the formation of detrimental mutations. The production of antibody diversity is dependent on activation-induced cytidine deaminase (AID), which promotes events that result in class switch recombination and point mutations in developing B leukocytes. These events occur when the genes for the variable regions of immunoglobulin are transcriptionally activated (discussed in [Teng and Papavasiliou 2007]). Also, gene activation has been associated with mutagenic events in UV-treated cells. Experiments using embryonic mouse cells engineered to transcribe the *Hprt* gene (Hypoxanthine-guanine phosphoribosyltransferase) in response to tetracycline treatment show that transcription promotes mutations (Hendriks et al. 2008). HPRT is part of the purine salvage pathway, and involves the conversion of hypoxanthine to guanosine monophosphate, a precursor to GTP. Loss of Hprt functional activity is detected by selection on 6-thioguanine, a genotoxic purine analog (Duan et al. 2004). These experiments combined gene activation with exposure to either UV, or an oxidative damaging agent. Results from this study showed an eightfold increase in the frequency of Hprt loss-of-function mutations, dependent on transcription of the *Hprt* gene prior to selection. The majority of transcription-mediated mutations induced by UV treatment were intragenic deletions in *Hprt* (Hendriks et al. 2008).

Transcription Mediates the Formation of Single-Stranded DNA and Secondary Structures that Are Mutagenic

One proposed model explaining transcription-mediated mutations is that the process of transcription generates single-stranded DNA, which increases the exposure of DNA bases to chemical damage (Fig. 3.3a). Interestingly, this idea was discussed as a possible explanation for the generation of "adaptive" mutations occurring in *E. coli* cells subject to prolonged carbon source starvation (Bridges et al. 1996; Davis 1989). Experiments that used a *lacZ* reversion assay in a bacterial system estimated a difference of almost three orders of magnitude in the rate of C-deamination in single-stranded DNA compared to double-stranded DNA at 37°C (Frederico et al. 1990). In mammalian cells, transcription mediates the formation of immunoglobulin variants with higher affinity for their antigens through the activity of activation induced deaminase (AID), which targets single-stranded DNA. Accessory proteins to AID target single-stranded DNA, and facilitate the deamination of cytosine to uracil. Uracil is then processed through the BER pathway, which includes cleavage by uracil DNA glycosylase (*ung*), generation of a DNA gap by an apurinic/apyrimidinic (AP) endonuclease, and subsequent replication and ligation. In addition, uracil may be processed by the mismatch repair system. Key elements that generate a wide spectrum of point mutations are the enlistment of error-prone polymerases (Pol θ and Pol η) that fill the gap generated during repair (reviewed in [Teng and Papavasiliou 2007]).

This model is supported by *E. coli* experiments in which transcriptional activation-dependent mutations induced by methylmethane sulfonate (a base alkylating agent that causes transitions at C residues and transversions at A residues [Wrzesinski et al. 2010]) mapped to the non-transcribed strand (Fix et al. 2008).

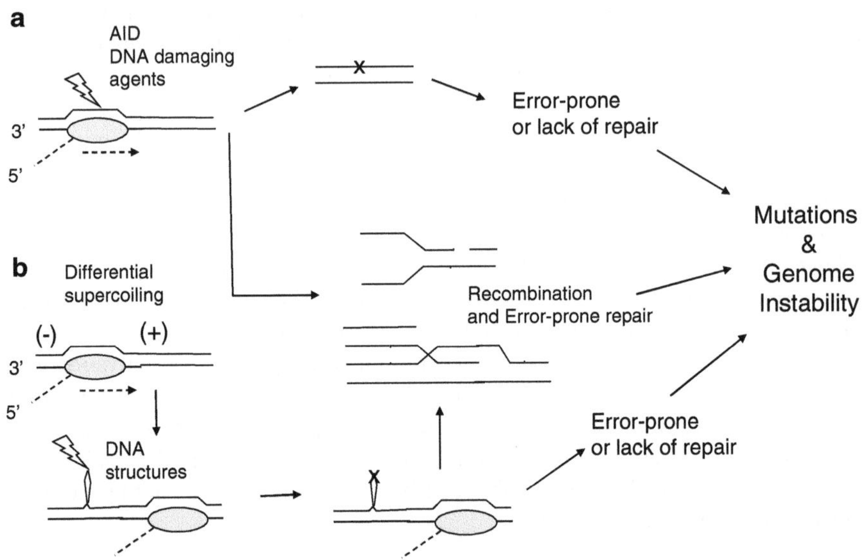

Fig. 3.3 The stress of transcription generates single-stranded DNA (**a**), which increases exposure to solvent or differential supercoiling (denoted by "+" and "–") and (**b**) mediates the formation of structures that make DNA more prone to lesions (lightning bolt) than double-stranded DNA. DNA lesions are processed, either by error-prone NER or BER. Single-stranded DNA and noncanonical structures mediate recombination. Lesion processing may generate single- and double-stranded gaps that are repaired by recombination and error-prone polymerases

Also, C to T mutations, induced by deamination, and G to T mutations, induced by reactive oxygen species, mapped more frequently to the non-transcribed strand in tRNA suppressor genes (Bockrath and Li 1998; Klapacz and Bhagwat 2005). These experiments used genetic backgrounds deficient in repair factors that target DNA lesions induced by methylation, deamination, or reactive oxygen species, suggesting that mutations were generated because of the lack of repair during subsequent DNA replication. This type of mutagenesis has also been observed in human genes (Skandalis et al. 1994). It is also possible that mutations arise from an error-prone BER pathway, analogous to the case presented for the generation of immunogenic variants in B cells (Teng and Papavasiliou 2007). In bacteria, evidence suggestive of mutagenic BER has been presented in the Gram-positive microorganism *B. subtilis* (Debora et al. 2011; Lopez-Olmos et al. 2012). It is also possible that the incomplete processing of lesions (increased at sites of high transcription) generates single- or double-stranded gaps that may also be repaired by recombination functions (illustrated in the middle of Fig. 3.3). Models of how recombination generates deletions and insertions have been presented elsewhere, and in the context of the production of mutations conferring escape from starvation stress (Galhardo et al. 2007; Hastings 2007; Roth et al. 2006).

The act of transcription has also been proposed to alter supercoiling conditions and trigger the formation of secondary structures in single-stranded DNA (Wright 2004).

Wright and coworkers support the view that in the wake of transcription, supercoiling stress placed on DNA forms single-stranded structures that prone G and C bases to lesions (Fig. 3.3b) (Wright 2000, 2004). The mutability of those bases is associated with the formation and stability of stem-loop structures that result from transcription-generated supercoiling, their position in the structure, and whether these bases are exposed to solvent conditions. Further, a metric called the mutability index forecasts the likelihood of base pairing in predicted structures, and the stability of those structures (Wright et al. 2003). In brief, the mutability index for a base is calculated by generating DNA sequence windows that include the base, and then folding those sequences using Zuker's mfold algorithm (Zuker 2003). The mutability index value is calculated by multiplying the percentage of observed folds in which the base is unpaired by the negative energy of folding of the most stable structure. Higher MI values can be thought of as an increase in the potential exposure of a given base to solvent within a transcription bubble, and thus a higher mutation potential. The types of damage that these residues are most frequently subject to are deamination and oxidative stress (Wright 2000, 2004).

This concept was examined in experiments that used defective *E. coli lacZ* alleles containing nonsense mutations at different nucleotide positions that varied in their mutability index values. Those experiments showed that the number of revertants at specific positions correlated with their mutability index. Also, the accumulation of mutations was dependent on both transcription activation and supercoiling state (Burkala et al. 2007; Schmidt et al. 2006). This model explains how lesions accumulate in the non-transcribed strand; however, it remains to be elucidated whether mutations are introduced during genome replication or through defective repair. The transcription-associated formation of stem-loop structures has been used to describe the acquisition of mutations in the p53 tumor suppressor gene, and to model how genetic variants in immunoglobulin genes are produced (Wright et al. 2002, 2004). An interesting implication of this model is that lesions are formed in the non-transcribed strand, and may require a round of replication to fix the mutation into the transcribed counterpart, which ultimately generates the altering phenotypic change. Conversely, recent experiments have presented evidence for the formation of mutations in the transcribed strand in the context of skin cells treated with UV to target CC residues. Such treatment results in cyclobutane pyrimidine dimers that may be deaminated and converted to UU bases. In the absence of repair, and upon two rounds of replication, this results in CC to TT transitions, a very prevalent mutation in the *p53* gene in embryonic stem cells [(Hendriks et al. 2008); discussed in (Helleday 2010)]. It is not clear how these mechanisms operate in the context of stressful conditions or replication-limiting conditions.

In summary the transcriptional process may mediate mutagenic events by exposing single-stranded DNA to solvent conditions, which would promote the formation of lesions. Mutagenic lesions are then the substrate for misprocessing or mutagenic repair (via BER or recombination). The involvement of stem-loop structures triggered by transcription and the differences in supercoiling state ahead of and behind the transcription elongation complex bias mutations to "hot spots" within genes with high mutability index values. Perhaps cells evolved such sites as substrates for increasing genetic diversity.

Collisions Between Transcription and Replication as a Source of Genomic Instability

In addition to point mutations, the formation of R-loops has also been shown to promote recombination events that lead to larger types of mutations (reviewed in [Aguilera 2002]). This process has been denoted as transcription-associated recombination (TAR). Experiments conducted in yeast provide evidence suggesting that transcription promotes recombination (see (Datta and Jinks-Robertson 1995; Kim et al. 2007), described above). These reports showed that the genetic components of recombination and error-prone replication influenced the appearance of transcription-dependent mutations. Subsequent experiments identified the THO complex as influencing recombination activity at highly transcribed regions (Luna et al. 2009). THO is a nuclear complex containing the factors Tho1, Hpr1, Mft1, and Thp2 (Chavez et al. 2000). The THO complex is associated with the transcription elongation machinery, and processes and exports mRNA out of the nucleus. Then, it is hypothesized that misprocessing of the nascent mRNA leads to the formation of extended R-loops (Gomez-Gonzalez and Aguilera 2009). Evidence supporting this concept has also been presented in experiments with strains that overproduce RNAases, or with defects in topoisomerase activity (Aguilera 2002). Overproduction of RNAase or changes in supercoiling influence the formation of RNA–DNA hybrids, and therefore reduces the formation of R-loops. When transcription elongation complexes encounter replication forks (Kim et al. 2007), such events collapse replication and recruit replication restart factors.

The idea that conflicts between transcription and replication produce genomic instability has also been extended to bacteria. Experiments in *B. subtilis* strains, altered in their gene orientation with respect to genome replication, showed increases in genome instability, particularly when transcription and replication collided head-on (Merrikh et al. 2011; Srivatsan et al. 2010). A key element in the production of genomic instability in those experiments was the activation of the SOS repair system, a well-known mutagenic pathway. In *E. coli*, the presence of transcription elongation factors, such as Mfd and DksA, prevents the generation of mutations or genome instability (Trautinger et al. 2005). Evidence for how the presence of transcription factors aids in restarting replication has been recently presented in vitro (Pomerantz and O'Donnell 2010), and in vivo in *E. coli*, without activating the SOS response (Tehranchi et al. 2010). Those experiments indicate that active or stalling transcription elongation complexes collapse the replication fork. Such events trigger the SOS response, which generates mutations. The role of Mfd and DksA is then to process or dissociate the stalled transcription complex before it encounters a replication fork.

One other way in which the process of transcription may engender genomic instability is by colliding with replication forks. Such collisions are processed by recombination functions that may generate rearrangements that lead to genome instability. Alternatively, error-prone polymerases recruited to sites of replication fork repair may also produce point mutations (Hastings et al. 2010). These events are promoted by the formation of DNA–RNA hybrids or stalled transcription complexes, and prevented by mRNA processing elements or transcription factors that dissociate the transcription elongation complex from the template strand, effectively preventing collisions.

Factors Interacting with Transcriptional Elongation Complexes May Promote Mutagenic Events

The previous two models may require one or two rounds of replication to fix mutations to the non-transcribed strand, or the formation of replication restart intermediates. In the context nongrowth conditions, or when genome replication is halted or inhibited, what is the effect of gene induction on mutagenesis? There are a few studies that approach this question, and they indicate that transcriptional activation produces mutagenic events in *B. subtilis* resting cells. The effect of gene induction on mutagenesis in *B. subtilis* resting cells was demonstrated by an experiment that used strains containing defective alleles for biosynthesis of two amino acids, and experimentally controlled gene induction for one of the two defective alleles. Test cells were maintained in conditions of starvation for two amino acids, in combination with transcriptional activation or repression. IPTG was used to experimentally control transcription. This experimental approach maintained cells under starvation conditions, even in the presence of mutations that restore growth in the absence of one amino acid. Those experiments showed that starving, nongrowing cells accumulated 10 times more revertants in the presence of IPTG than in its absence (Pybus et al. 2010). Because the cells tested in that study were subject to starvation for two amino acids, and revertants were detected only after supplying one of the growth-limiting amino acids, it is possible to speculate that the effect of transcriptional de-repression on the accumulation of mutations in *B. subtilis* cells do not require genome replication, collisions with replication forks, or the formation replication intermediates.

Experiments that examine mutagenesis in nongrowing conditions point to two proteins that act during transcriptional elongation and influence the accumulation of mutations in nongrowing cells in *B. subtilis* and *E. coli*. In *B. subtilis* Mfd, also known as transcription repair coupling factor (TRCF), was shown to influence stationary-phase mutagenesis. Those experiments measured the accumulation of mutations in cells subjected to prolonged starvation, and rates of mutation in actively growing cells in three different chromosomal auxotrophic markers in genetic backgrounds that differed in Mfd proficiency. Mfd increased mutagenesis in cells that were nongrowing, but it did not influence mutation rates in growing cultures (Ross et al. 2006). Subsequent experiments showed that proficiency of Mfd facilitated transcription of genes in cells experiencing stationary phase conditions (Martin et al. 2011; Pybus et al. 2010). It should be noted that defects in Mfd in *E.coli* did not influence the generation of suppressor mutations (Bridges 1995).

Table 3.1 Domains, functions, and proposed changes to amino acid residues in *E. coli* Mfd

Domain (N-C)[a]	Function	Positions[b]	Expected phenotype
D1a	UvrA recruitment	1-116	
D2	UvrA recruitment	117-216 H152A[c] R165A G166Q R181A	Defective in NER
D1b	UvrA recruitment	217-341	
D3	Flexible non-conserved	350-456	
D4	RNAP interaction	478-546 L499R	Defective in RNAP binding Deficient in TRCF functions
D5	Translocase	548-781 (571-804)	
D6	Translocase	782-990 (805-1012)	
D7	Handle	1001-1148 Deletion of domain	Defective in NER (see D2)

[a]Domains are listed from the N to C terminus of TRCF (as described in *E. coli*)
[b]Residue positions corresponding to specific domain
[c]Amino acid changes, proposed, or previously shown to inactivate a specific TRCF function. Data are adapted from Deaconescu et al. (Deaconescu et al. 2006)

However, *E. coli* NusA, a transcription antitermination factor, significantly increases mutations in cells experiencing starvation stress. This was examined in the context of mutations conferring lactose utilization and tetracycline resistance (Cohen and Walker 2010). Further experiments indicated that NusA did not influence the ability to transcribe the test genes in conditions of stress. In contrast to the models predicated on collisions between transcription and replication, mutagenesis studies in replication-limiting conditions suggest that some transcription factors mediate mutagenic processes.

The mechanisms by which these two transcription factors influence the generation of mutations in stressful conditions invoke error-prone repair, gratuitous repair, or the formation of transient phenotypes by transcriptional bypass. These three models are based on the enzymatic activities of Mfd and NusA. Mfd, also known as TRCF, was first described by Evelyn Witkin in *E. coli* in the context of UV mutagenesis (Witkin 1966). Subsequent experiments showed that Mfd recruits the nucleotide excision repair (NER or Uvr proteins), which recognizes and processes bulky DNA lesions, to the transcribed strand. This mechanism ensures that highly transcribed regions, or genes whose products are in high demand by the cell, are preferentially repaired. In the absence of Mfd, bulky lesions in bacteria are processed by the $UvrA_2B$ complex, which scans double-stranded DNA for distortions, with UvrBC generating incisions flanking the DNA lesion. These steps are followed by the activity of UvrD, a helicase, which generates a single-stranded gap, later filled by polymerase I. Mfd (TRCF) recruits UvrA to sites at which blocking lesions stall elongating RNAP complexes (Hanawalt and Spivak 2008). Structural and functional properties for Mfd (TRCF) are illustrated in Table 3.1. Structural and homology studies have described *E. coli* Mfd mutations, or regions that affect RNAP binding, translocase activity, and NER recruitment (UvrA recruitment).

The observations on *E. coli* Mfd functions, as well as *B. subtilis* experiments showing that the PolY1 error-prone DNA polymerase promotes mutagenesis in stationary phase cells, (Sung et al. 2003) lend support to the hypothesis of transcription-coupled, error-prone repair. This model argues that the gap generated during transcription-coupled repair is filled by an error-prone polymerase, and suggests that this type of mutagenesis targets the template strand of DNA. Alternatively, it is also possible to speculate that mutations may be generated by PolI in *B. subtilis*. Interestingly, *B. subtilis* PolI is effectively error-prone because it lacks 3′-5′ exonuclease activity (Duigou et al. 2005). Then, transcription-coupled repair may operate with high or low fidelity. Because stationary phase conditions are conducive to DNA lesions (Debora et al. 2011; Sung et al. 2003; Pedraza-Reyes and Yasbin 2004; Vidales et al. 2009), highly transcribed regions are likely to recruit transcription-coupled repair. DNA lesions present blocks to transcription, which are resolved by Mfd and the activity of error-prone DNA polymerases (Fig. 3.2). It should be noted that these types of mutagenesis are exclusive to stationary cells, because error-prone polymerases are highly expressed in those conditions (Duigou et al. 2004), and cells are under replication-limiting conditions. In these conditions, mutations at highly transcribed regions that confer a growth advantages are quickly selected, and increase over time during prolonged stress.

One other model by which gene activation may promote mutagenic events is that of transcriptional bypass and the formation of transient phenotypes. Transcriptional bypass happens when RNA polymerase (RNAP) synthesizes an mRNA molecule containing a mutated ribonucleotide residue. The erroneous synthesis of such messages is generated when DNA damage results in a miscoding lesion (e.g., deamination at a C base in the template DNA strand results in U, which, when read by RNAP, produces an A in the corresponding mRNA molecule). Alternatively, an erroneous message may be also generated by incorporating miscoding ribonucleotides during transcription, e.g., incorporating 8-oxo-rGTP (generated by oxidation of rGTP) opposite A bases in the template strand (Taddei et al. 1997). Incorporation of oxidized of ribonucleotides is controlled by MutT, and has been implicated in the formation of stationary phase mutagenesis (Vidales et al. 2009). A mutated mRNA may potentially generate an altered protein that confers a transient phenotype (Viswanathan et al. 1999a, b; Bridges 1999).

In the context of halted or limited replication conditions, a transient growth phenotype may license the cell back into growth and replicating conditions. The potential from the originating miscoding lesion in the template DNA strand is locked or introduced into the DNA in the absence of repair and subsequent replication (for reviews see [Holmquist 2002; Doetsch 2002]) or mutagenic repair (Debora et al. 2011). This idea is supported by experiments showing that (1) RNAP bypasses DNA lesions during transcription with different efficiencies (for reviews see [Saxowsky and Doetsch 2006]), (2) in vivo experiments that showed the formation of transient phenotypes in real time (Viswanathan et al. 1999a), and (3) this process is well conserved as it occurs in bacteria and eukaryotes (Bregeon et al. 2003; Charlet-Berguerand et al. 2006). Because this process is affected by RNAP processivity, factors influencing the transcription elongation process (Mfd, GreA, and NusA, to name a few) are speculated to affect transcriptional bypass (Svejstrup 2002).

The meeting of active transcription elongation complexes with DNA lesions may stall RNAP, or cause it to backtrack, which halts transcription and misaligns the active site of RNAP from the template strand and the 3′ of the nascent mRNA—the 3′ end of the mRNA may protrude out of the RNAP-DNA complex. Resuming active transcription without RNAP recycling may be accomplished by the activity of the Gre factors in *E. coli*, which process this event by cleaving the 3′ protruding end of the mRNA (Toulme et al. 2000). Interestingly, GreA (in *E. coli*) and transcription factors (in human cells) have been shown to influence transcription fidelity, rate, and the generation of transient phenotypes (Charlet-Berguerand et al. 2006; Borukhov et al. 2005; Gordon et al. 2009).

Stalled and backtracked elongation complexes may also be processed by Mfd. Evidence for this process has been documented by in vitro *E. coli* experiments showing that Mfd interacts with the upstream moiety of the transcription elongation complex, and, through its translocase activity, pushes the misaligned RNAP active site forward, potentially "kicking" it back into active transcription (Park et al. 2002). It is because of this translocase activity that Mfd has been speculated to influence transcriptional bypass. However, the role of Mfd in transcriptional bypass appears controversial. In *E. coli*, Mfd has been shown to prevent transcriptional bypass of miscoding lesions on the template strand by dissociating the RNAP elongation complex, and subsequent repair of the damage by the NER system (this would effectively promote new rounds of transcription initiation and elongation in the lesion-containing gene) (Bregeon et al. 2003). On the other hand, in vitro experiments using Hela cells extracts suggest that CSB, the Mfd functional homologue in humans, promotes transcriptional bypass, particularly at lesions generated by oxidative damage (Charlet-Berguerand et al. 2006; Bregeon et al. 2009). This discrepancy may be explained by the differences in gene size between eukaryotes and bacteria. In terms of energy conservation, in genes that require a long time to transcribe, it would be more favorable to resolve blocked transcription elongation complexes by "pushing" the RNAP back to active transcription at the lesion site, rather than assembling a new transcription complex, which would be required to start from the promoter region (discussed in [Hanawalt and Spivak 2008]).

NusA, a factor mediating transcription pausing, termination, and antitermination, is involved in synchronizing transcription and translation in bacteria. Recently, *E. coli* NusA has been shown to directly interact with DinB, an error-prone polymerase that protects cells from Nitrofurazone exposure, and speculated to mediate low fidelity transcription-coupled repair (Cohen et al. 2009; Cohen et al. 2010). This was demonstrated by genetic and biochemical experiments that examined cell sensitivity to nitrofurazone, which causes the formation of DNA adducts such as N^2-deoxyguanine. Those experiments also showed that NusA did not affect transcription of the gene under selection, or transcriptional bypass of DNA lesions similar to those caused by nitrofurazone. In the context of mutations that occur in conditions of prolonged starvation for a carbon source, cells carrying a temperature sensitive mutation in *nusA* showed severely decreased accumulation of stress-induced mutations; such defects were suppressed by the overexpression of DinB. These experiments support the idea that during stressful environmental conditions, DNA damage is repaired through an error-prone pathway (Cohen et al. 2010).

The observations concerning the promotion of mutagenic events, mediated by transcription factors that interact with elongating RNAP complexes, imply that cells detect DNA lesions located in the template strand, and that subsequent low-fidelity repair introduces mutations. However, other experimental evidence also suggests that gratuitous recruitment of Mfd, or its functional homologues, is triggered by DNA secondary structures in the non-template strand. This leads to a scenario in which mutations are generated in the template strand in the absence of DNA lesions, but are caused by secondary structures formed during transcription in the non-template strand (Tamkun 2007; Tornaletti 2009; Tornaletti et al. 2008). This and other models of how transcription activation promote mutations have been further discussed recently (Robleto et al. 2012).

Conclusions

In this chapter, we visited evidence and models that support the hypothesis that transcriptional activation of a gene promotes mutagenic events. The discussed models produce genetic diversity in cells experiencing nongrowth or stress. A key element in the transcription-dependent mutations is misrepair, or error-prone repair. The mutagenic events promoted by transcriptional activation may be generated by increasing exposure of single-stranded DNA to solvent conditions, by events that promote recombination events, or by transcription-coupled repair pathways that include error-prone replication.

During conditions in which replication is halted, transcriptional activation-dependent mutagenesis represents a cellular strategy to escape non-replicative conditions. In addition, it is possible that cells generate, through the process of transcriptional bypass of DNA lesions, transient phenotypes that confer growth advantages or escape from stress. The potential of the transient phenotype may be immortalized by subsequent DNA replication in the absence of repair. This process affords a non-replicating cell to "test" modified alleles for their ability to rescue such cell back into growth conditions (Holmquist 2002).

A novel and yet to be explored process that may influence the formation of mutations in actively growing or non-replicating conditions is that of reverse transcription. Retrotransposons are well-characterized insertion elements that cause gene knockouts and genome rearrangements. Similarly, retroviruses go through an RNA intermediate before incorporating into the host chromosome. As RNAPs are more error-prone than DNA polymerases, virus variants may be generated during the RNA intermediate and immortalized into DNA during the reverse transcription step. Such variants are then selected based on their virulence. Evidence for the influence of a reverse transcription step in the production of genetic diversity has been presented in the variation of immunoglobins (Teng and Papavasiliou 2007). It would be interesting to investigate whether bacterial eukaryotic cells generate allelic variants through a reverse transcription step, and whether such a process is restricted to DNA regions that are highly transcribed.

Ultimately, transcriptional activation-dependent mutagenesis produces genetic diversity and provides applied alternatives to engineer enhanced protein variants. On the other hand, the elements underlying the mechanisms that generate these types of mutations are conserved and implicated in the formation of mutations conferring antibiotic resistance in bacteria (Cohen and Walker 2010; Boshoff et al. 2003; Han et al. 2008) and the development of some cancers (Rodin et al. 2002). Hence, inhibition of transcription-mediated mutation by targeting the factors that drive its underlying mechanisms may be used as a disease therapy strategy.

Acknowledgments This manuscript was supported by Grants (NIH) GM07255 (CONACyT) 88482 and (NSF) MCB0843606 DBI1005223.

References

Aguilera A (2002) The connection between transcription and genomic instability. EMBO J 21:195–201

Bockrath R, Li BH (1998) Transcriptional mutagenesis and DNA strand asymmetrical mutations expressed in *Escherichia coli* under restrictive metabolic conditions. Mutat Res 422:351–355

Borukhov S, Lee J, Laptenko O (2005) Bacterial transcription elongation factors: new insights into molecular mechanism of action. Mol Microbiol 55:1315–1324

Boshoff HI, Reed MB, Barry CE 3rd, Mizrahi V (2003) DnaE2 polymerase contributes to in vivo survival and the emergence of drug resistance in *Mycobacterium tuberculosis*. Cell 113:183–193

Bregeon D, Doddridge ZA, You HJ, Weiss B, Doetsch PW (2003) Transcriptional mutagenesis induced by uracil and 8-oxoguanine in *Escherichia coli*. Mol Cell 12:959–970

Bregeon D, Peignon PA, Sarasin A (2009) Transcriptional mutagenesis induced by 8-oxoguanine in mammalian cells. PLoS Genet 5:e1000577

Bridges BA (1995) Starvation-associated mutation in *Escherichia coli* strains defective in transcription repair coupling factor. Mutat Res 329:49–56

Bridges BA (1999) Dirty transcripts from clean DNA. Science 284:62–63

Bridges BA, Sekiguchi M, Tajiri T (1996) Effect of mutY and mutM/fpg-1 mutations on starvation-associated mutation in *Escherichia coli*: implications for the role of 7,8-dihydro-8-oxoguanine. Mol Gen Genet 251:352–357

Brock RD (1971) Differential mutation of the beta-galactosidase gene of *Escherichia coli*. Mutat Res 11:181–186

Burkala E, Reimers JM, Schmidt KH, Davis N, Wei P, Wright BE (2007) Secondary structures as predictors of mutation potential in the *lacZ* gene of *Escherichia coli*. Microbiology 153:2180–2189

Charlet-Berguerand N, Feuerhahn S, Kong SE, Ziserman H, Conaway JW, Conaway R, Egly JM (2006) RNA polymerase II bypass of oxidative DNA damage is regulated by transcription elongation factors. EMBO J 25:5481–5491

Chavez S, Beilharz T, Rondon AG, Erdjument-Bromage H, Tempst P, Svejstrup JQ, Lithgow T, Aguilera A (2000) A protein complex containing Tho2, Hpr1, Mft1 and a novel protein, Thp2, connects transcription elongation with mitotic recombination in *Saccharomyces cerevisiae*. EMBO J 19:5824–5834

Cohen SE, Walker GC (2010) The transcription elongation factor NusA is required for stress-induced mutagenesis in *Escherichia coli*. Curr Biol 20:80–85

Cohen SE, Godoy VG, Walker GC (2009) Transcriptional modulator NusA interacts with translesion DNA polymerases in *Escherichia coli*. J Bacteriol 191:665–672

Cohen SE, Lewis CA, Mooney RA, Kohanski MA, Collins JJ, Landick R, Walker GC (2010) Roles for the transcription elongation factor NusA in both DNA repair and damage tolerance pathways in *Escherichia coli*. Proc Natl Acad Sci USA 107:15517–15522

Crozat E, Philippe N, Lenski RE, Geiselmann J, Schneider D (2005) Long-term experimental evolution in *Escherichia coli*. XII. DNA topology as a key target of selection. Genetics 169:523–532

Datta A, Jinks-Robertson S (1995) Association of increased spontaneous mutation rates with high levels of transcription in yeast. Science 268:1616–1619

Davis BD (1989) Transcriptional bias: a non-Lamarckian mechanism for substrate-induced mutations. Proc Natl Acad Sci USA 86:5005–5009

Deaconescu AM, Chambers AL, Smith AJ, Nickels BE, Hochschild A, Savery NJ, Darst SA (2006) Structural basis for bacterial transcription-coupled DNA repair. Cell 124:507–520

Debora BN, Vidales LE, Ramirez R, Ramirez M, Robleto EA, Yasbin RE, Pedraza-Reyes M (2011) Mismatch repair modulation of MutY activity drives *Bacillus subtilis* stationary-phase mutagenesis. J Bacteriol 193:236–245

Doetsch PW (2002) Translesion synthesis by RNA polymerases: occurrence and biological implications for transcriptional mutagenesis. Mutat Res 510:131–140

Duan J, Nilsson L, Lambert B (2004) Structural and functional analysis of mutations at the human hypoxanthine phosphoribosyl transferase (HPRT1) locus. Hum Mutat 23:599–611

Duigou S, Ehrlich SD, Noirot P, Noirot-Gros MF (2004) Distinctive genetic features exhibited by the Y-family DNA polymerases in *Bacillus subtilis*. Mol Microbiol 54:439–451

Duigou S, Ehrlich SD, Noirot P, Noirot-Gros MF (2005) DNA polymerase I acts in translesion synthesis mediated by the Y-polymerases in *Bacillus subtilis*. Mol Microbiol 57:678–690

Fix D, Canugovi C, Bhagwat AS (2008) Transcription increases methylmethane sulfonate-induced mutations in *alkB* strains of *Escherichia coli*. DNA Repair (Amst) 7:1289–1297

Frederico LA, Kunkel TA, Shaw BR (1990) A sensitive genetic assay for the detection of cytosine deamination: determination of rate constants and the activation energy. Biochemistry 29:2532–2537

Galhardo RS, Hastings PJ, Rosenberg SM (2007) Mutation as a stress response and the regulation of evolvability. Crit Rev Biochem Mol Biol 42:399–435

Gomez-Gonzalez B, Aguilera A (2009) R-loops do not accumulate in transcription-defective hpr1-101 mutants: implications for the functional role of THO/TREX. Nucleic Acids Res 37:4315–4321

Gordon AJ, Halliday JA, Blankschien MD, Burns PA, Yatagai F, Herman C (2009) Transcriptional infidelity promotes heritable phenotypic change in a bistable gene network. PLoS Biol 7:e44

Hall DW, Joseph SB (2010) A high frequency of beneficial mutations across multiple fitness components in *Saccharomyces cerevisiae*. Genetics 185:1397–1409

Han J, Sahin O, Barton YW, Zhang Q (2008) Key role of Mfd in the development of fluoroquinolone resistance in *Campylobacter jejuni*. PLoS Pathog 4:e1000083

Hanawalt PC, Spivak G (2008) Transcription-coupled DNA repair: two decades of progress and surprises. Nat Rev Mol Cell Biol 9:958–970

Hastings PJ (2007) Adaptive amplification. Crit Rev Biochem Mol Biol 42:271–283

Hastings PJ, Hersh MN, Thornton PC, Fonville NC, Slack A, Frisch RL, Ray MP, Harris RS, Leal SM, Rosenberg SM (2010) Competition of *Escherichia coli* DNA polymerases I, II and III with DNA Pol IV in stressed cells. PLoS One 5:e10862

Helleday T (2010) Mutagenesis: mutating a gene while reading it. Curr Biol 20:R57–58

Hendriks G, Calleja F, Vrieling H, Mullenders LH, Jansen JG, de Wind N (2008) Gene transcription increases DNA damage-induced mutagenesis in mammalian stem cells. DNA Repair (Amst) 7:1330–1339

Herman RK, Dworkin NB (1971) Effect of gene induction on the rate of mutagenesis by ICR-191 in *Escherichia coli*. J Bacteriol 106:543–550

Holmquist GP (2002) Cell-selfish modes of evolution and mutations directed after transcriptional bypass. Mutat Res 510:141–152

Hudson RE, Bergthorsson U, Ochman H (2003) Transcription increases multiple spontaneous point mutations in *Salmonella enterica*. Nucleic Acids Res 31:4517–4522

Kim N, Abdulovic AL, Gealy R, Lippert MJ, Jinks-Robertson S (2007) Transcription-associated mutagenesis in yeast is directly proportional to the level of gene expression and influenced by the direction of DNA replication. DNA Repair (Amst) 6:1285–1296

Kim H, Lee BS, Tomita M, Kanai A (2010) Transcription-associated mutagenesis increases protein sequence diversity more effectively than does random mutagenesis in *Escherichia coli*. PLoS One 5:e10567

Klapacz J, Bhagwat AS (2005) Transcription promotes guanine to thymine mutations in the non-transcribed strand of an *Escherichia coli* gene. DNA Repair (Amst) 4:806–813

Lopez-Olmos K, Hernandez MP, Contreras-Garduno JA, Robleto EA, Setlow P, Yasbin RE, Pedraza-Reyes M (2012) Roles of endonuclease V, uracil-DNA glycosylase, and mismatch repair in *Bacillus subtilis* DNA base-deamination-induced mutagenesis. J Bacteriol 194: 243–252

Luna R, Gonzalez-Aguilera C, Aguilera A (2009) Transcription at the proximity of the nuclear pore: a role for the THP1-SAC3-SUS1-CDC31 (THSC) complex. RNA Biol 6:145–148

Martin HA, Pedraza-Reyes M, Yasbin RE, Robleto EA (2011) Transcriptional de-repression and Mfd are mutagenic in stressed *Bacillus subtilis* cells. J Mol Microbiol Biotechnol 21:45–58

Merrikh H, Machon C, Grainger WH, Grossman AD, Soultanas P (2011) Co-directional replication-transcription conflicts lead to replication restart. Nature 470:554–557

Park JS, Marr MT, Roberts JW (2002) *E. coli* Transcription repair coupling factor (Mfd protein) rescues arrested complexes by promoting forward translocation. Cell 109:757–767

Pedraza-Reyes M, Yasbin RE (2004) Contribution of the mismatch DNA repair system to the generation of stationary-phase-induced mutants of *Bacillus subtilis*. J Bacteriol 186:6485–6491

Pomerantz RT, O'Donnell M (2010) Direct restart of a replication fork stalled by a head-on RNA polymerase. Science 327:590–592

Pybus C, Pedraza-Reyes M, Ross CA, Martin H, Ona K, Yasbin RE, Robleto E (2010) Transcription-associated mutation in *Bacillus subtilis* cells under stress. J Bacteriol 192:3321–3328

Reimers JM, Schmidt KH, Longacre A, Reschke DK, Wright BE (2004) Increased transcription rates correlate with increased reversion rates in *leuB* and *argH Escherichia coli* auxotrophs. Microbiology 150:1457–1466

Robleto EA, Yasbin R, Ross C, Pedraza-Reyes M (2007) Stationary phase mutagenesis in *B. subtilis*: a paradigm to study genetic diversity programs in cells under stress. Crit Rev Biochem Mol Biol 42:327–339

Robleto EA, Martin HA, Pedraza-Reyes M (2012) Mfd and transcriptional derepression cause genetic diversity in *Bacillus subtilis*. Front Biosci (Elite Ed) 4:1246–1254

Rodin SN, Rodin AS, Juhasz A, Holmquist GP (2002) Cancerous hyper-mutagenesis in *p53* genes is possibly associated with transcriptional bypass of DNA lesions. Mutat Res 510:153–168

Ross C, Pybus C, Pedraza-Reyes M, Sung HM, Yasbin RE, Robleto E (2006) Novel role of *mfd*: effects on stationary-phase mutagenesis in *Bacillus subtilis*. J Bacteriol 188:7512–7520

Roth JR, Kugelberg E, Reams AB, Kofoid E, Andersson DI (2006) Origin of mutations under selection: the adaptive mutation controversy. Annu Rev Microbiol 60:477–501

Rudner R, Murray A, Huda N (1999) Is there a link between mutation rates and the stringent response in *Bacillus subtilis*? Ann N Y Acad Sci 870:418–422

Saxowsky TT, Doetsch PW (2006) RNA polymerase encounters with DNA damage: transcription-coupled repair or transcriptional mutagenesis? Chem Rev 106:474–488

Schmidt KH, Reimers JM, Wright BE (2006) The effect of promoter strength, supercoiling and secondary structure on mutation rates in *Escherichia coli*. Mol Microbiol 60:1251–1261

Skandalis A, Ford BN, Glickman BW (1994) Strand bias in mutation involving 5-methylcytosine deamination in the human *hprt* gene. Mutat Res 314:21–26

Srivatsan A, Tehranchi A, MacAlpine DM, Wang JD (2010) Co-orientation of replication and transcription preserves genome integrity. PLoS Genet 6:e1000810

Sung HM, Yeamans G, Ross CA, Yasbin RE (2003) Roles of YqjH and YqjW, homologs of the *Escherichia coli* UmuC/DinB or Y superfamily of DNA polymerases, in stationary-phase mutagenesis and UV-induced mutagenesis of *Bacillus subtilis*. J Bacteriol 185:2153–2160

Svejstrup JQ (2002) Transcription repair coupling factor: a very pushy enzyme. Mol Cell 9:1151–1152

Taddei F, Hayakawa H, Bouton M, Cirinesi A, Matic I, Sekiguchi M, Radman M (1997) Counteraction by MutT protein of transcriptional errors caused by oxidative damage. Science 278:128–130

Tamkun JW (2007) Stalled polymerases and transcriptional regulation. Nat Genet 39:1421–1422

Tehranchi AK, Blankschien MD, Zhang Y, Halliday JA, Srivatsan A, Peng J, Herman C, Wang JD (2010) The transcription factor DksA prevents conflicts between DNA replication and transcription machinery. Cell 141:595–605

Teng G, Papavasiliou FN (2007) Immunoglobulin somatic hypermutation. Annu Rev Genet 41:107–120

Tornaletti S (2009) Transcriptional processing of G4 DNA. Mol Carcinog 48:326–335

Tornaletti S, Park-Snyder S, Hanawalt PC (2008) G4-forming sequences in the non-transcribed DNA strand pose blocks to T7 RNA polymerase and mammalian RNA polymerase II. J Biol Chem 283:12756–12762

Toulme F, Mosrin-Huaman C, Sparkowski J, Das A, Leng M, Rahmouni AR (2000) GreA and GreB proteins revive backtracked RNA polymerase in vivo by promoting transcript trimming. EMBO J 19:6853–6859

Trautinger BW, Jaktaji RP, Rusakova E, Lloyd RG (2005) RNA polymerase modulators and DNA repair activities resolve conflicts between DNA replication and transcription. Mol Cell 19:247–258

Vidales LE, Cardenas LC, Robleto E, Yasbin RE, Pedraza-Reyes M (2009) Defects in the error prevention oxidized guanine system potentiate stationary-phase mutagenesis in *Bacillus subtilis*. J Bacteriol 191:506–513

Viswanathan A, You HJ, Doetsch PW (1999a) Phenotypic change caused by transcriptional bypass of uracil in nondividing cells. Science 284:159–162

Viswanathan A, Liu J, Doetsch PW (1999b) *E. coli* RNA polymerase bypass of DNA base damage. Mutagenesis at the level of transcription. Ann N Y Acad Sci 870:386–388

Witkin EM (1966) Radiation-induced mutations and their repair. Science 152:1345–1353

Wright BE (1996) The effect of the stringent response on mutation rates in *Escherichia coli* K-12. Mol Microbiol 19:213–219

Wright BE (2000) A biochemical mechanism for nonrandom mutations and evolution. J Bacteriol 182:2993–3001

Wright BE (2004) Stress-directed adaptive mutations and evolution. Mol Microbiol 52:643–650

Wright BE, Minnick MF (1997) Reversion rates in a *leuB* auxotroph of *Escherichia coli* K-12 correlate with ppGpp levels during exponential growth. Microbiology 143(Pt 3):847–854

Wright BE, Reimers JM, Schmidt KH, Reschke DK (2002) Hypermutable bases in the *p53* cancer gene are at vulnerable positions in DNA secondary structures. Cancer Res 62:5641–5644

Wright BE, Reschke DK, Schmidt KH, Reimers JM, Knight W (2003) Predicting mutation frequencies in stem-loop structures of derepressed genes: implications for evolution. Mol Microbiol 48:429–441

Wright BE, Schmidt KH, Minnick MF (2004) Mechanisms by which transcription can regulate somatic hypermutation. Genes Immun 5:176–182

Wrzesinski M, Nieminuszczy J, Sikora A, Mielecki D, Chojnacka A, Kozlowski M, Krwawicz J, Grzesiuk E (2010) Contribution of transcription-coupled DNA repair to MMS-induced mutagenesis in *E. coli* strains deficient in functional AlkB protein. Mutat Res 688:19–27

Zuker M (2003) Mfold web server for nucleic acid folding and hybridization prediction. Nucleic Acids Res 31:3406–3415

Chapter 4
Transposon Mutagenesis in Disease, Drug Discovery, and Bacterial Evolution

Zhongge Zhang, Jing Wang, Maksim A. Shlykov, and Milton H. Saier Jr.

Abstract Transposons are diverse and ubiquitous in living organisms, where they function in gene expression regulation, chromosomal rearrangement, and acceleration of evolutionary change. Their potential uses in, and consequences to, human health are numerous. The smallest transposons, insertion sequence (IS) elements, are prevalent in prokaryotes. Recently, one such element, IS5, has been shown to activate expression of the *glpFK* operon in *E. coli* in a mechanistically well-defined Lamarckian fashion (Zhang and Saier, PLoS Genet 5:e1000689, 2009a; Mol Microbiol 74:29–43, 2009b). A recent publication by Wang and Wood (ISME J 5:1517–1525, 2011) has revealed that IS5 also mediates directed mutation of the *E. coli* flagellar master switch operon, *flhDC*, and unpublished evidence suggests that the same occurs in other *E. coli* operons. The question is no longer, "Darwin *or* Lamarck?"; we must now see biological evolution in terms of Darwin *and* Lamarck. We are at the threshold of a new era in biological research with a grasp of the genetic basis for Lamarckian evolution at hand.

Universal Transposons

General Features

Complex compound transposons and small insertion sequence (IS) elements have been found in diverse groups of organisms. The discovery and engineering of transposons active in vertebrate cells (Miskey et al. 2005) promoted research on

The most incomprehensible thing in the universe is that the universe is so comprehensible.
-Albert Einstein

Z. Zhang • J. Wang • M.A. Shlykov • M.H. Saier Jr. (✉)
Division of Biological Sciences, Department of Molecular Biology, University of California, 9500 Gilman Drive, San Diego, La Jolla, CA 92093-0116, USA
e-mail: zzhongge@ucsd.edu; missingly@yahoo.com; mshlykov@ucsd.edu; msaier@ucsd.edu

D. Mittelman (ed.), *Stress-Induced Mutagenesis*, DOI 10.1007/978-1-4614-6280-4_4,

oncogenes and tumor suppressors. These elements were first reported by Barbara McClintock in 1951 during a Cold Spring Harbor Symposium where she described her ground breaking work entitled *Chromosome Organization and Genetic Expression*. She noted that gene transposition requires the presence of an activator, and does not follow the laws of genetics as defined by Mendel. She also concluded that some transpositions affect more than one phenotype, and the creation of a new phenotype is sometimes reversible (McClintock 1951).

Today, transposition is defined as the relocation of specialized DNA elements within chromosomes or extra-chromosomal elements, mediated by a "transposase," the enzyme of transposition. Regions of DNA flanked by inverted repeats are both the binding sites and substrates of the transposase, and the DNA within the inverted repeats is referred to as transposable elements or "transposons." Transposition occurs isoenergetically, without the consumption of nucleoside triphosphates (Haren et al. 1999).

Transposons in Higher Eukaryotes

Retrotransposons: Introduction

Retrotransposons, a specialized class of transposons common in eukaryotes, have been important drivers of evolutionary change and diversification in higher organisms (Maksakova et al. 2008). They have been studied in detail in both animals and plants, although they have been, and continue to be, important for the evolution of many other organisms as well. They fall into several superfamilies, categorizations valuable for understanding their modes of action and for purposes of classification.

Accumulating evidence suggests that hybrid genetic dysfunctions accrue not only because of sequence divergence and incompatible alleles, but also because of numerous mechanistic disorders related to the maintenance of chromatin integrity (Michalak 2009). For example, it has been observed that hybridization in both plants and mammals, and probably many other types of eukaryotes, disrupts patterns of DNA methylation and imprinting. These epigenetic changes can result from the transcriptional activation and mobilization of transposable elements in hybrids, and are subject to small regulatory RNA control. These controls play roles in both suppressing transposable elements and epigenetic control (Maksakova et al. 2008; Michalak 2009).

Transposons and Epigenetics

Epigenetic changes associated with DNA methylation and histone modifications that lead to chromatin remodeling and regulation of gene expression underlie differentiation. They also result in the mis-programming of many types of developmental disorders, such as obesity, type 2 diabetes, cardiovascular diseases, and various metabolic syndromes (Junien et al. 2005). Converging data support the notion that individuals with these conditions often suffer from improper "epigenetic programming" during their fetal/postnatal development. This can result from inadequate nutrition and metabolic disturbances during fetal development and childhood. Such epigenetic disorders can be transmitted to the next generation(s). Susceptibility of epigenetic gene expression to altered environmental and internal conditions is well documented. Thus, a body of research suggests the participation of transposable elements in epigenetic program modification (Junien et al. 2005).

Retrotransposons in Plants

Tnt1 elements comprise a superfamily of LTR-retrotransposons distributed throughout much of the plant kingdom. The well-studied *Solanaceae* family provides an example. These elements represent excellent model systems for studying regulatory and evolutionary controls established between hosts and mobile genetic elements.

Tnt1 retrotransposons control their own activation, restricting expression to specific external environmental conditions. For example, the Tnt1A element is expressed in response to stress, and its activation by microbial factors is followed by gene amplification (Grandbastien et al. 2005). This demonstration revealed that factors produced by prokaryotic pathogens can generate genetic diversity in plants. This suggests a level of inter-domain communication not previously anticipated. It reveals how distinct interacting organisms within an ecological system can influence the evolutionary development of another organism.

There can be three-way regulatory interactions between plant cells, bacterial cells and transposable elements. The Tnt1A promoter has the potential of being activated by various biotic and abiotic stimuli as noted above. Several of these are specifically repressed in plants (e.g., tobacco), and are revealed only when the LTR promoter is placed in a heterologous context. Grandbastien et al. (2005) proposed that over evolutionary time, stimulus-specific repression might be established in order to minimize activation under conditions that might generate germinal transposition.

In addition to tight transcriptional control mechanisms, Tnt1A retrotransposons self-regulate their activities through the gradual generation of defective copies that have reduced transcriptional activity. Tnt1 retrotransposons found in various plant species are characterized by a high level of variability in the LTR sequences involved in transcription: they have evolved complexity by gaining new expression patterns. These patterns are frequently associated with responses to diverse stress conditions.

Tnt1A insertions that are associated with specific genic regions are initially favored. However, they may subsequently be counter-selected, while insertions in repetitive DNA are maintained. Still, amplification and loss of insertions may result from more significant events, as suggested by the large restructuring of Tnt1 populations observed in certain plants. The distribution of Tnt1 elements thus appears to be in dynamic flux, with amplification counterbalanced by loss of insertions.

Tnt1 insertional polymorphisms occur frequently, showing that the behavioral patterns of Tnt1 retrotransposons differ between host species, responding to differences in expression conditions during the environmental histories of the host species.

Human Retroviruses

Transposons, including retrotransposons and retroviruses, come in many varieties and in many different organismal types, from bacteria to plants and animals (see Zhang and Saier 2011; Zhang et al. 2011). In fact, endogenous retroviruses belong to the superfamily of transposable and retrotransposable genetic elements that represent about 8% of the human genome. These mobile genetic elements, and their numerous inactivated "junk" sequences, probably arose during early animal genome evolution; they represent nearly one half of human DNA. In some eukaryotes, including plants, these percentages are substantially higher. In fact, a significant part of this "unconventional genome" has retained potential activity, arguing against the use of the term "junk" for nonstructural gene regions of a eukaryotic genome.

Epigenetic control is involved in silencing many of these genetic elements. However, certain environmental factors, including the presence of viruses, are known to dysregulate their expression in a variety of cell types (Perron and Lang 2010). For example, embryonal cells with limited gene methylation are most susceptible to uncontrolled activation of these mobile genetic elements by viral infections and other chance events.

Certain viruses transactivate promoters in members of the human endogenous retroviral family type W (HERV-W) (Kurth and Bannert 2010). HERV-W RNA was first isolated from viral particles (e.g., Multiple Sclerosis-associated RetroViral element, MSRV) that had been associated with the evolution and prognosis of multiple sclerosis (MS) (Antony et al. 2011). HERV-W elements encode a powerful immuno-pathogenic envelope protein (ENV) that activates a pro-inflammatory/autoimmune cascade through interaction of the Toll-like receptor 4 located on immune cells. This ENV protein has repeatedly been detected in MS brain lesions and may be involved in this and other diseases. Epigenetic factors controlling HERV-W ENV protein expression thus prove critical for various neurological disorders (Perron and Lang 2010).

Transposon Use for Gene Therapy and Stem Cell Development

Effective gene therapy requires (1) accurate gene delivery into the relevant target cells, (2) long-term gene expression, and (3) minimal risk of secondary deleterious effects. The development of efficient and safe viral and nonviral vectors could greatly facilitate the development of clinical gene therapy protocols. However, to date, nonviral gene transfer approaches have typically resulted in limited stable gene transfer efficiencies in primary cells. The use of nonviral gene delivery approaches together with the use of the most recent transposon technologies may potentially overcome some of these limitations (Belay et al. 2011; VandenDriessche et al. 2009).

Large-scale genetic screens using mammalian cells have yielded stable gene expression in in vivo experiments using the mouse model system. Human clinical trials have been approved for the use of redirected T-cells for gene therapy of B-cell lymphoma. Additionally, induced pluripotent stem cells could theoretically be generated after genetic reprogramming, with transposons encoding reprogramming factors (VandenDriessche et al. 2009). These developments underscore the emerging potential of transposons for gene therapy applications and induced pluripotent stem cell generation for regenerative medicinal purposes (Belay et al. 2011).

Transposons for Cancer Gene Discovery

There is a major push in cancer research to the genes and signaling pathways that cause cancer. The hope is that this will lead to the identification of new drug targets that are amenable to molecular therapies. However, recent discoveries suggest that the cancer genome includes hundreds of infrequently mutated genes in addition to the few common cancer genes. It is important to establish whether the role of these genes in cancer is direct or indirect.

An approach currently used by the International Cancer Genome Consortium is to sequence the genomes of hundreds of tumors of each type of human cancer: 25,000 genomes in total (Copeland and Jenkins 2010). This expensive approach will undoubtedly yield information at the genomic, epigenomic and transcriptomic levels; however, by itself, it is not likely to allow identification of causal relationships.

A high-throughput method for cancer gene discovery in the mouse has involved retroviral insertional mutagenesis (RIM) (Kool and Berns 2009). Retroviruses can induce cancer as part of their normal viral cycles. For example, proviral DNA can integrate into the mouse genome to deregulate expression of an oncogene, or inactivate a tumor suppressor gene. The retroviral integration sites in tumors thus mark the locations of candidate cancer genes.

Transposon-based insertional mutagenesis (TIM) provides an alternative high-throughput approach for cancer gene discovery (Ivics and Izsvak 2010). Most transposable elements, excluding retrotransposons, use a "cut and paste" mechanism where the transposable element-encoded transposase catalyzes the excision of the transposon from an original location in the genome and promotes reintegration elsewhere. There are two types of transposons: autonomous transposons encode an active transposase and are thus capable of transposing on their own, while nonautonomous transposons lack a functional transposase, but retain the *cis*-acting DNA sequences that are necessary for transposition. Nonautonomous transposons are therefore active only when the transposase is supplied in *trans*. This, in principle, allows one to control the tissue in which TIM occurs by limiting where the transposase is expressed. Therefore, TIM can be used to selectively model many types of cancer (Copeland and Jenkins 2010).

Insertion Sequence Elements in Prokaryotes

Within the transposon superfamily are IS elements, the smallest transposons known. They generally encode only functions related to their mobility (Mahillon and Chandler 1998). More than 500 such elements have been identified in both prokaryotes and eukaryotes, and their length is usually less than 2.5 kb. In 1981, Esther Lederberg initiated the nomenclature system for IS elements by simply adding a single number to an IS element, i.e., IS1, IS2, IS3, etc. (Lederberg 1981).

A crucial aspect of transposition is the target site in the genome where integration takes place (Guynet et al. 2009). The majority of integration sites appear to be random, although exceptions have been observed. Some IS elements prefer to insert under the directive of a DNA structure such as bent DNA (Pribil and Haniford 2003), single-nucleotide mismatches (Yanagihara and Mizuuchi 2002), or specific short target sequences (Barabas et al. 2008).

The insertion sequence 5 (IS5) family is represented in both bacteria and archaea, with members that carry a single transposase gene. They are usually 800–1,700 bp long (Mahillon and Chandler 1998). The 1,195-base pair *E. coli* IS5 "hops" at increased rates under environmental stress conditions, as is true for many IS elements. The gene activation response provides the host bacterium with survival advantages during periods of nutrient depletion. It is capable of both precise and imprecise excision (Zhang et al. 2010).

Gene Activation by Insertion Sequences

Elevated neighboring gene expression is sometimes a consequence of the presence of an IS element such as IS1, IS2, or IS5 in *E. coli*. These ISs, when inserted at appropriate locations relative to the −10 hexamer of a promoter, can activate it, or create a novel promoter capable of enhancing the expression of the downstream genes (Galas and Chandler 1989). Some ISs have detectable −35 hexamers within the terminal IRs, as observed for IS2 (Szeverenyi et al. 1996), IS21 (Reimmann et al. 1989), IS30 (Dalrymple 1987), and IS257 (Leelaporn et al. 1994).

Four mechanisms of gene activation by transposable elements have been observed: (1) derepression, (2) creation of a completely new promoter, (3) formation of a novel hybrid promoter, and (4) activation of a silent native promoter. One well-studied example of gene activation by counteracting repression was conducted by Schnetz and Rak (1992) on IS5 enhancement of the β-glucoside (*bgl*) catabolic operon, which is one of several operons responsible for β-glucoside utilization in

E. coli. Insertion of IS5 at various positions, both upstream and downstream of the *bgl* promoter, resulted in enhancement of promoter activity (Schnetz and Rak 1992; Reynolds et al. 1981). Further studies revealed a most complex regulatory mechanism (Venkatesh et al. 2010). IS5 also activates the normally suppressed fucose/propanediol *fucAO* catabolic operon in *E. coli* in the presence of propanediol (Zhang et al. 2010; Chen et al. 1989; Hacking and Lin 1976). Such behavior allows bacteria to metabolize β-glucosides and propanediol as alternative nutrient sources, respectively.

A novel mechanism of transposon-mediated gene activation involving the *glpFK* operon in *E. coli* has been elucidated (Zhang and Saier 2009a, b). IS5 activates the *glpFK* operon in *crp* mutants lacking the cyclic AMP receptor protein. They thus utilize glycerol in the absence of Crp at rates that surpass the wild type rates. In this case, IS5 harbors unique sequences containing A-tracts and an IHF binding site, and these proved to be responsible for *glpFK* operon activation (Zhang and Saier 2009b). The IS5 hopping event exhibited all of the characteristics of directed mutation (Zhang and Saier 2009a), and the mechanism of this Lamarckian mutational event has been elucidated: binding of the glycerol repressor GlpR to the *glpFK* operators controls IS5 hopping to its target site upstream of the *glpFK* promoter (Zhang and Saier 2009a, b).

Unlike IS-promoted modes of gene activation, IS-mediated chromosomal rearrangements require the presence of two or more similar—if not identical—sequences in the genome. Such rearrangement may ultimately lead to the alteration of distal gene activities, and to genome plasticity. Two possible mechanisms may cause the rearrangement of chromosomal DNA by transposons: (1) indirectly by homologous recombination or (2) directly by an alternative transposition process (Gray 2000).

In another gene activating effect of IS5, a genomic rearrangement was produced by the presence of two IS5 elements (Zinser et al. 2003). First, an IS5 transpositional event inactivated the *cstA* gene (the product of which is possibly an oligopeptide permease (Schultz and Matin 1991)), and second, an inversion between this IS5 and another IS5 about 60 kb away activated the *ybeJ-gltJKL* four-gene operon (Zinser et al. 2003). Thus, gene activation by a single IS element can occur by a variety of mechanisms.

The Flagellar Master Switch

Microorganisms with flagella are able to seek favorable conditions and avoid detrimental environments, thus gaining a competitive advantage over other organisms (Macnab 1996). Motility and chemotaxis capabilities allow them to avoid starvation conditions and toxic compounds, and thus to survive under a wide variety of changing environmental conditions. Due to the high energy expenditure required for motility, and the possible elicitation of a strong immune response to flagellar antigens by host organisms, the expression of flagellar genes is highly regulated by a wide range of environmental parameters (see examples illustrated in Fig. 4.1).

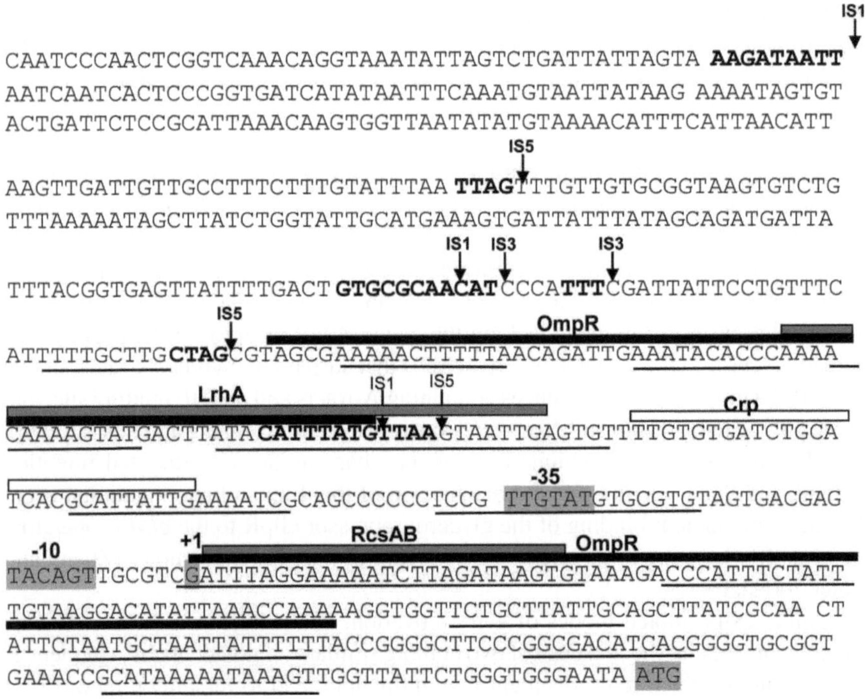

Fig. 4.1 The structure of the *flhDC* promoter region, showing the binding sites for three of the protein regulators as well as the insertion sites for IS elements. Modified from Fig. 4 in Barker et al. (2004). The −10 and −35 elements, transcription start site (+1), and the first codon of *flhD* are highlighted in *grey*. The H-NS binding sites are *underlined*. The IS insertion sites are *bolded*

Regulation of *E. coli* flagellar synthesis occurs via a three level hierarchy, in which expression of one gene at a given level requires the transcription of another gene, expressed at an earlier stage (Macnab 1996; Smith and Hoover 2009). At the top of the hierarchy is the *flhDC* master operon, which encodes a gene product, $FlhD_2C_2$. This heterotetrameric transcriptional regulator activates at least eight Class II flagellar promoters, and 39 non-flagellar promoters (Fernandez De Henestrosa et al. 2000; Stafford et al. 2005). Class II flagellar genes encode components of the flagellar protein export system, the flagellar basal body, and a minor RNA polymerase subunit σ factor, σ^F, which up-regulates the activity of Class III genes. These Class III genes encode the components of the flagellar filament (flagellin, FliC), as well as hook-associated, motor, and chemotaxis proteins. Class III genes are down-regulated by the anti-sigma factor, FlgM (Smith and Hoover 2009), which is retained inside the cell until synthesis of the flagellar basal body is complete (Hughes et al. 1993). Then FlgM is exported via this apparatus from the cytoplasm into the external medium, where it is inactive. This transport event frees σ^F for activation of Class III genes.

Numerous environmental signals control the expression of *flhDC*, (e.g., temperature, osmolarity, and pH, as well as carbon, nitrogen, phosphorous, and sulfur starvation) (Smith and Hoover 2009; Quan et al. 2002; Soutourina et al. 2002). Global regulatory proteins that influence *flhDC* expression include H-NS, the cyclic AMP receptor protein cAMP-CRP, the osmotic pressure sensor, EnvZ/OmpR, and the sulfur sensor, CysB (Quan et al. 2002; Bertin et al. 1994; Soutourina et al. 1999). In fact the *flhDC* operon is arguably sensitive to regulation by more agents and conditions than any other operon in *E. coli*, and the promoter region contains binding sites for numerous pleiotropic transcription factors in addition to those mentioned above (Wang and Wood 2011; Smith and Hoover 2009; Du et al. 2011; Krin et al. 2010; Lemke et al. 2009; Theodorou et al. 2011; Wada et al. 2011). In addition, the mRNA produced is stabilized by the presence of the RNA binding regulator, CsrA (Romeo 1998;Wei et al. 2000).

During nutrient depletion, bacteria may activate *flhDC* transcription, allowing them ultimately to move toward more nutrient rich environments. Activation of *flhDC* is often coordinated with growth stage; *E. coli* can undergo a "lifestyle" change from a motile planktonic existence to an adhesive-sedentary one when *E. coli* transitions from late-exponential to early-stationary phase (O'Toole et al. 2000; Hall-Stoodley et al. 2004). It is therefore not surprising that $FlhD_2C_2$ influences expression of numerous genes that have little or nothing to do with motility.

Effects of IS Insertions Upstream of the *flhDC* Operon on Motility and Gene Expression

It had previously been reported that transcriptional activation of the *flhDC* operon, encoding the master switch for flagellar biogenesis, can occur in response to the insertion of IS elements upstream of the promoter of this operon, accompanied by increased rates of motility and swarming (Barker et al. 2004; Gauger et al. 2007).

Table 4.1 presents the fast swimming mutants that we isolated that resulted from insertion of IS1, IS3, or IS5 into the upstream control region of the *flhDC* operon. Twelve independently isolated mutants had IS1 inserted in three different positions, in either the direct or reverse orientations. Another 12 mutants had IS3 inserted in two different positions, always in the reverse orientation, and 15 independently isolated mutants had IS5 inserted in two different positions. In one of these positions (−99) 3 of 12 mutants were of the direct orientation, while 9 were of the reverse orientation. The remaining three mutants (at position −318) were of the direct orientation. These results suggest that some positions accept the IS element in only one orientation (direct or reverse) to give rise to increased motility, while other positions allow insertion in either orientation to achieve *flhDC* operon activation. These IS insertions arose at frequencies that were many orders of magnitude higher than those of normal mutation rates. An additional mutant, IS5-169, with IS5 localized at position −169 was provided by the Barker Laboratory (Barker et al. 2004) and was transferred into the genetic background of BW25113. All strains examined in this study are therefore isogenic.

Table 4.1 Fast swimming insertion sequence (IS) mutants in the *flhDC* operon upstream regulatory region

Transposon	Number of mutants	Insertion site	Location	Orientation
IS1	3	CATTTATG	−107	Inverted
IS1	3	GTGCGCAA	−214	Inverted
IS1	6	AAGATAATT	−469	Direct
IS3	6	TTT	−199	Inverted
IS3	6	CAT	−206	Inverted
IS5	3	TTAA	−99	Direct
IS5	9	TTAA	−99	Inverted
IS5	3	TTAG	−318	Direct

Two selected flagellar genes in these mutants were fused to *lacZ* to monitor expression levels. First, fusions were made to the Class I *flhDC* operon, and second, fusions were made to *flhB*, a Class II gene encoding a component of the flagellar export pore/channel in the basal region of the flagellum. These fusions were constructed in *E. coli* K-12 strain BW25113 (Zhang and Saier 2009a, b; Zhang et al. 2010) in order to characterize fast swimming variants that were mutated by an IS element insertion upstream of the *flhDC* promoter. The insertional mutants had been isolated from soft agar plates consisting of 0.3% Fisher BioReagents Agar with 4 g/l Difco Nutrient Broth with or without 0.5% glucose or 1% glycerol. These insertional mutants were analyzed with respect to the effects of the IS insertions on gene expression and motility in soft agar.

All IS insertional mutations examined gave rise to increased *flhDC* expression under all conditions assayed. These conditions included minimal (M9) media with glucose or glycerol as the carbon source, and Luria Bertani (LB) or nutrient broth (NB), with or without glucose. The results obtained with the IS5 mutants in LB with or without glucose are presented in Fig. 4.3.

All strains showed best expression when grown in minimal or complex medium, with glucose strongly repressing expression in the wild type strain. However, all of the IS5 insertional mutants showed little or no glucose repression. Glycerol, which generally increased swarming motility, may have increased the cellular energy level, accounting for increased gene expression. Glucose, on the other hand, is known to mediate catabolite repression of flagellar synthesis in wild type cells (Polen et al. 2003; Silverman and Simon 1974). This observation is in contrast to an earlier report that claimed that in a close relative of *E. coli*, *Serratia marcescens*, glucose does not repress *flhDC* expression (Liu et al. 2000). In fact, the results observed by Liu et al. with *Serratia* resemble those we found with *E. coli* strains bearing upstream IS5. A broader study of motility in various enterobacteria revealed that some exhibit the phenomenon of glucose-mediated catabolite repression, while others do not (Lai et al. 1997). Interestingly, Kitamura et al. (1994) showed, surprisingly, that perturbations in phospholipid composition can also cause *flhDC* repression. This effect proved to involve the stationary phase/starvation sigma factor, σ^S (Uchiyama et al. 2010).

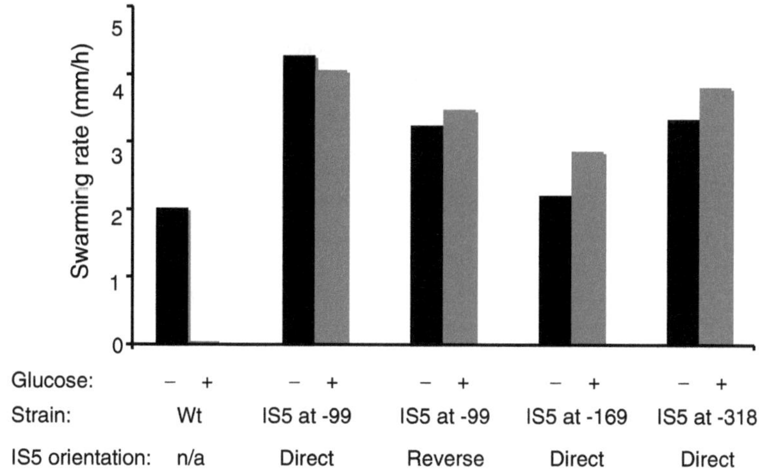

Fig. 4.2 Effects of IS5 at various locations and orientations on *E. coli* cell swarming in the absence or presence of glucose. Swarming plates containing 0.25% agar and 50% normal strength nutrient broth (NB) were made, as described previously (Zhang and Saier 2009a). When appropriate, glucose was added to the media at 0.5% (w/v). Overnight NB cultures were diluted with 50% normal strength NB to an OD_{600} of 1.0, and 1 μl of cells were inoculated onto the surfaces of semi-solid agar plates. The plates were incubated at 25°C. The diameters of the swarming zones were measured every hour until the swarming zones reached the plate margins. Linear lines for sizes of swarming zones versus time were drawn; slopes of these lines are a measure of the swarming rates (mm/h)

In the isogenic *E. coli* K12 BW25113 strains included in our study, motility and *flhDC* expression in wild type—but not insertional—mutants are clearly sensitive to glucose catabolite repression, as revealed in Fig. 4.2 and Table 4.2. When a downstream gene (*flhB* (Class II)) was examined, results in general were parallel to but stronger than those observed for *flhDC* as expected if $FlhD_2C_2$ is the master switch activator of the entire flagellar cascade (see Figs. 4.3 and 4.4 and Table 4.2). These results show that activation of *flhDC* expression is physiologically relevant, and suggests that this cascade effect is directly responsible for the observed effects of the IS5 insertions and the growth condition examined on swarming motility.

IS5 Mediates Lamarckian Directed Mutation of the Master Switch *flhDC* Operon

It is well established that mutation rates may be influenced by the environment, and, as noted above for IS5-mediated activation of the *glpFK* operon, these mutations can arise in a "directed" fashion by an established mechanism (Zhang and Saier 2009a, b; Zhang et al. 2010; Zhang and Saier 2010). Recently, Wang and Wood (2011) have provided compelling evidence that IS5 insertional mutations upstream of the *E. coli* K12 BW25113 *flhDC* operon occur in a directed fashion, similar to

Table 4.2 Effects of IS5 insertional mutations in the *flhDC* upstream region on swarming and expression of the *flhDC* and *flhB* genes of *E. coli*

	(A) Swarming		(B) *flhDC-lacZ* (Class I)	
	NB	NB+glc	NB	NB+glc
WT	2.0	0.04	24.3	4.6
-318	3.3	3.8	59.8	46.4
-169	2.2	2.9	47.9	38.9
-99 (D)	4.3	4.1	68.3	56.3
-99 (I) (inverted)	3.2	3.5	63.8	51.9
			(C) *flhB-lacZ* (Class II)	
WT			29.3	3.8
-318			360	251
-169			429	302
-99 (D)			483	395
-99 (I) (inverted)			454	344

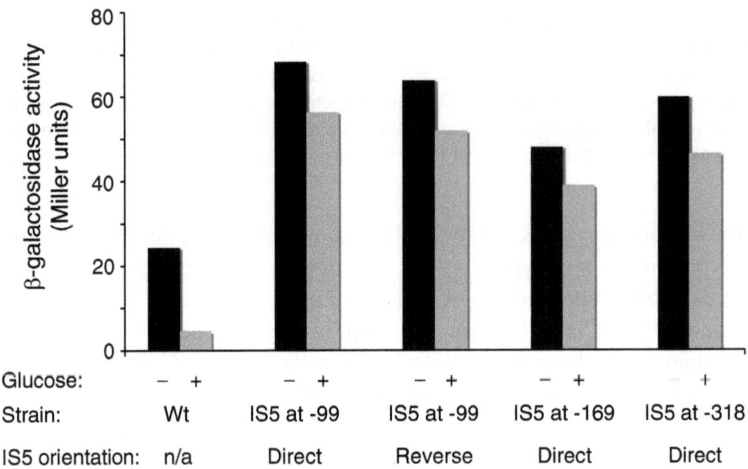

Fig. 4.3 Depiction of *flhDC* operon expression in response to various growth conditions in wild type and selected IS5 insertional mutants in the *flhDC* upstream region. All strains are isogenic (see also Table 4.2)

that observed for the *E. coli glp* operon. These investigators demonstrated that IS5 inserts into the upstream region of the *flhDC* operon at a specific site, in a manner that depends on whether or not the environment permits motility. As in the cases reported in section "Effects of IS Insertions Upstream of the *flhDC* Operon on Motility and Gene Expression," the IS5 insertions studied increased motility.

Although the mechanism was not fully defined, the results of Wang and Wood (2011) appear to establish a second example of directed mutation. IS5 inserts upstream of the *flhDC* promoter only when cells are grown in soft-agar plates that permit swarming motility. However, it does not insert in this region when cells are

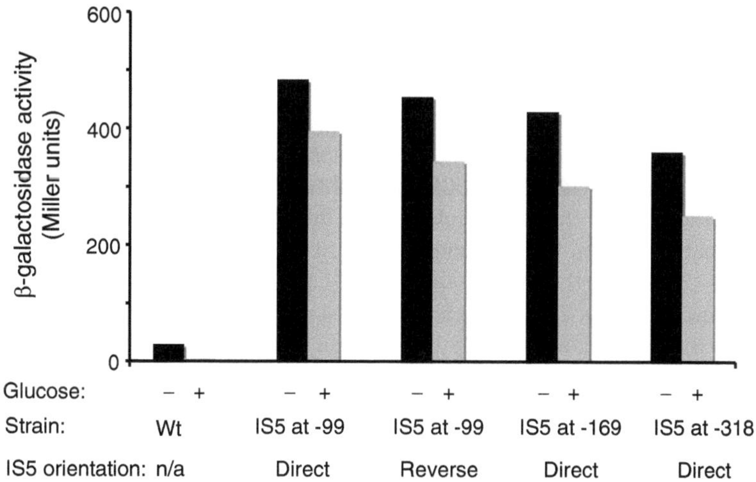

Fig. 4.4 Effects of IS5 insertional mutations in the *flhDC* upstream region on *flhB* gene (Class II) expression in response to various growth conditions. As observed for *flhDC* expression (Fig. 4.1), glucose always repressed the wild type but had little or no effect on the IS5 insertional mutants

grown on hard-agar plates, conditions that do not permit swarming motility. They also do not occur in planktonic cultures. Furthermore, only one IS5 insertional event was detected when the cells were grown on soft-agar plates. This fact contrasts with the situation described in "Effects of IS Insertions Upstream of the *flhDC* Operon on Motility and Gene Expression," above, probably reflecting differences in growth and selection conditions, since both research groups used the same *E. coli* K12 strain (BW25113). Regardless, the results reported indicated that the increased rate of insertion of IS5 into the *flhDC* upstream region under conditions that allowed swarming motility was not due to a general elevation of IS5 transposition throughout the whole genome.

Somewhat surprisingly, Wang and Wood (2011) also showed that highly motile cells with IS5 upstream of *flhDC* have greater biofilm forming capabilities. There also seemed to be a growth cost, probably due to the energetic burden of enhanced motility; these highly motile cells gave a reduced growth rate and a lower yield of cells in rich medium.

Functional flagella appeared to be required for IS5 insertion upstream of *flhD*, as there was no IS5 insertion upstream of *flhDC* in mutants defective for the *flhD*, *flgK*, and *motA* genes. The activating IS5 insertional mutations appeared to be relatively stable under various sets of normal physiological conditions, although IS5 is known to excise both precisely and imprecisely at low frequencies (Zhang et al. 2010). Additionally, the IS5 insertional mutations of Wang and Wood (2011) were shown to arise during biofilm formation. This could serve the function of creating genetic and phenotypic diversity. Hence, the cells appear to "sense" whether motility is feasible before a subpopulation undergoes a mutation to become hypermotile by a sensing mechanism that involves activation of the synthesis of the master transcription regulator, $FlhD_2C_2$.

Conclusions at a Glance

In his "Philosophie Zoologique," published over 200 years ago, Jean-Baptiste Lamarck suggested that evolution is driven primarily by non-randomly acquired, beneficial phenotypic change. Fifty years later, Darwin, in his "On the Origin of Species" instead provided evidence for his theory that random undirected change provided the material for natural selection. In this chapter, we see that certain observations that have been made recently better fit the Lamarckian paradigm. Koonin and Wolf (2009) noted that "the prokaryotic CRISPR-Cas system (Takeuchi et al. 2011; Terns and Terns 2011) of defense against mobile elements (Bhaya et al. 2011) seems to function via a bona fide Lamarckian mechanism, namely by integrating small segments of viral or plasmid DNA into specific loci in the host prokaryote genome and then utilizing the respective transcripts to destroy the cognate mobile element DNA (or RNA). They further noted that "a similar principle may be used in the piRNA branch of RNA interference, which is involved in defense against transposable elements in the animal germ line." They considered the possibility that horizontal gene transfer (HGT) may promote quasi-Lamarckian behavior in many, and perhaps all organisms, because in some cases, competence is inducible by stress, and because in some cases, the transfused genes confer a selective advantage. Thus, as noted for the IS5-activated *glpFK* and *flhDC* operons, tightly regulated stress-induced mutagenesis is promoted by the environmental stress conditions that the mutation relieves.

Many genetic mechanisms that generate diversity have been identified in microorganisms. Many chronic infectious diseases are characterized by the formation of biofilms that protect the bacteria from the host's adaptive immune response, as well as predation by phagocytic cells. Also, biofilms allow frequent horizontal gene transfer (HGT), promoting large-scale genome content changes (Ehrlich et al. 2010) if the biofilm is polyclonal (polymicrobial) in nature, as is usually the case. Evidence is available to support the conclusion that HGT enhances the production of novel strains with unique combinations of genic characteristics, and that production of these altered strains leads to persistence of infection. This concept of a population of bacteria undergoing mutagenesis to produce a "cloud" of similar strains to "confuse" and "overwhelm" the host's immune system parallels genetic diversity strategies used by viral and parasitic pathogens.

Transposons are ubiquitous and fall into several classes, each of which predominates in different types of organisms (see "Universal Transposons," above). Of great interest to eukaryotic molecular biologists are the retrotransposons and retroviruses that are particularly common in animals and plants, and may comprise up to 90% of a genome in both active and inactive states (see "Transposons in Higher Eukaryotes," above). These genetic elements alter epigenetic patterns of gene expression, respond in many cases to biotic and abiotic stress conditions, and activate or inactivate gene expression using dozens of different mechanisms. They are being used for gene therapy, stem cell development, and cancer gene discovery, tremendously facilitating basic research while providing practical solutions to some of our major health problems.

The smallest transposons found in nature are the prokaryotic IS elements ("Insertion Sequence Elements in Prokaryotes," above). There are many such elements, and they, like eukaryotic transposons, facilitate evolutionary change by gene activation or gene silencing, and by promoting major genomic rearrangements. In spite of extensive research regarding their "hopping" mechanisms, modes of action, and consequences, it is only now becoming apparent that they are vehicles of Lamarckian directed mutation (Zhang et al. 2011; Zhang and Saier 2010; Saier 2011).

Until 2011, the only published IS element-mediated example of directed mutation was IS5-promoted transcriptional regulation of the *E. coli glpFK* operon (Zhang and Saier 2009a, b; Zhang et al. 2011, 2010; Zhang and Saier 2010, 2011). Thus, while IS-related directed mutational events are only now becoming recognized, the general features of their mechanisms of gene activation are well established (see "Gene Activation by Insertion Sequences"). In fact, several such mechanisms are known. Related analyses, including the IS5-mediated activation of the cryptic *bgl* (β-glucoside utilization) operon in *E. coli*, revealed a degree of complexity almost unheard of in prokaryotes just a decade ago.

A major focus of this chapter is the new work appearing on IS control of the *E. coli* flagellar master switch operon, *flhDC*, which controls the hierarchical cascade of events that determine flagellar biogenesis and consequent motility in *E. coli*. Extensive studies in several laboratories have revealed that *flhDC* expression is controlled by numerous transcription factors and several IS elements. Our studies and those of Barker et al. (2004) and Gauger et al. (2007) have shown that three IS elements, IS1, IS3, and IS5 can activate *flhDC* expression, and downstream flagellar gene expression, in a probable causative hierarchical fashion ("The Flagellar Master Switch" and "Effects of IS Insertions Upstream of the *flhDC* Operon on Motility and Gene Expression"). These results now culminate with the recent and exciting results of Wang and Wood (2011; "IS5 Mediates Lamarckian Directed Mutation of the Master Switch *flhDC* Operon" above) showing that IS5-mediated *flhDC* activation occurs in a Lamarckian directed mutational process, reminiscent of the *glpFK* activation mechanism described, just 2 years earlier (Zhang and Saier 2009a, b). Surprisingly, the conditions Wang and Wood used resulted in the selective insertion of just one of the IS elements that we had studied, IS5, and coincidentally, it corresponds to the one we had independently studied in greatest detail (see Tables 4.1 and 4.2 as well as Figs. 4.1–4.3).

Where Are We Going Now?

We currently have two well-documented examples of IS5-mediated Lamarckian directed mutation. The *glpFK* and the *flhDC* operons of *E. coli* have been shown to be the targets of this type of mutation. Our current unpublished work indicates that the *fuc* regulon of *E. coli* is similarly subject to IS5-mediated directed mutation. The principle of directed mutation, now established with these two examples, may simply be representative. Many more examples are likely to be sought and discovered now that this principle is established. These exciting observations suggest that, just as Einstein showed that Newton's picture of the physical universe was incomplete, so is Darwin's view of biological evolution incomplete. The question is no longer "Darwin or Lamarck?" but "When Darwin?" and "When Lamarck?". It is an exciting time in biology.

Great biologists are led by the results of their research using model biological systems. They perform an experiment which tells them what the next experiment should be, which tells them what the next experiment should be, which tells them what the next experiment should be. They let Nature take the lead, and each experiment brings them closer to the truth. By this empirical approach, Nature reveals herself. Then, the scientific and nonscientific communities must become convinced of the veracity of their findings. When these discoveries go against established dogma, this task becomes most daunting.

Acknowledgments The research conducted in our laboratory was supported by NIH grant GM077402. We thank Carl Welliver for his assistance in the preparation of this chapter.

References

Antony JM, Deslauriers AM, Bhat RK, Ellestad KK, Power C (2011) Human endogenous retroviruses and multiple sclerosis: innocent bystanders or disease determinants? Biochim Biophys Acta 1812:162–176

Barabas O, Ronning DR, Guynet C, Hickman AB, Ton-Hoang B, Chandler M, Dyda F (2008) Mechanism of IS200/IS605 family DNA transposases: activation and transposon-directed target site selection. Cell 132:208–220

Barker CS, Pruss BM, Matsumura P (2004) Increased motility of Escherichia coli by insertion sequence element integration into the regulatory region of the flhD operon. J Bacteriol 186:7529–7537

Belay E, Dastidar S, Vandendriessche T, Chuah MK (2011) Transposon-mediated gene transfer into adult and induced pluripotent stem cells. Curr Gene Ther 11:406–413

Bertin P, Terao E, Lee EH, Lejeune P, Colson C, Danchin A, Collatz E (1994) The H-NS protein is involved in the biogenesis of flagella in Escherichia coli. J Bacteriol 176:5537–5540

Bhaya D, Davison M, Barrangou R (2011) CRISPR-Cas systems in bacteria and archaea: versatile small RNAs for adaptive defense and regulation. Annu Rev Genet 45:273–297

Chen YM, Lu Z, Lin EC (1989) Constitutive activation of the fucAO operon and silencing of the divergently transcribed fucPIK operon by an IS5 element in Escherichia coli mutants selected for growth on L-1,2-propanediol. J Bacteriol 171:6097–6105

Copeland NG, Jenkins NA (2010) Harnessing transposons for cancer gene discovery. Nat Rev Cancer 10:696–706

Dalrymple B (1987) Novel rearrangements of IS30 carrying plasmids leading to the reactivation of gene expression. Mol Gen Genet 207:413–420

Du H, Sheng X, Zhang H, Zou X, Ni B, Xu S, Zhu X, Xu H, Huang X (2011) RpoE may promote flagellar gene expression in Salmonella enterica serovar typhi under hyperosmotic stress. Curr Microbiol 62:492–500

Ehrlich GD, Ahmed A, Earl J, Hiller NL, Costerton JW, Stoodley P, Post JC, DeMeo P, Hu FZ (2010) The distributed genome hypothesis as a rubric for understanding evolution in situ during chronic bacterial biofilm infectious processes. FEMS Immunol Med Microbiol 59:269–279

Fernandez De Henestrosa AR, Ogi T, Aoyagi S, Chafin D, Hayes JJ, Ohmori H, Woodgate R (2000) Identification of additional genes belonging to the LexA regulon in Escherichia coli. Mol Microbiol 35:1560–1572

Galas D, Chandler M (1989) Bacterial insertion sequences. In: Berg D, Howe M (eds) Mob DNA. American Society for Microbiology, Washington, DC

Gauger EJ, Leatham MP, Mercado-Lubo R, Laux DC, Conway T, Cohen PS (2007) Role of motility and the flhDC Operon in Escherichia coli MG1655 colonization of the mouse intestine. Infect Immun 75:3315–3324

Grandbastien MA, Audeon C, Bonnivard E, Casacuberta JM, Chalhoub B, Costa AP, Le QH, Melayah D, Petit M, Poncet C et al (2005) Stress activation and genomic impact of Tnt1 retrotransposons in Solanaceae. Cytogenet Genome Res 110:229–241

Gray YH (2000) It takes two transposons to tango: transposable-element-mediated chromosomal rearrangements. Trends Genet 16:461–468

Guynet C, Achard A, Hoang BT, Barabas O, Hickman AB, Dyda F, Chandler M (2009) Resetting the site: redirecting integration of an insertion sequence in a predictable way. Mol Cell 34:612–619

Hacking AJ, Lin EC (1976) Disruption of the fucose pathway as a consequence of genetic adaptation to propanediol as a carbon source in Escherichia coli. J Bacteriol 126:1166–1172

Hall-Stoodley L, Costerton JW, Stoodley P (2004) Bacterial biofilms: from the natural environment to infectious diseases. Nat Rev Microbiol 2:95–108

Haren L, Ton-Hoang B, Chandler M (1999) Integrating DNA: transposases and retroviral integrases. Annu Rev Microbiol 53:245–281

Hughes KT, Gillen KL, Semon MJ, Karlinsey JE (1993) Sensing structural intermediates in bacterial flagellar assembly by export of a negative regulator. Science 262:1277–1280

Ivics Z, Izsvak Z (2010) The expanding universe of transposon technologies for gene and cell engineering. Mob DNA 1:25

Junien C, Gallou-Kabani C, Vige A, Gross MS (2005) Nutritionnal epigenomics: consequences of unbalanced diets on epigenetics processes of programming during lifespan and between generations. Ann Endocrinol (Paris) 66:2S19–2S28

Kitamura E, Nakayama Y, Matsuzaki H, Matsumoto K, Shibuya I (1994) Acidic-phospholipid deficiency represses the flagellar master operon through a novel regulatory region in Escherichia coli. Biosci Biotechnol Biochem 58:2305–2307

Kool J, Berns A (2009) High-throughput insertional mutagenesis screens in mice to identify oncogenic networks. Nat Rev Cancer 9:389–399

Koonin EV, Wolf YI (2009) Is evolution Darwinian or/and Lamarckian? Biol Direct 4:42

Krin E, Danchin A, Soutourina O (2010) RcsB plays a central role in H-NS-dependent regulation of motility and acid stress resistance in Escherichia coli. Res Microbiol 161:363–371

Kurth R, Bannert N (2010) Beneficial and detrimental effects of human endogenous retroviruses. Int J Cancer 126:306–314

Lai HC, Shu JC, Ang S, Lai MJ, Fruta B, Lin S, Lu KT, Ho SW (1997) Effect of glucose concentration on swimming motility in enterobacteria. Biochem Biophys Res Commun 231:692–695

Lederberg EM (1981) Plasmid reference center registry of transposon (Tn) allocations through July 1981. Gene 16:59–61

Leelaporn A, Firth N, Byrne ME, Roper E, Skurray RA (1994) Possible role of insertion sequence IS257 in dissemination and expression of high- and low-level trimethoprim resistance in staphylococci. Antimicrob Agents Chemother 38:2238–2244

Lemke JJ, Durfee T, Gourse RL (2009) DksA and ppGpp directly regulate transcription of the Escherichia coli flagellar cascade. Mol Microbiol 74:1368–1379

Liu JH, Lai MJ, Ang S, Shu JC, Soo PC, Horng YT, Yi WC, Lai HC, Luh KT, Ho SW et al (2000) Role of flhDC in the expression of the nuclease gene nucA, cell division and flagellar synthesis in Serratia marcescens. J Biomed Sci 7:475–483

Macnab R (1996) Flagella and motility. In: Neidhardt FEA (ed) Escherichia coli and Salmonella typhimurium: cellular and molecular biology. American Society for Microbiology, Washington, DC

Mahillon J, Chandler M (1998) Insertion sequences. Microbiol Mol Biol Rev 62:725–774
Maksakova IA, Mager DL, Reiss D (2008) Keeping active endogenous retroviral-like elements in check: the epigenetic perspective. Cell Mol Life Sci 65:3329–3347
McClintock B (1951) Chromosome organization and genic expression. Cold Spring Harb Symp Quant Biol 16:13–47
Michalak P (2009) Epigenetic, transposon and small RNA determinants of hybrid dysfunctions. Heredity (Edinb) 102:45–50
Miskey C, Izsvak Z, Kawakami K, Ivics Z (2005) DNA transposons in vertebrate functional genomics. Cell Mol Life Sci 62:629–641
O'Toole G, Kaplan HB, Kolter R (2000) Biofilm formation as microbial development. Annu Rev Microbiol 54:49–79
Perron H, Lang A (2010) The human endogenous retrovirus link between genes and environment in multiple sclerosis and in multifactorial diseases associating neuroinflammation. Clin Rev Allergy Immunol 39:51–61
Polen T, Rittmann D, Wendisch VF, Sahm H (2003) DNA microarray analyses of the long-term adaptive response of Escherichia coli to acetate and propionate. Appl Environ Microbiol 69:1759–1774
Pribil PA, Haniford DB (2003) Target DNA bending is an important specificity determinant in target site selection in Tn10 transposition. J Mol Biol 330:247–259
Quan JA, Schneider BL, Paulsen IT, Yamada M, Kredich NM, Saier MH Jr (2002) Regulation of carbon utilization by sulfur availability in Escherichia coli and Salmonella typhimurium. Microbiology 148:123–131
Reimmann C, Moore R, Little S, Savioz A, Willetts NS, Haas D (1989) Genetic structure, function and regulation of the transposable element IS21. Mol Gen Genet 215:416–424
Reynolds AE, Felton J, Wright A (1981) Insertion of DNA activates the cryptic bgl operon in E. coli K12. Nature 293:625–629
Romeo T (1998) Global regulation by the small RNA-binding protein CsrA and the non-coding RNA molecule CsrB. Mol Microbiol 29:1321–1330
Saier MH Jr (2011) Did adaptive and directed mutation evolve to accelerate stress-induced evolutionary change? J Mol Microbiol Biotechnol 21:5–7
Schnetz K, Rak B (1992) IS5: a mobile enhancer of transcription in Escherichia coli. Proc Natl Acad Sci USA 89:1244–1248
Schultz JE, Matin A (1991) Molecular and functional characterization of a carbon starvation gene of Escherichia coli. J Mol Biol 218:129–140
Silverman M, Simon M (1974) Characterization of Escherichia coli flagellar mutants that are insensitive to catabolite repression. J Bacteriol 120:1196–1203
Smith TG, Hoover TR (2009) Deciphering bacterial flagellar gene regulatory networks in the genomic era. Adv Appl Microbiol 67:257–295
Soutourina O, Kolb A, Krin E, Laurent-Winter C, Rimsky S, Danchin A, Bertin P (1999) Multiple control of flagellum biosynthesis in Escherichia coli: role of H-NS protein and the cyclic AMP-catabolite activator protein complex in transcription of the flhDC master operon. J Bacteriol 181:7500–7508
Soutourina OA, Krin E, Laurent-Winter C, Hommais F, Danchin A, Bertin PN (2002) Regulation of bacterial motility in response to low pH in Escherichia coli: the role of H-NS protein. Microbiology 148:1543–1551
Stafford GP, Ogi T, Hughes C (2005) Binding and transcriptional activation of non-flagellar genes by the Escherichia coli flagellar master regulator FlhD2C2. Microbiology 151:1779–1788
Szeverenyi I, Bodoky T, Olasz F (1996) Isolation, characterization and transposition of an (IS2)2 intermediate. Mol Gen Genet 251:281–289
Takeuchi N, Wolf YI, Makarova KS, Koonin EV (2011) Nature and intensity of selection pressure on CRISPR-associated genes. J Bacteriol 194:1216–1225
Terns MP, Terns RM (2011) CRISPR-based adaptive immune systems. Curr Opin Microbiol 14:321–327

Theodorou MC, Theodorou EC, Kyriakidis DA (2011) Involvement of AtoSC two-component system in Escherichia coli flagellar regulon. Amino Acids 43:833–844

Uchiyama J, Nobue Y, Zhao H, Matsuzaki H, Nagahama H, Matsuoka S, Matsumoto K, Hara H (2010) Involvement of sigmaS accumulation in repression of the flhDC operon in acidic phospholipid-deficient mutants of Escherichia coli. Microbiology 156:1650–1660

VandenDriessche T, Ivics Z, Izsvak Z, Chuah MK (2009) Emerging potential of transposons for gene therapy and generation of induced pluripotent stem cells. Blood 114:1461–1468

Venkatesh GR, Kembou Koungni FC, Paukner A, Stratmann T, Blissenbach B, Schnetz K (2010) BglJ-RcsB heterodimers relieve repression of the Escherichia coli bgl operon by H-NS. J Bacteriol 192:6456–6464

Wada T, Morizane T, Abo T, Tominaga A, Inoue-Tanaka K, Kutsukake K (2011) EAL domain protein YdiV acts as an anti-FlhD4C2 factor responsible for nutritional control of the flagellar regulon in Salmonella enterica Serovar Typhimurium. J Bacteriol 193:1600–1611

Wang X, Wood TK (2011) IS5 inserts upstream of the master motility operon flhDC in a quasi-Lamarckian way. ISME J 5:1517–1525

Wei B, Shin S, LaPorte D, Wolfe AJ, Romeo T (2000) Global regulatory mutations in csrA and rpoS cause severe central carbon stress in Escherichia coli in the presence of acetate. J Bacteriol 182:1632–1640

Yanagihara K, Mizuuchi K (2002) Mismatch-targeted transposition of Mu: a new strategy to map genetic polymorphism. Proc Natl Acad Sci USA 99:11317–11321

Zhang Z, Saier MH Jr (2009a) A novel mechanism of transposon-mediated gene activation. PLoS Genet 5:e1000689

Zhang Z, Saier MH Jr (2009b) A mechanism of transposon-mediated directed mutation. Mol Microbiol 74:29–43

Zhang Z, Saier MH Jr (2010) Transposon-mediated directed mutation. Nova Acta Leopoldina 111:19–23

Zhang Z, Saier MH Jr (2011) Transposon-mediated adaptive and directed mutations and their potential evolutionary benefits. J Mol Microbiol Biotechnol 21:59–70

Zhang Z, Yen MR, Saier MH Jr (2010) Precise excision of IS5 from the intergenic region between the fucPIK and the fucAO operons and mutational control of fucPIK operon expression in Escherichia coli. J Bacteriol 192:2013–2019

Zhang Z, Yen MR, Saier MH Jr (2011) A new direction for directed mutation. Trends Evol Biol 3:13–21

Zinser ER, Schneider D, Blot M, Kolter R (2003) Bacterial evolution through the selective loss of beneficial Genes. Trade-offs in expression involving two loci. Genetics 164:1271–1277

Chapter 5
Hsp90 as a Capacitor of Both Genetic and Epigenetic Changes in the Genome During Cancer Progression and Evolution

Xiangyi Lu, Luan Wang, Vincent E. Sollars, Mark D. Garfinkel, and Douglas M. Ruden

Abstract In this chapter, we focus on the role of the chaperone protein Hsp90 as a capacitor for morphological variation that is released during times of stress. Hsp90 helps to fold numerous client proteins, which constitute a veritable "who's who" of important signaling molecules, such as Akt, Raf, Src, chromatin-modifying proteins, nuclear hormone receptors, and kinetochore assembly proteins. We first review evidence that Hsp90 functions trans-generationally as a capacitor for morphological variation via both genetic and epigenetic means: in the former by revealing cryptic genetic variation and in the latter by generating heritable epialleles. Then we discuss two mechanisms by which altered Hsp90 function can mutate DNA: transposon mobilization, and chromosomal aneuploidy. Next, we hypothesize how beneficial cryptic epigenetic variation might be stabilized, or locked in place, by the directed DNA-level mutation of epigenetically assimilated epialleles. Finally, we describe how Hsp90 functions intra-generationally within an organism's lifetime by releasing cryptic phenotypic variation during development in a stressful environment, and how this can be hijacked during the progression of diseases such as cancer.

X. Lu
Institute of Environmental Health Sciences, Wayne State University, Detroit, MI 48201, USA

L. Wang • D.M. Ruden (✉)
Institute of Environmental Health Sciences, Wayne State University, Detroit, MI 48201, USA

Department of Obstetrics and Gynecology, C. S. Mott Center for Human Growth and Development, Wayne State University, Detroit, MI 48201, USA
e-mail: douglasr@wayne.edu

V.E. Sollars
Department of Biochemistry and Microbiology, Joan C. Edwards School of Medicine, Marshall University, One John Marshall Drive, Huntington, WV 25755, USA

M.D. Garfinkel
Department of Biological Sciences, University of Alabama in Huntsville, Huntsville, AL 35899, USA

D. Mittelman (ed.), *Stress-Induced Mutagenesis*, DOI 10.1007/978-1-4614-6280-4_5,

These granules may be called gemmules. They are collected from all parts of the system to constitute the sexual elements, and their development in the next generation forms the new being; but they are likewise capable of transmission in a dormant state to future generations and may then be developed.

The Variation of Plants and Animals under Domestication Darwin (1868) (Darwin 1883).

Epigenetics is Back! Hsp90 and Phenotypic Variation Pigliucci (2003).

Introduction

The first of the above quotes appeared in Darwin's book The Variation of Animals and Plants under Domestication (Darwin 1883), which was published 9 years after his On the Origin of Species (Darwin 1859). Darwin proposed that "gemmules" were secreted by the organs of the body after adapting to a stressful environment and, via transport through the blood, accumulated in the germ cells. Through these gemmules, Darwin proposed a mechanism to explain Lamarck's theory for the inheritance of acquired characteristics, which was published almost 60 years earlier in 1809 (Lamarck 1809). However, to put Darwin's reliance on Lamarckian inheritance in historical context, Darwin's gemmule hypothesis was made in ignorance of Mendel's pioneering investigations of "particulate" heritable units (i.e., genes), which did not become the organizing principle of genetics until their rediscovery in the early 1900s. The synthesis of Darwinian evolutionary theory with Mendelism subsequently focused evolution almost entirely on genetic mechanisms (Mendel 1865). Consequently, Lamarck's acquired characteristics hypothesis, Darwin's gemmule hypothesis, and other pre-Mendelian hypotheses for nongenetic means of inheritance have been largely forgotten or ignored in the past century.

However, research in the past couple of decades has shown that Darwin's notional gemmules have many interesting similarities to nuclear hormones, some of which, such as glucocorticoids, are induced by stress, travel via the blood, and influence both intra-generational somatic properties and possibly even trans-generational germ line processes (Ruden et al. 2008). Furthermore, endocrine disruptors such as diethylstibesterol (DES), which is more potent than the hormone estrogen in binding to and activating the estrogen receptor, have been shown to increase the incidence of cervical cancer in women for at least two generations (reviewed in Ruden et al. 2005a). Therefore, in this chapter we revisit modern hypotheses that often have striking similarities to the Darwin's gemmule hypothesis. As evolutionary biologist Massimo Pigliucci stated in the second of the above quotes, "Epigenetics is back!" (Pigliucci 2003).

The rediscovery of Mendel's laws of genetic inheritance was the beginning of what is called the "Modern Synthesis." In 1942 (Huxley 1942), a book with this title was written by Huxley, compiling the prevailing evolutionary theories of the time. (Huxley's book has recently been republished with a forward and historical perspective by Pigliucci (Huxley 1942)). The major theme of Huxley's Modern Synthesis is that species can survive stressful environments by selection acting upon existing genetic variation for adaptations that increase survival and reproductive fitness in the stressful conditions. Separate from the Huxley reprint edition, Pigliucci and colleagues updated the Modern Synthesis for the twenty-first century, naming the successor the "Extended Synthesis" (Pigliucci et al. 2010). Pigliucci's synthesis incorporates new knowledge such as epigenetics and evolutionary-developmental biology (evo-devo). (For a thorough description of evo-devo, we recommend a wonderful book by Sean B. Carroll called Endless Forms Most Beautiful, published in 2005 Carroll 2005). Around the same time as Pigliucci's book on this topic, we published a review (Ruden 2011), in which we called the update of Huxley's Modern Synthesis the "(New) New Synthesis" and the "Ultramodern Synthesis," although Pigliucci's term has been more widely accepted by the evolutionary community (Pigliucci et al. 2010).

The focus of this chapter is on the role of Hsp90 in facilitating both genetic and epigenetic changes in the genome during stress, cancer progression, and its role in evolutionary processes. Hsp90 is a chaperone protein that is required for the proper folding of over 200 signaling molecules (Taipale et al. 2010; Dezwaan and Freeman 2008; Whitesell and Lindquist 2005). We discuss how Hsp90 functions as a capacitor for morphological evolution via both genetic (Rutherford et al. 2007; Queitsch et al. 2002) and epigenetic means (Ruden et al. 2003; Sollars et al. 2003). A capacitor, in the biological sense, is something that holds back cryptic (i.e., hidden) phenotypic variation (Rutherford and Lindquist 1998). An electrical capacitor stores charge, and then releases the charge in a sudden burst, but an evolutionary or genetic capacitor holds back phenotypic variation in a normal environment, and then reveals the new phenotypes in a stressful environment. Genetic capacitors have also been called "adaptively inducible canalizers" by Meikeljohn and Hartl (2002) and "Waddington's widget" in a review from our laboratory (Ruden et al. 2003), a term that has been adopted by some researchers in the ecology and zoology community (Kohler et al. 2009). In this chapter we develop the concept of how Hsp90 might function as both a genetic and epigenetic capacitor for evolution and for disease progression.

> One Ring to rule them all, One Ring to find them,
> One Ring to bring them all and in the darkness bind them.
> The Lord of the Rings (Tolkien 1954).

Hsp90 as a Genetic Capacitor for Morphological Evolution

Hsp90 is an evolutionarily conserved protein that forms a ring-like dimer when it binds and facilitates the proper folding of its client proteins. Hsp90 has been called a "master regulator of master regulators" (Lu et al. 2011) and the "hub of protein homeostasis" (Taipale et al. 2010). Here, with apologies to the estate of J.R.R. Tolkien and his many fans, we call Hsp90 the "One Ring to rule them all," because it regulates the function of many signaling pathways. The Hsp90 client list, or "Ring-bearers" to continue the analogy, includes proteins such as Akt, Raf, and Src

(Taipale et al. 2010; Trepel et al. 2010), as well as several chromatin remodeling proteins such as Trithorax (Tariq et al. 2009; Ruden and Lu 2008) and Smyd3 (Ruden et al. 2005a; Hamamoto et al. 2004), and a kinetochore assembly protein (Chen et al. 2012). To push this analogy further, like the infamous ring in Tolkien's The Lord of the Rings (Tolkien 1954), in the "darkness" of stress, Hsp90 literally binds them: that is, Hsp90 forms stable, non-covalent associations with its clients as they become partially unfolded under stressful conditions such as heat shock or nutritional deprivation. Even though Hsp90 makes up as much as 5% of the cellular protein, during stressful conditions Hsp90 can be functionally inactivated as it is titrated by dimerizing or multimerizing with its partially unfolded "stressed" client proteins. While Hsp90 likely plays a critical role in beneficial evolutionary processes, as with the "One Ring," it also has dark powers to promote the rapid evolution of pre-malignant cancer cells to metastasis and drug resistance. (Note, however, that in Tolkien's work, the One Ring was made by Sauron, who imbued it with his hate, malice, and desire for subjugation and domination of weaker peoples. Neither he nor the One Ring had any redeeming qualities whatsoever. This contrasts severely with the biology of Hsp90. The normal function of Hsp90, under normal conditions, is to be a chaperone for numerous signaling proteins, facilitating their normal activities. This is benevolent, not malicious).

Rutherford and Lindquist, in their classic paper "Hsp90 as a Capacitor for Morphological Evolution" (Rutherford and Lindquist 1998), proposed that Hsp90 is a key component of a system that allows organisms to adapt to stressful environments. They showed that *Drosophila melanogaster* laboratory strains that have reduced Hsp90 function due to heterozygosity for several different mutations in the Hsp83 gene displayed a variety of morphological abnormalities at low frequencies in a least seven different body parts, including the eyes, wings, and legs (Rutherford and Lindquist 1998). After several generations of selection for these abnormal phenotypes, eventually nearly 100% of the flies had the morphologically abnormal phenotype, even when Hsp90 activity was fully restored by adding back the wild-type Hsp83 gene. Similar results were obtained when Hsp90 function was inhibited by its potent and specific drug inhibitor geldanamycin (Rutherford and Lindquist 1998). They called Hsp90 a "capacitor" for morphological variation because inhibition of Hsp90 activity resulted in depletion of developmental buffering, akin to the electrical charge depletion by an electrical capacitor, so that a variety of morphological abnormalities would appear in the offspring (Rutherford and Lindquist 1998).

What Rutherford and Lindquist (1998) did in 1998, as recognized by McLaren (McLaren 1999) and our laboratory (Ruden et al. 2003) in later reviews, essentially was to repeat and update classical experiments done by Waddington some 50 years earlier. Charles Waddington is often called "the father of epigenetics" partly because he popularized the word epigenetics, at least as it applies to development. He showed that stressing *D. melanogaster* larvae, such as by heat shock or by the (at the time) commonly used anesthetic diethyl ether, can cause a small percentage of adults to emerge that have a novel phenotype, such as missing crossveins in the wing (i.e., the crossveinless phenotype) (Waddington 1953) or developing enlarged, wing-like halteres (i.e., the Ultrabithorax phenotype) (Waddington 1956). Waddington (1942) defined epigenetics as "the interaction of the genome with the environment," but more recent definitions of epigenetics specify cellular biochemical mechanisms such as histone modifications, small RNAs, and even prions (reviewed in Jablonka and Lamb 2005). As with Rutherford and Lindquist's experiments (Rutherford and Lindquist 1998), Waddington observed that after more than 20 generations of heat-shock and re-selection, eventually 100% of the offspring had the crossveinless or Ultrabithorax phenotype, depending on which phenotype had initially been selected. As mentioned above, the novel phenotypes were stabilized so that environmental stress was no longer necessary for the offspring to have these phenotypes. How this might involve Hsp90 is discussed below.

In Waddington's classic publication on this subject, he called the process of selecting for the stress-induced phenotypes over several generations "genetic assimilation" (Waddington 1942). We note, on this the 70th anniversary year of Waddington's paper, that he did not shy away from invoking Larmarckian-sounding mechanisms to explain "genetic assimilation." Indeed he titled the paper "Assimilation of Acquired Characteristics" (Waddington 1942), a clear allusion to Lamarck's theory. Waddington described the "normal" (i.e., wild-type) phenotype as being "canalized"—to use another of his famous coinages—by genetic variation in the starting population such that all of the flies in the population have normal crossveins or halteres. As an analogy, which remains popular among biologists who study epigenetics, he described the normal phenotype as having arisen in developmental time as if it was a ball that had rolled to the bottom of a potential energy well or channel, carved through a potentially complex shape that Waddington called an "epigenetic landscape" (reviewed in Waddington 1974).

In the epigenetic landscape, the walls of the channel represent developmentally unfavorable states while the deepest channel, down which the ball (i.e., cell type, tissue or organ anlagen, or entire organism) normally rolls, represents normal development. Stress, according to Waddington, can cause the topology of the "epigenetic landscape" to change so that the ball travels down an alternate valley, which leads toward a new environment-induced phenotype. In stressful environments, according to Waddington, the ball can roll down numerous routes, depending on the type and strength of the stress; the variability of the developmental outcome (and therefore of the phenotype) is increased by the stress. To illustrate the shifting landscape, Waddington drew numerous strings to represent the genetic variation that can be tightened or loosened by stress or environmental factors. In certain situations, as discussed later, the new stress-induced phenotypes have a selective advantage to the organism, but this is not always the case, as in Waddington's experiments. The strings that reshape Waddington's epigenetic landscape resemble the modern concept of interacting proteins in complex signaling and metabolic pathways (reviewed in Waddington 1974).

In modern molecular terms, what is likely happening in a shifting epigenetic landscape is that several signaling molecules become partially inactivated by stress. In a population, allelic variation would ensure that some variants of signaling molecules would be more susceptible to stress than other signaling molecules (e.g., they denature at a lower temperature). In stressful conditions, the Hsp90 pool could become titrated by binding to the partially denatured proteins. The remaining Hsp90 proteins would be insufficient in number relative to its many native-folded client-signaling proteins, with the result that they would become inactivated due to the lack of available Hsp90 protein.

Meikeljohn and Hartl, in a critique of Rutherford and Lindquist's capacitor hypothesis (Rutherford and Lindquist 1998), argued against the theory that Hsp90 can function as a "capacitor," which they renamed an "adaptively inducible canalizer" (Meiklejohn and Hartl 2002). "Adaptively inducible canalizer," according to Meikeljohn and Hartl, is a more descriptive term that can serve as an alternative to Rutherford and Lindquist's more ambiguous word "capacitor" (Rutherford and Lindquist 1998). According to Rutherford and Lindquist's theory, but using Meikeljohn and Hartl's (and Waddington's) terminology, Hsp90 "canalizes" the ground state of the normal phenotype, but it is "adaptively inducible" by stress so that beneficial phenotypes are generated that can be eventually re-canalized to a new route. In other words, Waddington's ball can roll down an alternative valley that is made deeper than the original valley by tightening and loosening a few strings. In other words, the topography of the epigenetic landscape is controlled by the environment.

One of the main criticisms of the capacitor model by Meikeljohn, Hartl, and other evolutionary biologists, and of Waddington's assimilation experiments in general, is that the vast majority of abnormal phenotypes that are induced by the stresses used in laboratory experiments are not beneficial to survival, but rather are often deleterious to survival. How, for instance, do the crossveinless or Ultrabithorax phenotypes in Waddington's experiments help a fly to survive heat stress or ether exposure? The answer, at least for these examples, is that they do not. However, Waddington's response to similar criticism in the 1950s was that he was not proposing that genetic assimilation is how evolution necessarily happens; his goal was only to show that assimilation could be a possible evolutionary mechanism (reviewed in Ruden et al. 2003). Waddington left it to future generations of evolutionary biologists to determine whether genetic assimilation is a valid evolutionary mechanism.

Despite his demonstration of genetic assimilation, Waddington's results were disregarded by many evolutionary biologists because, in fact, the genetically assimilated phenotypes probably decrease fitness in many environments. Fortunately, since Waddington performed his initial assimilation experiments, a couple of examples of beneficial genetic assimilation have been discovered. For example, the monkey flower Mimulus produces more trichomes when their leaves are partially eaten by insects (or, in the laboratory, when holes are produced with a hole punch) (Scoville et al. 2011; Holeski 2007). The trichomes are sticky protrusions that protect the leaves from getting further eaten by insects. An interesting aspect of the trichome is that the offspring of plants with trichomes also produce these structures, even without having their leaves eaten by insects. Since many generations of genetic assimilation are not required for the offspring to immediately (i.e., in one generation) have the extra-trichome phenotype, this has been proposed to be an example of trans-generational assimilation of a beneficial metastable epiallele (Scoville et al. 2011; Holeski 2007), a phenomenon described in the next section.

Another example of trans-generational epigenetic inheritance of a stress-induced beneficial adaptation, this time in an animal, is that several species of the water flea Daphnia develop thick “helmets” with spiky protrusions on their heads when fish-conditioned water is added to the culture medium (Oda et al. 2011; Laforsch and Tollrian 2004; Brooks 1965). In the absence of fish-conditioned water, Daphnia develop more-rounded heads that lack sharp spikes and thickened helmets. The sharp helmet spikes are thought to make it difficult for small fish to eat Daphnia animals because the sharp spikes would jab them in the mouths when they bite them. Spike induction in Daphnia by developmental exposure to fish water is an example of trans-generational inheritance because the offspring of fish-water-exposed Daphnia also have a spike on their thick helmets. Neither the water-diffusible signal in the fish water that induces spike formation in Daphnia nor the developmental mechanism of spike formation is known. However, Daphnia is a simple model system, recently adopted by the USA. National Institutes of Health (NIH) as a biomedically useful model organism (www.nih.gov/science/models/daphnia/), and by the USA. Environmental Protection Agency (EPA) as an ecotoxicogenomics model organism, which can be used to better understand the evolutionary mechanisms for genotype–environment interactions and beneficial trans-generational epigenetic processes. In the next section, we describe a trans-generational epigenetic system in *D. melanogaster* that we developed in our laboratory.

Hsp90 as an Epigenetic Capacitor for Morphological Evolution

When Rutherford and Lindquist’s “Hsp90 as a Capacitor for Morphological Evolution” paper came out in 1998 (Rutherford and Lindquist 1998), two of us (V.E.S. and D.M.R.) thought that it might explain some unusual observations in our laboratory (Sollars et al. 2003). We were searching for enhancers of an abnormal eye phenotype caused by mutations in the Krüppel (Kr) zinc-finger transcription factor that is required for development of the middle portion of the fly embryo (i.e., the posterior thoracic and anterior abdominal segments) (Sollars et al. 2003). The allele we were using, Krüppel-Irregular facets-1 (Kr^{If-1}) causes the Kr protein to be ectopically expressed in the eye imaginal disks, and this causes the eyes to be small and irregularly shaped (Sollars et al. 2003). In our genetic screen for enhancers of Kr^{If-1}, we identified alleles of the Hsp90 gene (called Hsp83 in Drosophila) as dominant enhancers of the eye phenotype with very peculiar, and, at the time, unprecedented, properties (Sollars et al. 2003). Because of its importance to the field of trans-generational epigenetic inheritance, we describe the Kr^{If-1} results in detail below.

Maternal heterozygosity for an Hsp83 loss-of-function mutant allele causes a dramatic phenotype in the offspring: ectopic large bristle outgrowths (ELBOs) protruding from the eyes of adult *D. melanogaster* (Sollars et al. 2003) (Fig. 5.1). We originally thought that they were the proximal parts of limb-like appendages growing out from the eyes, but we named them for their most obvious characteristic—clusters of large

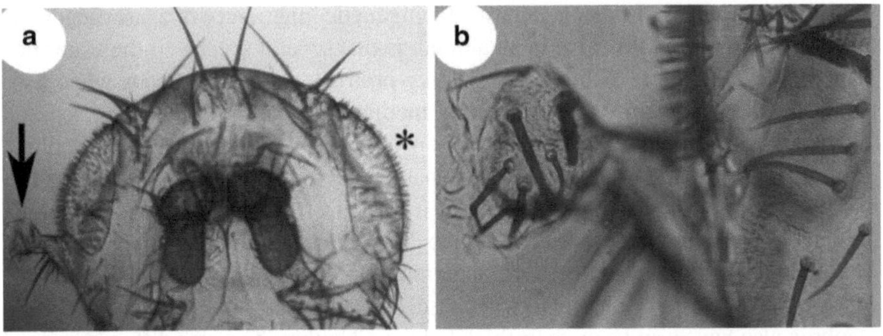

Fig. 5.1 Ectopic large bristle outgrowths (ELBOs). Scanning electron micrograph of eyes from $Kr^{If-1}/+$ flies with ELBOs. (**a**) Fly head with one normal eye (*) and one ELBO eye (*arrow*). (**b**) Enlargement of ELBO with large bristles

bristles growing out the eyes (Sollars et al. 2003). The unusual aspect of the ELBO phenotype is that it only occurs if the mother was missing one copy of the Hsp83 gene; ELBOs were not produced in the progeny of the father was missing one copy of Hsp83. Furthermore, the ELBOs are produced at equivalent frequencies in the progeny regardless of whether they inherited their mothers' Hsp83 mutation (Sollars et al. 2003). Since this phenotype appeared to be due to new mutations, we called them dominant Maternal-Effect Kr^{If-1} mutations (i.e., Me(Kr^{If-1})) (Sollars et al. 2003).

However, the most unusual aspect of ELBOs is that the flies with ELBOs, regardless of their Hsp83 genotype, when mated together, produced offspring with even higher frequencies of ELBOs than their parents' generation, even in a whole-genome-isogenized strain with no genetic variation (Sollars et al. 2003). In fact the ELBO frequency went up continuously from generation to generation until it plateaued after 9–10 generations at 60–70% penetration (Sollars et al. 2003). This was true regardless of whether the founding maternal ancestors were mutated for Hsp90 or were fed geldanamycin, a potent and specific inhibitor of Hsp90 that binds to the ATP site and inactivates the protein. Geldanamycin (GA) is the Hsp90 inhibitor used by Lindquist and colleagues in their genetic assimilation experiments (Rutherford and Lindquist 1998). Our remarkable and novel finding was that it is sufficient for Hsp90 expression or activity to be reduced in a single maternal ancestor in order for phenotypic effects to persist through several generations and be visible in a distant descendent (Sollars et al. 2003).

Since our ELBO results were published in 2003, they have been replicated in several laboratories (Gangaraju et al. 2011; Matyash et al. 2004). Additionally, new genetic inducers of ELBOs in the Kr^{If-1} background have been identified, such as piwi and hopscotch (hop) (Gangaraju et al. 2011). Hop is an Hsp90 co-chaperone, so it is not unexpected that hop mutant alleles could induce ELBOs. However, when Lin's laboratory identified piwi mutations as inducers of ELBOs, a possible mechanism for some aspects of the trans-generational epigenetic phenotype was newly revealed (Gangaraju et al. 2011) (Fig. 5.2). Their paper was the first to show that the Piwi protein is a client of Hsp90, and that inactivation of Hsp90 causes Piwi activity to be correspondingly reduced.

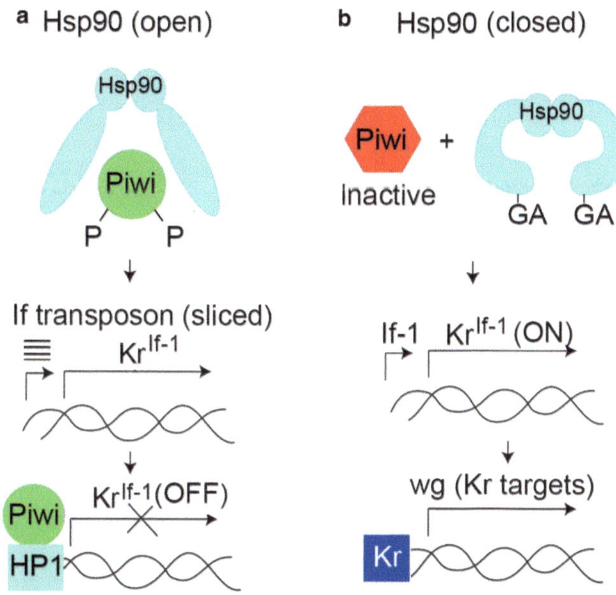

Fig. 5.2 Hsp90 as a genetic and epigenetic capacitor for phenotypic variation. (**a**) Hsp90 binds to Piwi and this leads to its activation by phosphorylation. *Left*, active Piwi slices transposon RNA into small fragments, and the Piwi-RNA forms a complex to repress transcription of the transposons. *Right*, the Kr^{If-1} locus is hypothesized to possess a repeat sequence in the promoter that gets sliced by active Piwi. Repression of the repeat at this locus leads to repression of Krüppel (Kr) expression in the female germ line. (**b**) In the absence of Hsp90, Piwi is not phosphorylated and is not active. Krüppel expression is ectopically present in many tissues, such as the eye imaginal disks, where it leads to the formation of ectopic large bristle outgrowths. However, we do not think that this model is sufficient to explain some of the genetic data (see text)

Piwi is an attractive candidate for a critical role in the ELBO phenotype because its function is to repress expression of transposons by RNA-directed chromatin remodeling (Gangaraju et al. 2011). Piwi is in the Argonaute family of siRNA-producing enzymes, and is thought to slice transposon RNA and recruit inhibitory chromatin remodeling factors (Fig. 5.2). As mentioned above, the Kr^{If-1} irregular facet phenotype is caused by ectopic expression of Kr protein in the eye imaginal disks. This ectopic expression is most likely caused by insertion of an as-yet uncharacterized repeat into the regulatory region of the Kr gene. Therefore, what Lin's laboratory showed is that Piwi mutations cause more expression of Kr in the eye imaginal disks, and this increased expression causes the ELBO phenotype (Gangaraju et al. 2011) (Fig. 5.2). Piwi is discussed further in the next section.

Figure 5.3 shows a model for how stress might change the epigenetic reprogramming of a developmental gene. In the absence of stress, Hsp90 is a chaperone for the histone 3 lysine 4 (H3K4) histone methyltransferase Trithorax (Tariq et al. 2009), which leads to the activation of genes (Fig. 5.3a). We propose that stressful environments functionally inactivate Hsp90 and this leads to the inactivation of Trithorax (Fig. 5.3a). Since both Trithorax-group (TrxG) protein complexes and Polycomb-group (PcG)

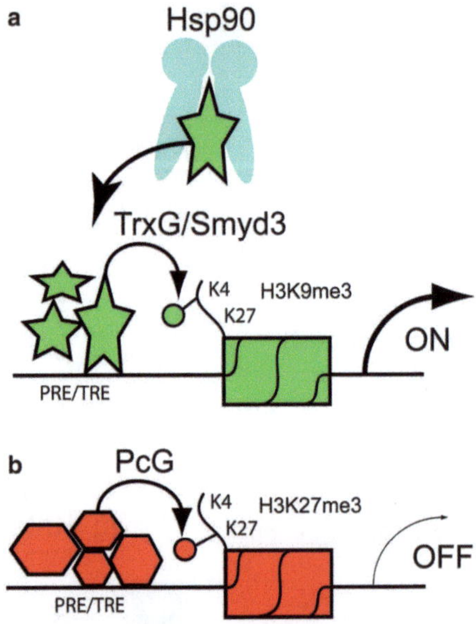

Fig. 5.3 Model for Hsp90 acting as a capacitor for histone modifications. (**a**) In a non-stressed cell, Hsp90 binds to the H3K4 histone methyltransferase (Trx or Smyd3) and activates the enzymatic activity. Chromatin of PRE/TREs carrying this active histone mark is open for transcriptional activation. (**b**) In a stressed cell, Hsp90 is inactive and Trx is not activated. Consequently, PRE/TREs are occupied by PcG complex proteins, one of which is a H3K27 histone methyltransferase (PcG). Chromatin thus modified by this inactive histone mark leads to transcriptional repression

protein complexes bind to the same DNA regulatory elements in target genes, called PREs/TREs (Polycomb response elements/Trithorax response elements), inactivation of Trithorax can cause Polycomb complex proteins to replace them. Whereas Trx codes for a transcription-activating H3K4 histone methyltransferase, the PcG complex contains a subunit that codes for a transcription-inhibitory histone 3 lysine 27 (H3K27) histone methyltransferase activity. This epigenetic switch would then cause a gene that is normally programmed to be "on" by the H3K4 methylation (catalyzed by TrxG) to be switched "off" by the H3K27 methylation activity of PcG (Fig. 5.3b). This epigenetic switch has been discussed in a recent review from our laboratory (Ruden and Lu 2008) (Fig. 5.3).

Hsp90 and Transposon Mobilization by Piwi

As mentioned above, Gangaraju and colleagues (2011) in Lin's laboratory used the ELBO trans-generational epigenetic system to show that Piwi also suppresses epigenetic variation in a process that requires Hsp90, and they propose that the combined genetic and epigenetic increase in diversity in times of stress can dramatically increase phenotypic variation that can be selected. Gangaraju and colleagues (2011) show that

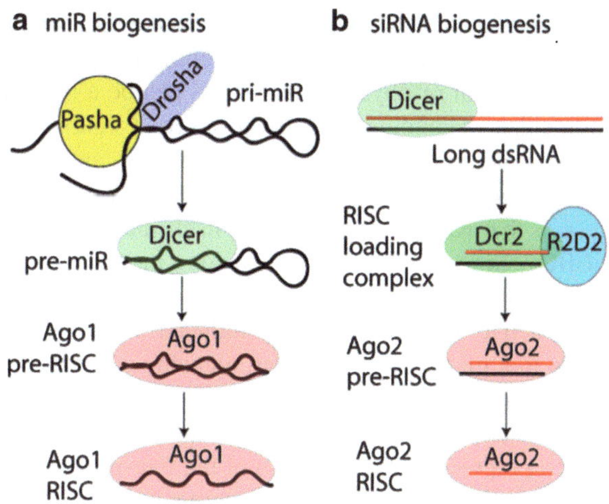

Fig. 5.4 Biogenesis of miR and siRNAs. (**a**) The pri-miR is encoded from a genomic locus with an imperfect inverted repeat (*bulges*). Cleavage by Drosha generates the pre-miR, and cleavage by Dicer and transfer to Ago1 forms the pre-RISC complex. (**b**) In siRNA biogenesis, dsRNA is cleaved by Dicer and loaded into the RISC loading complex. Piwi is in the Argonaute (Ago) family and produces Pi RNAs in Piwi RISC complexes (not shown)

combined maternal and zygotic loss of Piwi causes the ELBO phenotype in the F1 offspring, and that Hsp90 activates Piwi by binding to it and allowing it to be phosphorylated at as-yet unknown sites by as-yet unidentified kinases. They also show that flies selected for restored Piwi function can be either epigenetically or genetically "canalized" (buffered) for the ELBO phenotype (Gangaraju et al. 2011), similar to what we showed in our original paper (Sollars et al. 2003). Furthermore, the Lin laboratory showed, as we did (Sollars et al. 2003), that Wg (Wingless/Wnt), is a downstream target of Kr that is apparently ectopically expressed in the eye imaginal disks. Our lab later determined that Wg is not actually expressed in the eye imaginal disks, but rather in the hemocytes, which adhere to the eye imaginal disks, and are likely involved in the tissue reengineering that generates the ELBOs (Ruden et al. 2003).

Piwi prevents transposon mobilization in the male germ line by "slicing" (chopping up) transposon RNAs (Zambon et al. 2006; Brennecke et al. 2007) (Fig. 5.4). Specchia and colleagues (2010) showed that Hsp90 and Piwi prevent phenotypic variation by suppressing transposon mobilization in the male germ line. In this manner, Specchia and colleagues (2010) propose that Hsp90 is an "adaptively-inducible canalizer" that can increase genetic variation in times of stress, and this newly induced genetic variation can be genetically assimilated over many generations to increase the fitness of the organism. In other words, Specchia and colleagues propose that Piwi normally acts to prevent transposon mobilization (Specchia et al. 2010). But when environmental stress inactivates Piwi, the resulting de-repression of transposon mobility could cause new transposon insertion mutations and chromosomal rearrangements, thereby creating new genetic variability—including adaptive "hopeful monsters" that can be selected upon. "Hopeful monsters" are, according to Gould in his 1980 book A Panda's Thumb, Goldschmidt's term for "major structural transitions [that] can occur rapidly without a smooth series of intermediate stages" (Gould 1980) (more recently reviewed in Chouard 2010). Major changes caused by stress-induced transposon mobilization is not a new idea—Barbara McClintock introduced the idea of "genomic stress" in her 1983 Nobel Prize lecture, in which she proposed that transposon mobilization in times of environmental or physiological stress can be a last-ditch effort for an organism to reorganize its genome in an effort to save a sinking ship (McClintock 1984).

What is unique to the work of Gangaraju and colleagues (2011), in addition to the exciting finding that Piwi is involved in this trans-generational epigenetic process, is that Kr mRNA expression itself increases in the heads of the flies with the ELBOs. Since Kr^{If-1} is a spontaneous mutation that is presumably caused by a transposon insertion, an attractive hypothesis is that, when Hsp90 activity is abundant, active Piwi-RISC (RNA induced silencing complexes) silences the Kr^{If-1} allele in a manner similar to its silencing of transposons (Fig. 5.2). Piwi-RISC is thought to silence transposons in two sequential steps: first by slicing the transposon transcripts, and then by interacting with HP1 (heterochromatin protein 1) and directly repressing transcription of the transposons (Fig. 5.2a). If Kr^{If-1} is caused by a transposon insertion, then recruitment of Piwi-RISC and HP1 to the Kr^{If-1} promoter can at least partly repress expression of Kr in the eye imaginal disks and prevent ELBO formation (Fig. 5.2a). If Hsp90 is inactivated, either by the specific inhibitor geldanamycin or by mutating its gene, Hsp83, then Piwi-susceptible transposons are expressed and mobilized, and the Kr^{If-1} is de-repressed and ectopic Kr protein in the eye imaginal disks is produced in sufficient abundance to cause the ELBO phenotype (Fig. 5.2b).

While this is an attractive hypothesis to explain how inactive Piwi might cause transposon mobilization, we believe that a more complex mechanism must be responsible to explain the trans-generational epigenetic phenomena. We found, for instance, that Kr is expressed in the eye imaginal disk indistinguishably in $Kr^{If-1}/+$ flies regardless of whether ELBOs were present or not. Indeed, since Kr^{If-1}/Kr^{If-1} flies survive and have almost no eye tissue, increasing the expression of Kr mRNA, as shown by Gangaraju and colleagues (2011), should only cause the eyes to be smaller and not induce ELBOs.

We propose instead that ELBOs are induced by a more complex mechanism. For example, Piwi might be involved in DNA looping or some other topological conformation such as localization of PREs or insulators to "Polycomb bodies" which are distinct nuclear loci where Polycomb-group protein complexes aggregate (Saurin et al. 1998). These complexes, we propose, are what are selected for and stabilized in the trans-generational selection experiments. In our original paper, for instance, we found that the strongest inducer of the ELBO phenotype was maternal deficiency of the TrxG gene verthandi (vtd), which was recently found to encode the Rad21 subunit of Drosophila cohesin (Hallson et al. 2008). This supports our hypothesis because Rad21-cohesin was found to associate with the CTCF sites at the mouse IGF2-H19 DMR (differentially methylated region) and siRNA knockdown of Rad21-cohesin causes both disruption of large DNA loops and loss-of-imprinting at this locus (Stedman et al. 2008). Also, Piwi has been shown to be required for long-range interactions at Drosophila PREs, which control long-term expression or repression of the Hox genes by the TrxG and PcG genes, respectively (Grimaud et al. 2006).

Based on these results, we propose an alternative model that Piwi, perhaps in a complex with the Vtd-encoded cohesin protein, forms an intra-chromosomal loop, inter-chromosomal ectopic synapsis, or some other large-scale chromatin structure among various Kr target genes, such as Wg or the Kr locus itself, and this looping somehow prevents Kr from activating expression of the target genes. When Piwi is inactive, the loop is broken and Kr can now activate its target genes and induce ELBOs. This mechanism can be tested, for instance, by mutating the various domains of Piwi, such as the slicing catalytic function, which should not affect chromatin looping, or the HP1-interaction site, which might affect it.

In summary, the results of Gangaraju and colleagues (2011) provide evidence for a more general model of Piwi functioning as an "adaptively-inducible canalizer" to both suppress transposon-mediated mutagenesis and stabilize metastable epialleles. In other words, the environment is feeding back onto the genome more than was ever thought to occur in the past. There are several other trans-generational epigenetic systems in Drosophila, and in a recent review we proposed that they might all involve Hsp90 (Ruden and Lu 2008). In the next section we describe a trans-generational epigenetic system in plants that goes far beyond what has been done in Drosophila.

Trans-generational Epigenetic Inheritance in Plants

Cholot and colleagues took the trans-generational epigenetic experiments in Drosophila several steps further by replicating similar experiments in *Arabidopsis thaliana* on a genome-wide scale (Roux et al. 2011). They started with a strain that has very little DNA methylation, which is possible because the RNA-induced transcriptional silencing systems in plants are mostly redundant with the DNA methylation systems (Roux et al. 2011). We note that Drosophila, but not other insects, such as honeybees (Wang et al. 2006), appear to have lost their DNA methylation systems millions of years ago (although whether Drosophila is completely incapable of DNA methylation is both controversial and unresolved, see Schaefer and Lyko 2010; Phalke et al. 2009).

Plants with little DNA methylation were sickly and had a variety of low-penetrance phenotypic abnormalities, such as the height of the plant and the degree of branching. This range of phenotypes resembles those of flies with reduced Hsp90 activity noticed by Rutherford and Lindquist, as described above (Rutherford and Lindquist 1998). What Cholot and colleagues then did was to reintroduce the missing gene that caused severe reduction of DNA methylation, and studied what happened on a genome-wide level (whole-genome analyses of DNA are discussed in the last section of this chapter).

Cholot and colleagues identified three classes of epialleles: class 1 restored DNA methylation immediately, and included sites such as repeat loci that use RNA-directed DNA methylation systems; class 2 showed slow restoration of methylation that required two or three generations; and in class 3 methylation was never restored properly or took more than three generations for restoration (Roux et al. 2011).

Class 1 epialleles are the least interesting because they are hard-wired by genetic systems to be continuously methylated. However, as discussed in their paper, class 2 and 3 epialleles have profound implications (Roux et al. 2011). For example, since class 2 and class 3 epialleles require several generations for restoration, they can be considered metastable epialleles that could be subjected to either natural or artificial selection and contribute to evolution. This is remarkable because, in theory, one could select for a variety of phenotypes, such as tall or short or multi-branched or poorly branched, in an isogenized Arabidopsis strain devoid of DNA-level heterozygosity or polymorphism that has been "epigenetically erased" and restored, such as by what we call here the "Cholot process." We describe a similar process as "epigenetic assimilation" in a review that we wrote several years prior to Cholot's publications in this area (Garfinkel et al. 2004).

If the Cholot process succeeds in generating novel phenotypes of Arabidopsis in the absence of genetic variation, it would be interesting to determine the stability of the resulting metastable epiphenotypes. Jablonka and Lamb (2005), in their groundbreaking book "Evolution in Four Dimensions," proposed, years before Cholot's experiments, that a species without any genetic variation could evolve into many dramatically different phenotypes by purely epigenetic means. As discussed in the next section, we hypothesize that, after many generations, mutations are directed to the metastable epialleles in the epigenetically assimilated strains that strengthen their phenotypes to complete stability and full penetrance. If these ideas prove to be correct and generalize to other plant species, the Cholot epigenetic selection process, followed by directed mutations to fix the new selected phenotypes, could revolutionize plant breeding and generate potentially any novel phenotype that is important to the agricultural industry.

DES and Trans-generational Epigenetics in Humans: The Role of Hsp90

Between 1947 and 1971, over one million American women were exposed to DES when they took the drug to reduce the risk of miscarriage (Li et al. 2003a). Further human exposure to DES was caused via the contaminated meat products due to the feeding of this drug to beef cattle and other livestock to accelerate their growth (Knight et al. 1980). This was likely a significant source of exposure in humans: in 1971 alone, an estimated 27,600 kg of DES were used in livestock feed lots (Knight et al. 1980). Studies found that DES was present at biologically relevant levels in beef, and DES is metabolically very stable (Knight et al. 1980). Daughters of mothers exposed to DES during the first trimester often have a rare form of cancer, clear-cell adenocarcinoma (CCAC), as well as developmental abnormalities in the structure of the uterus, cervix, or vagina (Li et al. 2003a).

The adverse effects of DES on reproductive system development and cancer were confirmed in studies with laboratory animals exposed to DES during pregnancy. However, the mechanisms remain poorly understood, and the exact risk of

DES exposure in humans is still not clear. A recent study of DES exposure in humans from Robboy's laboratory concludes, "The findings support an association between in utero DES exposure and high-grade squamous neoplasia, although a role for more intensive screening among DES-exposed women in the production of this excess could not be completely ruled out" (Hatch et al. 2001).

DES induces developmental abnormalities in the reproductive tract of rodents, such as mice (McLachlan et al. 1980; Newbold et al. 1990; Iguchi and Takasugi 1987), rats (Branham et al. 1988; Medlock et al. 1992), and hamsters (Leavitt et al. 1981; Gilloteaux and Steggles 1985) via perinatal exposure (i.e., exposure within the first 3 months after birth). When neonatal mice are exposed to DES, ~90% of the mice develop uterine cancer by 18 months of age (Newbold et al. 1990). The DES exposure in these studies was 2 mg/pup/day for the first five days of life, which is comparable to the human exposures (Newbold et al. 1990). Most strikingly, DES affects uterine development in a trans-generational manner in the daughters and the granddaughters of DES-exposed pregnant mice (Newbold et al. 1998, 2000; Turusov et al. 1992; Walker and Haven 1997). Therefore, even though DES has not been used commercially in the USA since the early 1970s, these results suggest that another generation of humans is at risk due to DES exposure in their grandmothers, or even possibly great grandmothers, who may have been exposed between 1947 and 1971. It is now time to begin studies on the possible effects of DES on the granddaughters (and perhaps even great granddaughters) of women exposed in the mid-twentieth century.

Although the mechanisms for the trans-generational effects of DES exposure have not yet been fully determined, the effects of neonatal exposure to DES on uterine cancer in the same generation have been shown to be mediated, at least in part, by altering the CpG methylation pattern of key uterine cancer genes in mice. Altered DNA methylation by DES is probably a gene-specific phenomenon because the CpG methylation of the promoters of the HoxA10 and HoxA11 genes are not altered by neonatal exposure to DES, even though DES dramatically down-regulates expression of these genes (Li et al. 2001). Also, Li and colleagues have recently shown that neonatal exposure of DES in mice, using the above conditions, induces persistent increases in c-fos mRNA expression and hypomethylation of specific enhancer binding sites (Li et al. 2003a, b). It possible that the trans-generational effects of DES exposure on uterine development are explained by CpG methylation alterations in many of the key uterine cancer genes during gametogenesis.

Hsp90 and the Generation of Chromosomal Aneuploidies

So far, we have shown that Hsp90 is a capacitor for both genetic and epigenetic variation. Recently, Chen and colleagues showed that Hsp90 inactivation by stress potentiates rapid cellular adaptation through induction of aneuploidy (Chen et al. 2012). Aneuploidy is defined as a departure in chromosomal number from normal diploidy (in somatic cells) or from normal haploidy (in germ cells). Aneuploidy may be observed on an evolutionary timescale—especially in plants, where whole genome duplications and selective chromosome loss commonly distinguish closely related species. Aneuploidy is also a hallmark of cancer. Recent studies have shown that aneuploidies are a form of genome-scale changes that can confer adaptive phenotypes under diverse stressful environments.

Chen and colleagues showed in Chap. 1 of this book that the Hsp90 inhibitor geldanamycin is the best inducer of aneuploidy in the budding yeast *Saccharomyces cerevisiae*. The chaperone function of Hsp90 has an evolutionarily conserved role in kinetochore assembly. The kinetochore is the centromeric region of the chromosome that attaches to the microtubule spindle during mitosis. Reduced Hsp90 activity thus destabilizes the chromosomal attachment to the mitotic spindle, which would facilitate non-disjunction or metaphase chromosome loss. Chen and colleagues used a heterozygous genetic mutation on chromosome XV that causes sensitivity to a drug, which can be suppressed by adding an additional copy of the wild type gene via duplication the non-mutated chromosome XV. These findings demonstrate that aneuploidy is another type of stress-induced mutation and reveals a new role for Hsp90 as a capacitor for genetic variation.

In the next couple of sections, we present a couple of speculative models, which we published previously, demonstrating how Hsp90 might function as a capacitor for genetic variation via trinucleotide repeat expansion and contraction. Since the publication of our hypothesis, many of these ideas have subsequently been validated, at least in part.

Hsp90 and Its Role in the Expansion and Contraction of Trinucleotide Repeats

In the previous sections, we proposed that metastable epialleles could be epigenetically assimilated and then subsequently mutated to stabilize the assimilated phenotype. Several laboratories have proposed that directing mutations to metastable epialleles can occur via the high mutagenic potential of methylated DNA. For example, Whitelaw and colleagues (Rakyan and Whitelaw 2003) have shown that 5-methylcytosine (5mC), such as is found in retrotransposons, is approximately ten times more mutagenic than unmethylated cytosine, such as is found in CpG islands, which are present in over half of mammalian promoter regions. DNA methylation in retrotransposons has been shown to inhibit expression of the RNA and thereby prevent transposition and insertion into new loci. The reason for the increased mutagenicity of 5mC over cytosine is that 5mC can spontaneously deaminate to T, whereas unmethylated cytosine can spontaneously deaminate to U; the base-excision DNA repair pathway more efficiently repairs the latter (David 2012).

In 2008, we proposed the Epigenetic Directed Genetic Error (EDGE) hypothesis to explain how epigenetically assimilated metastable epialleles can be mutated so that they have a dramatic effect on an organism's phenotype (Ruden et al. 2005b). Our EDGE hypothesis was inspired by the findings of Fondon and Garner, who in 2005 discovered that several morphological characteristics of dogs involve expansion and contraction of trinucleotide repeats in the coding regions of genes (Fondon and Garner 2004). For example, expansion of the polyalanine repeat in the Runx-2 gene from 10 amino acids to 11 amino acids presumably causes this gene to be misexpressed which causes the English bull terrier's nose to slope further downward (Fondon and Garner 2004).

In our EDGE hypothesis, we proposed that the Runx-2 gene originally was a metastable epiallele in which the polyalanine codon-containing DNA repeat was methylated. The Runx-2 repeat, and repeats in general, encode the amino-terminal nonstructured portion of the gene, which includes the CpG island, a CG rich region in the promoter, the transcription start site, and the first exon. Therefore, methylation of the Runx-2 repeat could cause the gene to repressed by DNA-methylation directed transcriptional repression. The EDGE process could then direct mutations to the methylated repeat via expansion (or, in other cases, contraction; see Ruden et al. 2005b) of the repeat. In the case of Runx-2, for reasons not understood (see below), the expanded repeat has a similar effect to DNA methylation (but stronger) and causes the downward-sloping nose phenotype. The reason that we proposed that trinucleotide repeats in coding regions are being especially susceptible to the EDGE model is that such repeats are estimated to be 100,000 times more mutagenic than the genome in general, at least when tested in bacteria (Moxon et al. 2006).

If the EDGE hypothesis is correct, then how might DNA methylation induce repeat expansion and contraction? Here, we update the EDGE hypothesis by giving it a more mechanistic name, the miRDERR (miR-Directed Epigenetic Repression of RCGs) hypothesis (Fig. 5.5), where miR are micro-RNAs, and RCGs are repeat-containing genes, such as the genes with trinucleotide repeats discussed above. The miRDERR hypothesis provides a mechanism to explain the EDGE phenomenon that we suggested to explain rapid morphological evolution in dogs. The miRDERR hypothesis proposes that miRs can direct repressive chromatin marks to repeated sequences in RCGs. Here, we present an observation that supports the miRDERR hypothesis: the human APC tumor suppressor gene has three mutation hotspots at CpG dinucleotides near three of the coding repeats (Fig. 5.6).

Why do we think that this supports the miRDERR hypothesis? First, a major question is, "Why do small changes in the repeat lengths of RCGs encoding developmental proteins in dogs, such as the Runx-2 gene, affect developmental processes?" We speculate that RCGs are targets for either the siRNA (small interfering RNA) or miR pathways, and, consequently, this decreases the steady-state mRNA levels of the RCGs (Fig. 5.5). The reason for this, we believe, is that the repeats themselves are a type of highly repetitive sequences that are expressed at high levels, and the complementary sequences are also expressed at high levels as anti-sense RNAs or miRs. In the case of anti-sense RNA, double-stranded RNA can form, and this can be processed to make siRNA. In the case of miRs, they can target the RCG mRNA for translational repression and direct or indirect degradation (Fig. 5.5). For example, mRNAs that are translation-blocked by miRs are often degraded (Matranga and Zamore 2007; Zamore 2007). In order to show that siRNAs or miRs are targeting RCGs for degradation, they analyzed published microarray data (Rehwinkel et al. 2006).

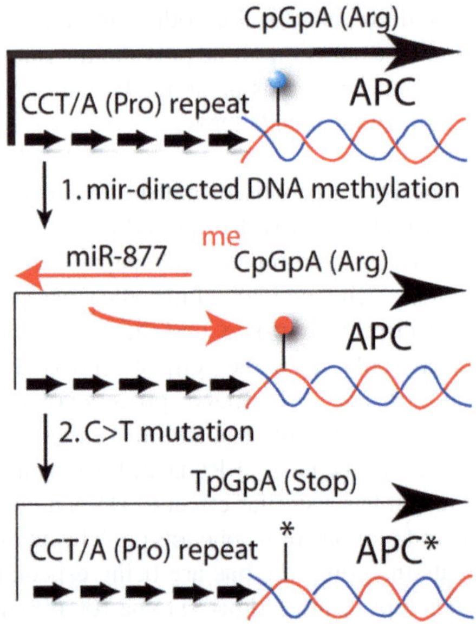

Fig. 5.5 The miRDERR Hypothesis (miR Directed Epigenetic Repression of RCGs). (**a**) First (1), miR-877 binds to the CCU/A repeat in nascent transcripts of the APC mRNA, thereby directing repressive chromatin and DNA methylation (*red dot*) near the repeat. This leads to a decrease in APC mRNA levels (*thin arrow*). (**b**) Second (2), Since methylated cytosines are mutagenic, C→T mutations increase in frequency near the repeat. This might explain the largest APC mutation hotspot [CGA (Arg) to TGA (Stop)] in colon cancer. Blue dot, unmethylated CpG. *, mutated CpG

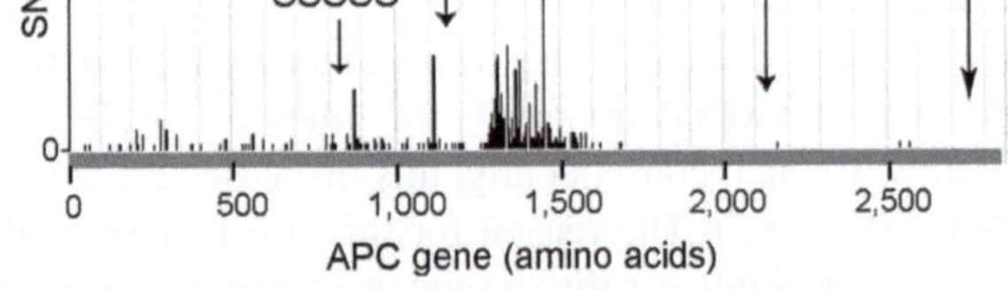

Fig. 5.6 APC hotspots are adjacent to repeats. *Top*, COSMIC diagram of APC mutations found in 8,324 tumor samples analyzed; 1,744 were simple mutations. The locations of the five n4 repeats are shown. The indels (insertion-deletions) and protein domains are also shown

The human APC tumor suppressor gene has mutation hotspots at CpG dinucleotides near coding repeats (Fig. 5.6). The miRDERR hypothesis predicts that mutations will be directed at or near the repeats in RCGs during cancer progression. To test this hypothesis, we performed a bioinformatic analysis of the somatic mutations in the APC gene that are annotated in the COSMIC database (Catalog of Somatic Mutations in Cancer). We chose APC because this is a commonly mutated gene in colorectal cancer (there are 1,174 simple mutations in 8,324 analyzed samples), and because APC contains several nocpg_n4 repeats (Fig. 5.6). Interestingly, the three largest mutation hotspots in APC are in the closest CpG to three of the nocpg_n4 repeats, SSSSS, EEEE, and PPPPP. For example, the largest hotspot is at amino acid 1,450 [CGA (Arg) to TGA (Stop)]. It is the CpG closest to the five-proline (PPPPP) repeat at amino acids X-Y, a distance of only seven codons (Fig. 5.6). Similarly, the hotspot at amino acid 876 [CGA (Arg) to TGA (Stop)] is near the five-serine (SSSSS) repeat located at amino acids Q-R, or 39 amino acids away. Finally, the hotspot at amino acid 1,114 [CGA (Arg) to TGA (Stop)] is also 39 codons away from the short repeat coding for the four glutamic acid (EEEE) residues at amino acids J-K (Fig. 5.6). As discussed below, one possibility is that the PPPPP repeat directs methylation to nearby CpGs through a miR directed epigenetic mechanism, such as proposed in the miRDERR hypothesis.

MicroRNAs (miRs) are transcribed from a class of noncoding genes whose final transcriptional product is a ~22–25 nucleotide functional RNA molecule. miRs play important roles in the regulation of target genes by binding to complementary regions of mRNA to repress their translation or regulate degradation through conserved complexes (Bartel and Chen 2004; Grimson et al. 2007; Jones-Rhoades et al. 2006) (Fig. 5.4). Recent reports have suggested that miRs may play roles in human cancers (Chang and Mendell 2007; Dews et al. 2006; Kent and Mendell 2006; O'Donnell et al. 2005) and during normal development (Cohen et al. 2007; Cohen and Brennecke 2006; Cohen et al. 2006; Sandmann and Cohen 2007; Stark et al. 2008). We speculate that homopolymer amino acid coding repeats are likely targets for miRs because long repeats often have high "purity," i.e., the same codon repeated consecutively several times (Ruden, Zeeberg, Chen, Rasouli, Page, Garfinkel, and Jamison, unpublished results), and are therefore low complexity repetitive sequences. Many miRs target repeated sequences, so this is an attractive hypothesis.

If the miRDERR hypothesis is correct, then what are the miRs that target the repeats in the APC gene? The database for identifying potential miR targets is called miRBase (Griffiths-Jones 2006; Griffiths-Jones et al. 2006) (http://microrna.sanger.ac.uk). We used miRBase to identify the best human miRs that have significant homology to the repeat sequences by entering the repeat sequence into miRBase and searching for high scoring human miRs. We found that human miR hsa-miR-877 has a high score (12/15 match complementary to the mRNA) for the PPPPP repeat in the human APC gene and is likely a target. Whether the human miR hs-miR-877 targets the PPPPP repeat in the APC gene in vivo, leading to the truncation of the APC protein that is often observed in human cancers will require experimental validation in the laboratory.

> False facts are highly injurious to the progress of science, for they often endure long; but false views, if supported by some evidence, do little harm, for everyone takes a salutary pleasure in proving their falseness.
>
> The Descent of Man, and Selection in Relation to Sex, p. 385 (Darwin 1871).

Conclusions and Future Studies

In this chapter, we discussed the role of Hsp90 and stress in regulating evolution by genetic and epigenetic mechanisms. As discussed earlier, many evolutionary biologists have disputed the importance of Hsp90 as a capacitor for morphological evolution. However, the original paper of Rutherford and Lindquist in 1998 (Rutherford and Lindquist 1998) has stood up very well, and the role of Hsp90 as an adaptively inducible canalizer is gaining more and more prominence among evolutionary biologists (Pigliucci et al. 2010). We believe that studies of Hsp90's capacitor function will continue to penetrate into the evolutionary biology community, and will eventually be seen as a cornerstone for understanding gene–environment interactions. Time will tell whether our optimism for the role of Hsp90 as a capacitor for evolution is justified.

Acknowledgments This work was supported by NIH grants ES012933 and ES02183 to D.M.R., and DK071073 to X.L.

References

Bartel DP, Chen CZ (2004) Micromanagers of gene expression: the potentially widespread inf. uence of metazoan microRNAs. Nat Rev Genet 5(5):396–400
Branham WS et al (1988) Uterine abnormalities in rats exposed neonatally to diethylstilbestrol, ethynylestradiol, or clomiphene citrate. Toxicology 51(2–3):201–212
Brennecke J et al (2007) Discrete small RNA-generating loci as master regulators of transposon activity in Drosophila. Cell 128(6):1089–1103
Brooks JL (1965) Predation and relative helmet size in cyclomorphic daphnia. Proc Natl Acad Sci USA 53(1):119–126
Carroll SB (2005) Endless forms most beautiful: the new science of evo devo and the making of the animal kingdom, 1st edn. Norton, New York, p 350, xi
Chang TC, Mendell JT (2007) microRNAs in vertebrate physiology and human disease. Annu Rev Genomics Hum Genet 8:215–239
Chen G et al (2012) Hsp90 stress potentiates rapid cellular adaptation through induction of aneuploidy. Nature 482(7384):246–250
Chouard T (2010) Evolution: revenge of the hopeful monster. Nature 463(7283):864–867
Cohen SM, Brennecke J (2006) Developmental biology. Mixed messages in early development. Science 312(5770):65–66
Cohen SM, Brennecke J, Stark A (2006) Denoising feedback loops by thresholding—a new role for microRNAs. Genes Dev 20(20):2769–2772
Cohen A et al (2007) Alterations in microRNA expression profiles reveal a novel pathway for estrogen regulation. Endocrinology 149(4):1687–1696
Darwin C (1859) The origin of the species. Suriano G (ed) Random House
Darwin C (1871) The descent of man, and selection in relation to sex. D. Appleton and company, New York
Darwin C (1883) Variation in animals and plants under domestication. Appleton and Co., New York
David R (2012) DNA repair: how chromosomes find their "soul mate. Nat Rev Mol Cell Biol 13(5):281

Dews M et al (2006) Augmentation of tumor angiogenesis by a Myc-activated microRNA cluster. Nat Genet 38(9):1060–1065
Dezwaan DC, Freeman BC (2008) HSP90: the Rosetta stone for cellular protein dynamics? Cell Cycle 7(8):1006–1012
Fondon JW, Garner HR (2004) Molecular origins of rapid and continuous morphological evolution. Proc Natl Acad Sci USA 101(52):18058–18063
Gangaraju VK et al (2011) Drosophila Piwi functions in Hsp90-mediated suppression of phenotypic variation. Nat Genet 43(2):153–158
Garfinkel MD et al (2004) Multigenerational selection and detection of altered histone acetylation and methylation patterns: toward a quantitative epigenetics in Drosophila. Methods Mol Biol 287:151–168
Gilloteaux J, Steggles AW (1985) Endometrium cell surface abnormalities in the Syrian hamster as a result of in utero exposure to diethylstilbestrol. Scan Electron Microsc (Pt 1):303–309.
Gould SJ (1980) The panda's thumb: more reflections in natural history, 1st edn. Norton, New York, 343
Griffiths-Jones S (2006) miRBase: the microRNA sequence database. Methods Mol Biol 342:129–138
Griffiths-Jones S et al (2006) miRBase: microRNA sequences, targets and gene nomenclature. Nucleic Acids Res 34(Database Issue):D140–D144
Grimaud C et al (2006) RNAi components are required for nuclear clustering of Polycomb group response elements. Cell 124(5):957–971
Grimson A et al (2007) MicroRNA targeting specificity in mammals: determinants beyond seed pairing. Mol Cell 27(1):91–105
Hallson G et al (2008) The Drosophila cohesin subunit Rad21 is a trithorax group (trxG) protein. Proc Natl Acad Sci USA 105(34):12405–12410
Hamamoto R et al (2004) SMYD3 encodes a histone methyltransferase involved in the proliferation of cancer cells. Nat Cell Biol 6(8):731–740
Hatch EE et al (2001) Incidence of squamous neoplasia of the cervix and vagina in women exposed prenatally to diethylstilbestrol (United States). Canc Causes Contr 12(9):837–845
Holeski LM (2007) Within and between generation phenotypic plasticity in trichome density of Mimulus guttatus. J Evol Biol 20(6):2092–2100
Huxley J (1942) Evolution, the modern synthesis. G. Allen & Unwin Ltd., London, p 645
Iguchi T, Takasugi N (1987) Postnatal development of uterine abnormalities in mice exposed to DES in utero. Biol Neonate 52(2):97–103
Jablonka E, Lamb MJ (2005) Evolution in four dimensions: genetic, epigenetic, behavioral, and symbolic variation in the history of life and mind. MIT Press, Cambridge, MA, p 462, x
Jones-Rhoades MW, Bartel DP, Bartel B (2006) MicroRNAS and their regulatory roles in plants. Annu Rev Plant Biol 57:19–53
Kent OA, Mendell JT (2006) A small piece in the cancer puzzle: microRNAs as tumor suppressors and oncogenes. Oncogene 25(46):6188–6196
Knight WA 3rd et al (1980) Steroid hormone receptors in the management of human breast cancer. Ann Clin Res 12(5):202–207
Kohler HR et al (2009) Snail phenotypic variation and stress proteins: do different heat response strategies contribute to Waddington's widget in field populations? J Exp Zool Part B Mol Dev Evol 312(2):136–147
Laforsch C, Tollrian R (2004) Embryological aspects of inducible morphological defenses in Daphnia. J Morphol 262(3):701–707
Lamarck JBP (1809) Zoological philosophy. Chicago Press, Chicago
Leavitt WW, Evans RW, Hendry WJ 3rd (1981) Etiology of DES-induced uterine tumors in the Syrian hamster. Adv Exp Med Biol 138:63–86
Li S et al (2001) Promoter CpG methylation of Hox-a10 and Hox-a11 in mouse uterus not altered upon neonatal diethylstilbestrol exposure. Mol Carcinog 32(4):213–219
Li S et al (2003a) Environmental exposure, DNA methylation, and gene regulation: lessons from diethylstilbesterol-induced cancers. Ann N Y Acad Sci 983:161–169

Li S et al (2003b) Neonatal diethylstilbestrol exposure induces persistent elevation of c-fos expression and hypomethylation in its exon-4 in mouse uterus. Mol Carcinog 38(2):78–84

Lu X et al (2011) Hsp90 inhibitors and drug resistance in cancer: the potential benefits of combination therapies of Hsp90 inhibitors and other anti-cancer drugs. Biochem Pharmacol 83(8):995–1004

Matranga C, Zamore PD (2007) Small silencing RNAs. Curr Biol 17(18):R789–R793

Matyash A, Chung HR, Jackle H (2004) Genome-wide mapping of in vivo targets of the Drosophila transcription factor Kruppel. J Biol Chem 279(29):30689–30696

McClintock B (1984) The significance of responses of the genome to challenge. Science 226(4676):792–801

McLachlan JA, Newbold RR, Bullock BC (1980) Long-term effects on the female mouse genital tract associated with prenatal exposure to diethylstilbestrol. Cancer Res 40(11):3988–3999

McLaren A (1999) Too late for the midwife toad: stress, variability and Hsp90. Trends Genet TIG 15(5):169–171

Medlock KL, Branham WS, Sheehan DM (1992) Long-term effects of postnatal exposure to diethylstilbestrol on uterine estrogen receptor and growth. J Steroid Biochem Mol Biol 42(1):23–28

Meiklejohn CD, Hartl DL (2002) A single mode of canalization. Trends Ecol Evol 17(10):468–473

Mendel G (1865) Experiments in plant hybridisation. Harvard University Press, Boston

Moxon R, Bayliss C, Hood D (2006) Bacterial contingency loci: the role of simple sequence DNA repeats in bacterial adaptation. Annu Rev Genet 40:307–333

Newbold RR, Bullock BC, McLachlan JA (1990) Uterine adenocarcinoma in mice following developmental treatment with estrogens: a model for hormonal carcinogenesis. Cancer Res 50(23):7677–7681

Newbold RR et al (1998) Increased tumors but uncompromised fertility in the female descendants of mice exposed developmentally to diethylstilbestrol. Carcinogenesis 19(9):1655–1663

Newbold RR et al (2000) Proliferative lesions and reproductive tract tumors in male descendants of mice exposed developmentally to diethylstilbestrol. Carcinogenesis 21(7):1355–1363

O'Donnell KA et al (2005) c-Myc-regulated microRNAs modulate E2F1 expression. Nature 435(7043):839–843

Oda S et al (2011) Morphological changes in Daphnia galeata induced by a crustacean terpenoid hormone and its analog. Environ Toxicol Chem/SETAC 30(1):232–238

Phalke S et al (2009) Retrotransposon silencing and telomere integrity in somatic cells of Drosophila depends on the cytosine-5 methyltransferase DNMT2. Nat Genet 41(6):696–702

Pigliucci M (2003) Epigenetics is back! Hsp90 and phenotypic variation. Cell Cycle 2(1):34–35

Pigliucci M, Müller G, Konrad Lorenz Institute for Evolution and Cognition Research (2010) Evolution, the extended synthesis. MIT Press, Cambridge, MA, p 495, viii

Queitsch C, Sangster TA, Lindquist S (2002) Hsp90 as a capacitor of phenotypic variation. Nature 417(6889):618–624

Rakyan V, Whitelaw E (2003) Transgenerational epigenetic inheritance. Curr Biol 13(1):R6

Rehwinkel J et al (2006) Genome-wide analysis of mRNAs regulated by Drosha and Argonaute proteins in Drosophila melanogaster. Mol Cell Biol 26(8):2965–2975

Roux F et al (2011) Genome-wide epigenetic perturbation jump-starts patterns of heritable variation found in nature. Genetics 188(4):1015–1017

Ruden DM (2011) The (new) new synthesis and epigenetic capacitors of morphological evolution. Nat Genet 43(2):88–89

Ruden DM, Lu X (2008) Hsp90 affecting chromatin remodeling might explain transgenerational epigenetic inheritance in Drosophila. Curr Genom 9(7):500–508

Ruden DM et al (2003) Waddington's widget: Hsp90 and the inheritance of acquired characters. Semin Cell Dev Biol 14(5):301–310

Ruden DM et al (2005a) Hsp90 and environmental impacts on epigenetic states: a model for the trans-generational effects of diethylstibesterol on uterine development and cancer. Human Mol Genet 14(1):R149–R155

Ruden DM et al (2005b) Epigenetic regulation of trinucleotide repeat expansions and contractions and the "biased embryos" hypothesis for rapid morphological evolution. Curr Genom 6(3):145–155

Ruden DM et al (2008) The EDGE hypothesis: epigenetically directed genetic errors in repeat-containing proteins (RCPs) involved in evolution, neuroendocrine signaling, and cancer. Front Neuroendocrinol 29(3):428–444
Rutherford SL, Lindquist S (1998) Hsp90 as a capacitor for morphological evolution. Nature 396(6709):336–342
Rutherford S, Hirate Y, Swalla BJ (2007) The Hsp90 capacitor, developmental remodeling, and evolution: the robustness of gene networks and the curious evolvability of metamorphosis. Crit Rev Biochem Mol Biol 42(5):355–372
Sandmann T, Cohen SM (2007) Identification of Novel Drosophila melanogaster MicroRNAs. PLoS One 2(11):e1265
Saurin AJ et al (1998) The human polycomb group complex associates with pericentromeric heterochromatin to form a novel nuclear domain. J Cell Biol 142(4):887–898
Schaefer M, Lyko F (2010) Lack of evidence for DNA methylation of Invader4 retroelements in Drosophila and implications for Dnmt2-mediated epigenetic regulation. Nat Genet 42(11): 920–921, author reply 921
Scoville AG et al (2011) Differential regulation of a MYB transcription factor is correlated with transgenerational epigenetic inheritance of trichome density in Mimulus guttatus. New Phytol 191(1):251–263
Sollars V et al (2003) Evidence for an epigenetic mechanism by which Hsp90 acts as a capacitor for morphological evolution. Nat Genet 33(1):70–74
Specchia V et al (2010) Hsp90 prevents phenotypic variation by suppressing the mutagenic activity of transposons. Nature 463(7281):662–665
Stark A et al (2008) A single Hox locus in Drosophila produces functional microRNAs from opposite DNA strands. Genes Dev 22(1):8–13
Stedman W et al (2008) Cohesins localize with CTCF at the KSHV latency control region and at cellular c-myc and H19/Igf2 insulators. EMBO J 27(4):654–666
Taipale M, Jarosz DF, Lindquist S (2010) HSP90 at the hub of protein homeostasis: emerging mechanistic insights. Nature reviews. Mol Cell Biol 11(7):515–528
Tariq M et al (2009) Trithorax requires Hsp90 for maintenance of active chromatin at sites of gene expression. Proc Nat Acad Sci USA 106(4):1157–1162
Tolkien JRR (1954) The lord of the rings. Allen & Unwin, London
Trepel J et al (2010) Targeting the dynamic HSP90 complex in cancer. Nat Rev Cancer 10(8):537–549
Turusov VS et al (1992) Occurrence of tumours in the descendants of CBA male mice prenatally treated with diethylstilbestrol. Int J Cancer 50(1):131–135
Waddington CH (1942) Canalization of development and the inheritance of acquired characters. Nature 150:563–565
Waddington CH (1953) Genetic assimilation of an acquired character. Evolution 7:118–126
Waddington CH (1956) Genetic assimilation of the bithorax complex. Evolution 10:1–13
Waddington CH (1974) A catastrophe theory of evolution. Ann NY Acad Sci 231(1):32–42
Walker BE, Haven MI (1997) Intensity of multigenerational carcinogenesis from diethylstilbestrol in mice. Carcinogenesis 18(4):791–793
Wang Y et al (2006) Functional CpG methylation system in a social insect. Science 314(5799):645–647
Whitesell L, Lindquist SL (2005) HSP90 and the chaperoning of cancer. Nat Rev Cancer 5(10):761–772
Zambon RA, Vakharia VN, Wu LP (2006) RNAi is an antiviral immune response against a dsRNA virus in Drosophila melanogaster. Cell Microbiol 8(5):880–889
Zamore PD (2007) RNA silencing: genomic defence with a slice of pi. Nature 446(7138):864–865

Chapter 6
Inheritance of Stress-Induced Epigenetic Changes Mediated by the ATF-2 Family of Transcription Factors

Ki-Hyeon Seong, Toshio Maekawa, and Shunsuke Ishii

Abstract The inheritance of stress effects has attracted much attention of late, partly because it resembles the inheritance of acquired characteristics. Accumulating evidence suggests that various stress effects may be inherited, including environmental stresses and responses to adverse nutritional conditions. Although factors such as epigenetic regulation have been implicated, the precise underlying mechanisms are obscure. The ATF-2 family of transcription factors is directly phosphorylated by stress-activated protein kinases, such as p38, in response to various stresses. In the absence of stress, they contribute to the formation of heterochromatin, which is rich in histone H3 Lys-9 trimethylation, and is a typical epigenetic marker that can be inherited. Disrupted heterochromatin, resulting from phosphorylation of ATF-2 and its related proteins, can also be inherited. This review summarizes current data on the inheritance of stress-induced epigenetic changes mediated by the ATF-2 family of transcription factors.

Introduction

Much attention has been paid to the inheritance of stress effects that do not induce changes in DNA sequences. This is because this phenomenon partly resembles the inheritance of acquired characteristics, which was originally proposed by Lamarck (1809) and has been denied historically. The process is well known in plants; for example, maize plants inherit color changes induced by environmental stresses—such as changes in light levels and temperature (Mikula 1995)—although the underlying mechanism is unknown. The expression of certain genes induced by UV light in *A. thaliana* is also inherited (Molinier et al. 2006). Almost 20 years ago, based on

K.-H. Seong • T. Maekawa • S. Ishii (✉)
Laboratory of Molecular Genetics, RIKEN Tsukuba Institute,
3-1-1 Koyadai, Tsukuba, Ibaraki 305-0074, Japan
e-mail: sishii@rtc.riken.jp

D. Mittelman (ed.), *Stress-Induced Mutagenesis*, DOI 10.1007/978-1-4614-6280-4_6,

epidemiological studies, David Barker proposed that babies with a low birth weight are more susceptible to lifestyle-related diseases, such as diabetes and heart failure, when they become adults (Barker et al. 1993). This hypothesis is now known as the DOHaD (developmental origin of health and diseases) hypothesis, which suggests that diseases originate from unbalanced nutrition of the fetus in utero and infants, and are then "transmitted" to the next generation (Waterland and Michels 2007). Recent experiments in rats and mice indicate that changes in gene expression in the pancreas and liver caused by high-fat diets and low-protein diets, respectively, are inherited by offspring (Ng et al. 2010; Carone et al. 2010). Another major question is whether the effects of psychological stress can be inherited. When rat pups are separated from their mother, this psychological stress is maintained and the pups show abnormal behavior into adulthood (Francis et al. 1999). However, it is not yet clear whether this effect is transmitted to the next generation.

Recent progress on epigenetic regulation strongly suggests that various chemical modifications of DNA and histones may play a key role in the inheritance of stress effects without involving changes in the DNA. In particular, recent studies show that the stress-responsive ATF-2 family of transcription factors can modulate the epigenetic status of heterochromatin (Seong et al. 2011; Maekawa et al. 2010b), and have opened the door to understanding the mechanisms underlying stress-induced epigenetic changes. Here, we review the role of the ATF-2 family of transcription factors in stress-induced epigenetic changes.

Stress and the ATF-2 Family of Transcription Factors

Over 20 years ago, the specific DNA element CRE (cAMP response element), which is localized in the promoter region of neuronal genes, was identified as mediating transcriptional induction by increasing intracellular cAMP levels (Montminy et al. 1986). Multiple groups tried to clone the transcription factor that binds to CRE, resulting in the identification of CREB (Gonzalez et al. 1989) and CRE-BP1 (Maekawa et al. 1989). Both contain a B-ZIP-type DNA-binding domain, which consists of a leucine zipper structure and a basic amino acid cluster; however, they have no amino acid sequence similarity. Both CREB and CRE-BP1 bind CRE as dimers formed via the leucine zipper structures like a scissors. CREB is directly phosphorylated by protein kinase A, which is activated by an increase in intracellular cAMP levels; phosphorylated CREB then activates a group of target genes containing CRE (Gonzalez and Montminy, 1989). The transcriptional co-activator CBP then binds to phosphorylated CREB and mediates transcription (Chrivia et al. 1993). CRE was also identified as a specific DNA element that mediates transcriptional activation by the adenoviral E1A oncoprotein, which is unable to bind to CRE directly (Lee et al. 1987). The search for specific transcription factors that mediate E1A-dependent transcription involved identifying factors that bind to CRE. Multiple factors—so-called activating transcription factors (ATFs)—were identified (Hai et al. 1989). Although all ATFs contain the B-ZIP-type DNA-binding domain,

ATF-2 was found to be identical to CRE-BP1. ATF-2/CRE-BP1 interacts with E1A and mediates E1A-dependent transcriptional activation (Liu and Green 1990). Although ATF-2 constitutively activates transcription as a homodimer or heterodimer with c-Jun in the absence of E1A, or without an increase in cAMP levels (Matsuda et al. 1991), the physiological role of ATF-2 was long unknown.

The discovery of stress-activated protein kinases (SAPKs) such as Jun N-terminal protein kinase (JNK) and p38 opened the door to further understanding the physiological role of ATF-2. SAPKs belong to the mitogen-activated protein kinase (MAPK) family, which is activated via MAPKKK and MAPKK in response to various stimuli, including growth factors, inflammatory cytokines, and stress (Chang and Karin 2001). Vertebrate ATF-2 possesses a transcriptional activation domain in its N-proximal region, which consists of a metal finger structure and two phosphorylation sites for JNK and p38 (Gupta et al. 1995; Nagadoi et al. 1999). In response to various stimuli, ATF-2 is phosphorylated by JNK or p38 at both Thr-69 and Thr-71 (van Dam et al. 1997; Livingstone et al. 1997), which causes transcriptional activation via subsequent interactions with the co-activator CBP, which possesses histone acetyltransferase (HAT) activity (Sano et al. 1998) (Fig. 6.1a). Thus, the transcriptional activation domain of ATF-2 acts as a stress-responsive element. Vertebrates express three members of the ATF-2 subfamily: ATF-2, CRE-BPa, and ATF-7 (originally called ATF-a), which show significant homology in their DNA-binding and transcriptional activation domains (Nomura et al. 1993; Gaire et al. 1990) (Fig. 6.1b). They contain a metal finger structure and two p38 phosphorylation sites within the N-terminal activation domain, and a B-ZIP-type DNA-binding domain within the C-terminal region. ATF-2 and ATF-7 are ubiquitously expressed, with the highest expression levels observed in the brain, whereas CRE-BPa shows more limited tissue expression. *Drosophila* has only one homolog of ATF-2, which contains two p38 phosphorylation sites and the B-ZIP-type DNA-binding domain within the N- and C-terminal regions, respectively (Sano et al. 2005) (Fig. 6.1c). ATF-2 and CRE-BPa are phosphorylated by either JNK or p38, but ATF-7 and *Drosophila* ATF-2 (dATF-2) are phosphorylated only by p38.

Various types of stress induce phosphorylation of ATF-2 by JNK or p38, which then activates transcription of a group of target genes. Typical exogenous types of stress that activate the p38-ATF-2 pathway include environmental stress, such as heat shock (HS) and osmotic stress, ultraviolet (UV) light, and hypoxia (Craig et al. 2004) (Table 6.1). In response to these stresses, the target genes encoding chaperones and apoptosis-regulation genes are induced. Another type of exogenous stress is infection by pathogens, which activates ATF-2 via p38 (Han et al. 1998; Kim et al. 2002). For example, lipopolysaccharide (LPS), present in the outer membrane of gram-negative bacteria, CpG DNA, or mycoplasmal macrophage-activating lipopeptide-2 kD (MALP-2), activates the p38-ATF-2 pathway via toll-like receptors (Hirose et al. 2009). Typical endogenous stimuli that activate the p38-ATF-2 pathway include inflammatory cytokines (Brinkman et al. 1999) such as TNF-α, and IL1, the levels of which are increased by psychological stress (Black 2002) or pathogen infection. Reactive oxygen species (ROS), which are induced during oxidative

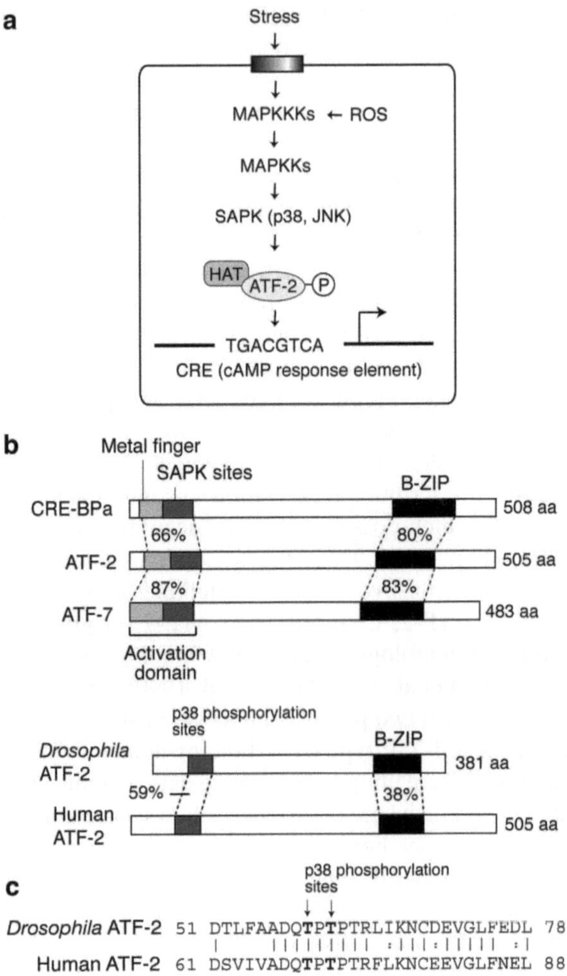

Fig. 6.1 The role of ATF-2 in stress-induced responses and the three members of the vertebrate ATF-2 family. (**a**) ATF-2 is activated by SAPKs via MAPKKK and MAPKK in response to various stresses, and subsequently activates transcription by interacting with the co-activator, CBP (which possesses histone acetyltransferase (HAT) activity), after binding to CRE. (**b**) Domain structure of ATF-2 and homology between the three members of the vertebrate ATF-2 family. ATF-2 contains a B-ZIP-type DNA-binding domain and a *trans*-activation domain, which consists of a metal finger structure and SAPK phosphorylation sites. (**c**) Homology between *Drosophila* ATF-2 and human ATF-2. Two p38 phosphorylation sites are conserved between human and *Drosophila* ATF-2

phosphorylation in the mitochondria to generate ATP, are also a source of endogenous stress that activates the p38-ATF-2 pathway (Benhar et al. 2001). These stresses induce the expression of immunity-related, neuronal, and metabolism-related genes (Table 6.1).

Table 6.1 Different types of stress that regulate ATF-2 family transcription factors

Stress or condition	Type of stress	Target genes	Related diseases	Related phenomena
Heat, Osmotic shock UV light, Hypoxia	Environmental stresses	Chaperones Apoptosis		Adaptation, evolution
Ligands of TLRs	Pathogen infection	Immunity	Allergy etc.	Memory of innate immunity?
Inflammatory cytokines (TNF-α, IL-1 etc.)	Psychological stress & pathogen infection	Neuronal	Mental diseases	Inheritance of mental diseases?
ROS	Nutritional stress	Metabolism	Diabetes etc.	DOHaD hypothesis

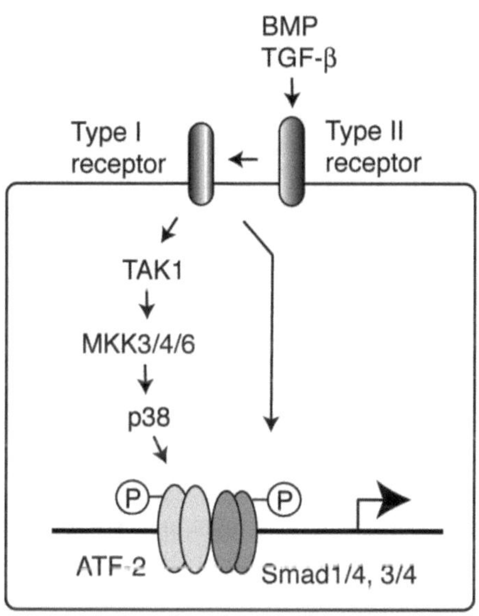

Fig. 6.2 Role of ATF-2 in the BMP/TGF-β signaling pathway. ATF-2 is phosphorylated via the TAK1-p38 pathway in response to BMP/TGF-β. However, BMP/TGF-β induces phosphorylation of the Smad1 and Smad3 transcription factors, which leads to nuclear entry of the Smad1/4 or Smad3/4 complexes. ATF-2 interacts with the Smad complex on the target genes to synergistically activate transcription

Several growth factors also activate ATF-2 to induce transcription. BMP (bone morphogenetic protein) and TGF-β (transforming growth factor β) bind to type II receptors and lead to the activation of type I receptors (Sano et al. 1999) (Fig. 6.2). The transcription factors, Smad1/3, are phosphorylated by type I receptors and form hetero-oligomers with Smad4, which then enter the nucleus to activate transcription. At the same time, TGF-β-activated kinase (TAK1), a member of the MAPKKK family, is activated by type I receptors to induce ATF-2 phosphorylation via p38. Some BMP/TGF-β target genes have binding sites for both Smads and ATF-2, which synergistically activate transcription.

Epigenetic Changes Mediated by *Drosophila* ATF-2 or Vertebrate ATF-7 in Response to Stress

Chromatin is classified as either heterochromatin or euchromatin, in which transcription is inactive or active, respectively (Fig. 6.3a). Heterochromatin is located close to the centromere or telomere and is enriched in histone H3 Lys-9 trimethylation (H3K9me3), to which heterochromatin protein 1 (HP1) binds via its chromodomain (Grewal and Moazed 2003). This epigenetic marker (H3K9me3) is conserved during mitosis and meiosis, and is transmitted to the next generation. A recent genome-wide analysis indicates that heterochromatin structures containing H3K9me3 are not only distributed throughout the pericentromeric and peritelomeric regions, but also throughout the euchromatin region (Barski et al. 2007). Studies on the mechanism of heterochromatin formation began in the fission yeast *Schizosaccharomyces pombe* (Volpe et al. 2002). The components of the RNA interference (RNAi) machinery, which generates small inhibitory RNA (siRNA) from double-stranded RNA (ds-RNA) to degrade mRNA, were identified as the regulators of heterochromatin formation (Grewal and Elgin 2007). Heterochromatin is enriched in repetitive sequences, which undergo low levels of transcription to form double-stranded RNAs. The RNAi machinery is thought to produce siRNA from these ds-RNAs, which then recruit histone methyltransferase to generate H3K9me3.

In addition to the RNAi machinery, the ATF-2 homolog in *S. pombe*, Atf1, is also independently involved in heterochromatin formation (Jia et al. 2004). Atf1 possesses the B-ZIP-type DNA-binding domain and is phosphorylated by the stress-activated kinase, Sty1, in response to various types of stress. We used *Drosophila* PEV (position effect variegation) to examine the role of ATF-2 in higher organisms (Henikoff 1990). In the w^{m4} line, the *white* gene, which is involved in biosynthesis of the red eye pigment, is located close to the pericentromeric heterochromatin on the X chromosome due to a large gene inversion. The *white* gene is silenced by heterochromatin, such that the w^{m4} line exhibits the mottled eye phenotype. This PEV assay has been used to genetically identify many regulators of heterochromatin. When a heterozygous or homozygous *dATF-2* mutation was introduced into the w^{m4} line, silencing of the *white* gene was abrogated in a dose-dependent manner (Fig. 6.3b), indicating an essential role for dATF-2 in heterochromatin formation (Seong et al. 2011). When the *dATF-2* mutation was combined with a mutation of the RNAi machinery (Spindle-E, encoding a DEAD-motif RNA helicase), the effect of the two mutations was additive (Fig. 6.3b), indicating that dATF-2 and the RNAi machinery independently regulate heterochromatin formation.

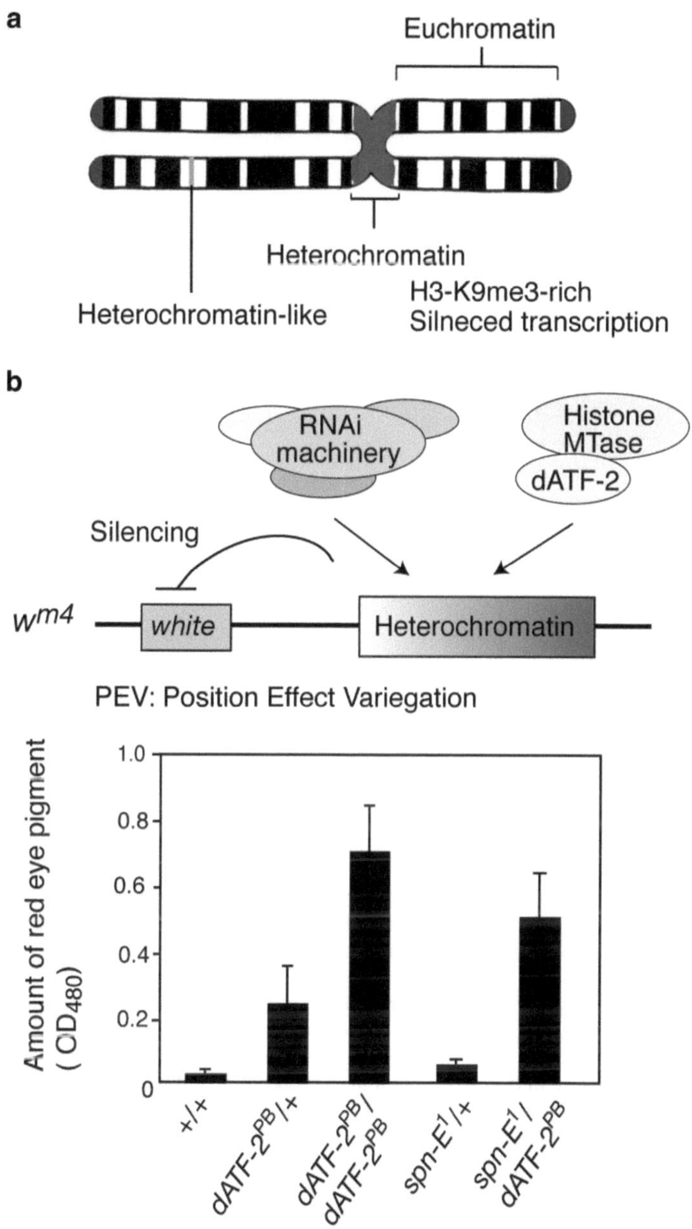

Fig. 6.3 Independent roles of ATF-2 and the RNAi machinery during heterochromatin formation in *Drosophila*. (**a**) Typical heterochromatin, which is enriched in H3K9me3, is located in regions close to centromeres and telomeres. In addition, similar heterochromatin-like structures are localized in the euchromatin region. (**b**) In the w^{m4} line, the *white* gene is located close to centromeric heterochromatin on the X chromosome due to a large gene inversion, which causes silencing of the *white* gene (*upper panel*). In the w^{m4} line carrying a *dATF-2* heterozygous or homozygous mutation, heterochromatin is disrupted, resulting in *white* gene induction in a dose-dependent manner (*lower panel*). The effect of mutating *Spn-E*, a component of the RNAi machinery, on *white* gene silencing was additive to that induced by the *dATF-2* mutation, indicating that the two pathways independently contribute to heterochromatin formation. This is a modification of Fig. 1 first published by Seong et al. (2011)

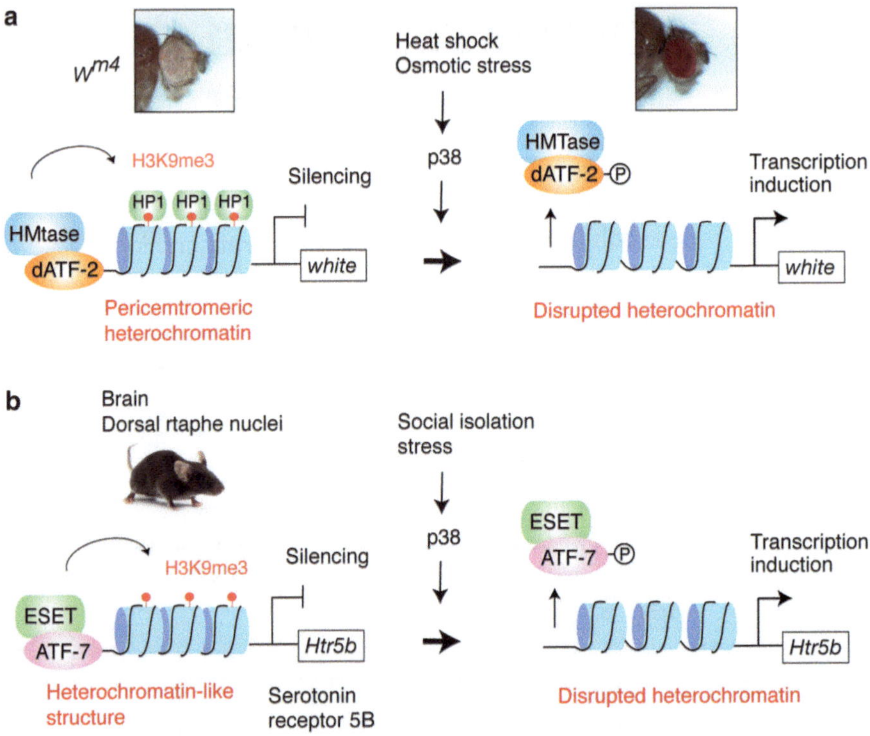

Fig. 6.4 ATF-2-dependent disruption of heterochromatin induced by stress. (**a**) Heat shock (HS) or osmotic stress disrupts pericentromeric heterochromatin via dATF-2. dATF-2 is required to form and maintain pericentromeric heterochromatin. When flies are exposed to heat shock or osmotic stress during early embryogenesis or at the larval stage, dATF-2 is phosphorylated by p38 and is released from the heterochromatin. This disrupts the heterochromatin and *white* gene transcription. (**b**) Social isolation stress disrupts heterochromatin-like structures on the *Htr5b* gene via ATF-7. In the dorsal raphe nuclei within the mouse brain, ATF-7 binds to the *Htr5b* gene and silences its transcription by recruiting the H3K9 trimethyltransferase ESET complex. Social isolation stress induces ATF-7 phosphorylation via p38, which then leads to the release of ATF-7 from the *Htr5b* gene, resulting in the disruption of the heterochromatin-like structures and *Htr5b* gene transcription

When the w^{m4} line was exposed to HS stress (37 °C for 1 h) at various time points during development, subsequent HS immediately after egg-laying significantly abrogated *white* gene silencing. Since the process of heterochromatin formation is divided into two steps (an establishment stage during early development, and a subsequent maintenance stage) (Hall et al. 2002), this suggests that the establishment stage is more sensitive to stress. Upon HS stress, dATF-2 is phosphorylated by p38, which is then released from the heterochromatin. This causes a decrease in the levels of H3K9me3 and disrupts the heterochromatin structure (Fig. 6.4a). When flies were exposed to osmotic stress, which induces dATF-2 phosphorylation more efficiently than HS stress, at a later developmental stage (the larval stage) by feeding with medium containing 0.3 M NaCl, a similar disruption of heterochromatin was observed. This suggests that stronger stresses may even disrupt heterochromatin at later developmental stages.

Of the three vertebrate ATF-2 subfamily members, only ATF-7 is involved in epigenetic silencing. ATF-7 interacts with mAM (mouse ATFa-associated Modulator), a component of the H3K9 trimethyltransferase ESET (SETDB1) complex (De Graeve et al. 2000; Wang et al. 2003). *Atf-7* mutant mice are viable but show abnormal behavior (Maekawa et al. 2010b). The mice exhibit increased marble-burying behavior compared with wild-type mice, which is a form of anxiety-related behavior. Furthermore, they display a significant increase in acoustic startle responses, and lower levels of pre-pulse inhibition (PPI) of the acoustic startle response. These phenotypes are related to the neurotransmitter serotonin, which is present at high levels in the dorsal raphe nuclei. Increased levels of serotonin receptor 5B (*Htr5b*) mRNA are detected in mutant dorsal raphe nuclei. ATF-7 directly binds to the *Htr5b* gene, and represses its transcription by forming heterochromatin-like structures via recruitment of histone H3K9 trimethyltransferase ESET (Fig. 6.4b, left). Interestingly, when wild-type mice are exposed to stress associated with social isolation, they exhibit various abnormal behaviors (Rodgers and Cole 1993), some of which resemble to those of *Atf-7* mutant mice. Usually, four to five mice are reared in one cage; mice reared alone for more than 1 month show evidence of psychological stress. When mice are exposed to social isolation stress, p38 and ATF-7 are phosphorylated and the *Htr5b* gene transcription is induced in the dorsal raphe nuclei (Fig. 6.4b, right) (Maekawa et al. 2010b). The mechanism by which isolation stress induces p38 activation is unknown, although one possibility is that peripheral levels of an inflammatory cytokine, such as TNF-α, are increased by psychological stress (Black 2002) and the cytokine enters the brain to activate the p38-ATF-7 pathway.

Inheritance of Stress-Induced Epigenetic Changes

To investigate the inheritance of stress-disrupted heterochromatin, male w^{m4} flies (G1(H)) were HS-stressed during early embryogenesis and then crossed with non-stressed female flies. The offspring (G2(HN)) showed slightly, but significantly, increased levels of *white* gene expression (Fig. 6.5a), indicating that the effect of HS stress on the father had been transmitted to the children. A similar pattern of inheritance was observed when female flies were used, indicating that the effects of HS stress are paternally and maternally inherited. When w^{m4} flies were fed medium containing 0.3 M NaCl during the larval stage, the osmotic stress-disrupted heterochromatin was also transmitted to the next generation. Thus, heterochromatin disrupted by various stresses can be inherited.

To test the inheritance of the stress-induced disrupted heterochromatin over multiple generations, the offspring (G2(HN)) from HS-exposed male flies were crossed with non-stressed flies. The resulting offspring (G3(HNN)) did not show higher

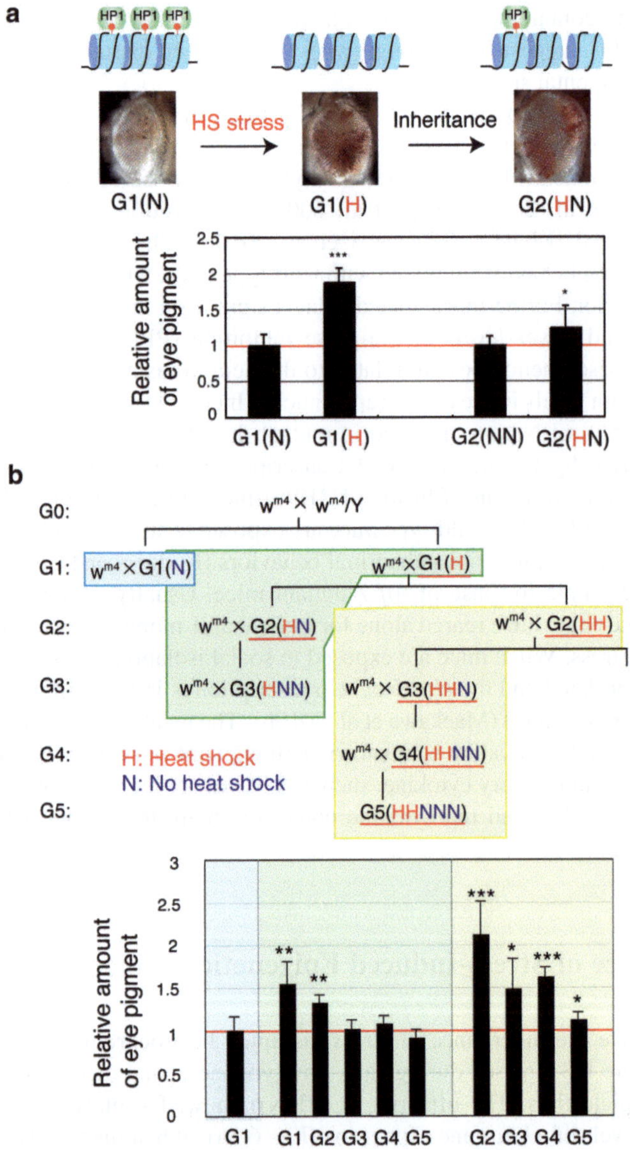

Fig. 6.5 Inheritance of stress-induced, dATF-2-dependent epigenetic changes over multiple generations. (**a**) When w^{m4} flies were exposed to HS stress during early embryogenesis, the levels of red eye pigment (G1(H)) were up-regulated compared with those in non-stressed flies (G1(N)). When these flies were then crossed with non-stressed flies, the offspring exhibited up-regulated levels of red eye pigment (G2(HN)). This figure has been modified from the original Fig. 6a published by Seong et al. (2011). (**b**) (*Green line* and *box*) When G1 generation flies were exposed to HS stress, the effects (shown by the upregulation of the *red* eye pigment) were transmitted to G2 generation flies but not to successive generations. (*Yellow line* and *box*) When both G1 and G2 generation flies were exposed to HS stress, the effects were observed in the G3, G4, and G5 generations. ***$P<0.001$; **$P<0.01$; *$P<0.05$. This figure is a modified version of Fig. 7 first published by Seong et al. (2011)

levels of *white* gene expression (Fig. 6.5b), indicating that the effect of HS stress on the parents had been transmitted to the children but not to the grandchildren. When G1- and G2-generation flies were exposed to HS stress, upregulation of *white* gene expression was observed in successive G3, G4, and G5 generations; thus, stress-disrupted heterochromatin is not stably inherited, although the strength and frequency of the stress may affect the number of generations over which the disrupted heterochromatin is transmitted.

Physiological Role of ATF-2-Related Proteins

To understand the physiological role of stress-induced and ATF-2-mediated epigenetic changes, it is useful to know the phenotype caused by mutations of the ATF-2 family of transcription factors, such as that shown by the *Atf-7* knockout mice described above. *Atf-2* and *Cre-bpa* null mutant mice show defects in placental and lung development, respectively (Maekawa et al. 1999, 2010a). Hypoxia plays an important role in determining the common developmental features observed in these tissues (Giaccia et al. 2004). PDGF receptor-α was identified as the direct target gene for ATF-2 during placental development. *Atf-2* hypomorphic mutants exhibit chondrodysplasia (Reimold et al. 1996), suggesting a role in bone formation. ATF-2 is also a critical tumor suppressor. Mice heterozygous for *Atf-2* are highly prone to mammary tumors, in which two apoptosis regulators, *Maspin* and *Gadd45α*, are up-regulated (Maekawa et al. 2007, 2008). Skin tumors are also generated in keratinocyte-specific *Atf-2* knockout mice (Bhoumik et al. 2008). A metabolic role for dATF-2 is suggested by the reduced glyceroneogenesis observed in the fat bodies of *dATF-2* knockdown flies (Okamura et al. 2007). The critical role of ATF-2 family transcription factors is demonstrated not only by the abnormal behavior of *Atf-7* mutant mice (Maekawa et al. 2010b), but also by the changes in sleep and locomotion in *dATF-2* knockdown flies (Shimizu et al. 2008). A large volume of biochemical data indicates that the p38 pathway functions in the toll-like receptor signaling pathway, but information obtained using mutants of ATF-2-related proteins is limited. The *C. elegans Atf-7* mutant also shows decreased resistance to pathogens (Shivers et al. 2010). Thus, the ATF-2 family of transcription factors is involved in many different biological phenomena. It would be interesting to examine whether any of those phenomena are epigenetically regulated by stress-induced and ATF-2-depenednt epigenetic changes (Table 6.2).

Mechanism of Inheritance of Stress-Disrupted Heterochromatin

Recent data suggest a critical role for ATF-2 family transcription factors in stress-induced epigenetic changes. However, the mechanism(s) underlying the inheritance of epigenetic change remains unknown. To be inherited, stress-induced

Table 6.2 Phenotype of ATF-2 family member mutants

Type of defect	Phenotype	Gene and mutant	References
Developmental defect	Placental defect	Mouse *Atf-2* null mutant	Maekawa et al. (1999)
	Lung defect	Mouse *Cre-bpa* mutant	Maekawa et al. (2010a)
	Chondrodysplasia	Mouse *Atf-2* hypomorphic mutant	Reimold et al. (1996)
Cancer	Mammary tumor	Mouse *Atf-2* heterozygous mutant	Maekawa et al. (2007)
	Skin tumor	Mouse *Atf-2* keratinocyte-specific mutant	Bhoumik et al. (2008)
Metabolism	Reduced glyceroneogenesis	*Drosophila Atf-2* knockdown mutant	Okamura et al. (2007)
Behavior	Increased acoustic startle response	Mouse Atf-7 mutant	Maekawa et al. (2010b)
	Sleep and locomotion	*Drosophila Atf-2* knockdown mutant	Shimizu et al. (2008)
Innate immunity	Decreased resistance to pathogen	*C. elegans Atf-7* mutant	Shivers et al. (2010)

epigenetic changes must occur in germ cells. In *Drosophila* germ cells, HS induced changes in dATF-2 localization on heterochromatin and subsequent heterochromatin disruption (Seong et al. 2011). ATF-7 is also expressed in mouse germ cells and is phosphorylated in response to LPS stimulation (our own unpublished data). Stress-disrupted heterochromatin in germ cells may be transmitted to sperm, zygotes, and early embryos; however, it is not known how heterochromatin is marked during these processes. During spermatogenesis, chromatin is opened by histone variants and histone acetylation, whereas histones are replaced by protamines in the spermatids (Gaucher et al. 2010). Furthermore, after fertilization, DNA methylation in the zygotic pro-nucleus is removed (Mayer et al. 2000), and the open chromatin structure is maintained from the zygote to the inner cell mass (ICM) of the blastocyst embryo. Typical heterochromatin structures start to form immediately after differentiation from the ICM. However, some evidence suggests possible markers for heterochromatin during spermatogenesis and early development. Mature sperm still contain small amounts of histone and histone variants (Bench et al. 1996; Wu et al. 2008), which are enriched in heterochromatin regions (van der Heijden et al. 2006). DNA methylation is evident in sperm DNA, and is not completely removed in the zygotes after fertilization. There is also some evidence indicating that certain RNA species are transmitted to zygotes via sperm (Rassoulzadegan et al. 2006). Since ds-RNAs are critical for heterochromatin formation, RNA may be an important marker for heterochromatin. Heterochromatin may be reset and maintained during germ cell development or early development using this putative marking system. If stress severely disrupts

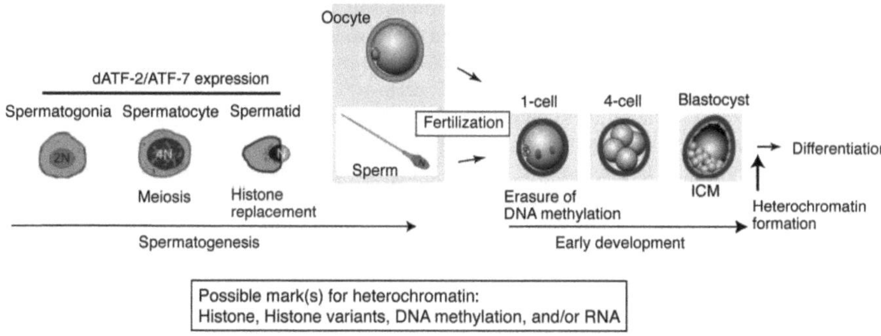

Fig. 6.6 How is the heterochromatin structure inherited through meiosis, fertilization, and early development? dATF-2 and ATF-7 are expressed in *Drosophila* and mouse germ cells, respectively. In spermatids, histones are replaced by protamines, while DNA methylation of the zygotic paternal genome is removed immediately after fertilization. Typical heterochromatin structures are formed after starting differentiation from the ICM. The candidate markers for heterochromatin during spermatogenesis, fertilization, and early development are histones, histone variants, DNA methylation, and/or RNA

heterochromatin in germ cells, these systems may not be able to completely reset the heterochromatin, so that partially disrupted heterochromatin might be transmitted to the next generation. Further studies are needed if we are to fully understand the mechanism(s) underlying the inheritance of epigenetic changes (Fig. 6.6).

References

Barker DJ, Gluckman PD, Godfrey KM, Harding JE, Owens JA, Robinson JS (1993) Fetal nutrition and cardiovascular disease in adult life. Lancet 341:938–941

Barski A, Cuddapah S, Cui K, Roh TY, Schones DE, Wang Z, Wei G, Chepelev I, Zhao K (2007) High-resolution profiling of histone methylations in the human genome. Cell 129:823–837

Bench GS, Friz AM, Corzett MH, Morse DH, Balhorn R (1996) DNA and total protamine masses in individual sperm from fertile mammalian subjects. Cytometry 23:263–271

Benhar M, Dalyot I, Engelberg D, Levitzki A (2001) Enhanced ROS production in oncogenically transformed cells potentiates c-Jun N-terminal kinase and p38 mitogen-activated protein kinase activation and sensitization to genotoxic stress. Mol Cell Biol 21:6913–6926

Bhoumik A, Fichtman B, Derossi C, Breitwieser W, Kluger HM, Davis S, Subtil A, Meltzer P, Krajewski S, Jones N, Ronai Z (2008) Suppressor role of activating transcription factor 2 (ATF-2) in skin cancer. Proc Natl Acad Sci USA 105:1674–1679

Black PH (2002) Stress and the inflammatory response: a review of neurogenic inflammation. Brain Behav Immun 16:622–653

Brinkman BM, Telliez JB, Schievella AR, Lin LL, Goldfeld AE (1999) Engagement of tumor necrosis factor (TNF) receptor 1 leads to ATF-2- and p38 mitogen-activated protein kinase-dependent TNF-α gene expression. J Biol Chem 274:30882–30886

Carone BR, Fauquier L, Habib N, Shea JM, Hart CE, Li R, Bock C, Li C, Gu H, Zamore PD, Meissner A, Weng Z, Hofmann HA, Friedman N, Rando OJ (2010) Paternally induced transgenerational environmental reprogramming of metabolic gene expression in mammals. Cell 143:1084–1096

Chang L, Karin M (2001) Mammalian MAP kinase signalling cascades. Nature 410:37–40

Chrivia JC, Kwok RP, Lamb N, Hagiwara M, Montminy MR, Goodman RH (1993) Phosphorylated CREB binds specifically to the nuclear protein CBP. Nature 365:855–859
Craig CR, Fink JL, Yagi Y, Ip YT, Cagan RL (2004) A Drosophila p38 orthologue is required for environmental stress responses. EMBO Rep 5:1058–1063
De Graeve F, Bahr A, Chatton B, Kedinger C (2000) A murine ATFa-associated factor with transcriptional repressing activity. Oncogene 19:1807–1819
Francis D, Diorio J, Liu D, Meaney MJ (1999) Nongenomic transmission across generations of maternal behavior and stress responses in the rat. Science 286:1155–1158
Gaire M, Chatton B, Kedinger C (1990) Isolation and characterization of two novel, closely related ATF cDNA clones from HeLa cells. Nucleic Acids Res 18:3467–3473
Gaucher J, Reynoird N, Montellier E, Boussouar F, Rousseaux S, Khochbin S (2010) From meiosis to postmeiotic events: the secrets of histone disappearance. FEBS J 277:599–604
Giaccia AJ, Simon MC, Johnson R (2004) The biology of hypoxia: the role of oxygen sensing in development, normal function, and disease. Genes Dev 18:2183–2194
Gonzalez GA, Montminy MR (1989) Cyclic AMP stimulates somatostatin gene transcription by phosphorylation of CREB at serine 133. Cell 59:675–680
Gonzalez GA, Yamamoto KK, Fischer WH, Karr D, Menzel P, Biggs W 3rd, Vale WW, Montminy MR (1989) A cluster of phosphorylation sites on the cyclic AMP-regulated nuclear factor CREB predicted by its sequence. Nature 337:749–752
Grewal SI, Elgin SC (2007) Transcription and RNA interference in the formation of heterochromatin. Nature 447:399–406
Grewal SI, Moazed D (2003) Heterochromatin and epigenetic control of gene expression. Science 301:798–802
Gupta S, Campbell D, Dérijard B, Davis RJ (1995) Transcription factor ATF2 regulation by the JNK signal transduction pathway. Science 267:389–393
Hai TW, Liu F, Coukos WJ, Green MR (1989) Transcription factor ATF cDNA clones: an extensive family of leucine zipper proteins able to selectively form DNA-binding heterodimers. Genes Dev 3:2083–2090
Hall IM, Shankaranarayana GD, Noma K, Ayoub N, Cohen A, Grewal SI (2002) Establishment and maintenance of a heterochromatin domain. Science 297:2232–2237
Han ZS, Enslen H, Hu X, Meng X, Wu IH, Barrett T, Davis RJ, Ip YT (1998) A conserved p38 mitogen-activated protein kinase pathway regulates Drosophila immunity gene expression. Mol Cell Biol 18:3527–3539
Henikoff S (1990) Position-effect variegation after 60 years. Trends Genet 6:422–426
Hirose N, Maekawa T, Shinagawa T, Ishii S (2009) ATF-2 regulates lipopolysaccharide-induced transcription in macrophage cells. Biochem Biophys Res Commun 385:72–77
Jia S, Noma K, Grewal SI (2004) RNAi-independent heterochromatin nucleation by the stress-activated ATF/CREB family proteins. Science 304:1971–1976
Kim DH, Feinbaum R, Alloing G, Emerson FE, Garsin DA, Inoue H, Tanaka-Hino M, Hisamoto N, Matsumoto K, Tan MW, Ausubel FM (2002) A conserved p38 MAP kinase pathway in *Caenorhabditis elegans* innate immunity. Science 297:623–626
Lamarck J-B (1809) Philosophie zoologique, ou exposition des considérations relatives à l'histoire naturelle des animaux. Dentu, Paris
Lee KA, Hai TY, SivaRaman L, Thimmappaya B, Hurst HC, Jones NC, Green MR (1987) A cellular protein, activating transcription factor, activates transcription of multiple E1A-inducible adenovirus early promoters. Proc Natl Acad Sci USA 84:8355–8359
Liu F, Green MR (1990) A specific member of the ATF transcription factor family can mediate transcription activation by the adenovirus E1a protein. Cell 61:1217–1224
Livingstone C, Patel G, Jones N (1997) ATF-2 contains a phosphorylation-dependent transcriptional activation domain. EMBO J 14:1785–1797
Maekawa T, Bernier F, Sato M, Nomura S, Singh M, Inoue Y, Tokunaga T, Imai H, Yokoyama M, Reimold A, Glimcher LH, Ishii S (1999) Mouse ATF-2 null mutants display features of severe type of meconium aspiration syndrome. J Biol Chem 274:17813–17819
Maekawa T, Jin W, Ishii S (2010a) The role of ATF-2 family transcription factors in adipocyte differentiation: anti-obesity effects of p38 inhibitors. Mol Cell Biol 30:613–625

Maekawa T, Kim S, Nakai D, Makino C, Takagi T, Ogura H, Yamada K, Chatton B, Ishii S (2010b) Social isolation stress induces ATF-7 phosphorylation and impairs silencing of the 5-HT 5B receptor gene. EMBO J 29:196–208

Maekawa T, Sakura H, Kanei-Ishii C, Sudo T, Yoshimura T, Fujisawa J, Yoshida M, Ishii S (1989) Leucine zipper structure of the protein CRE-BP1 binding to the cyclic AMP response element in brain. EMBO J 8:2023–2028

Maekawa T, Sano Y, Shinagawa T, Sakuma T, Nomura S, Licht JD, Ishii S (2008) ATF-2 controls transcription of *Maspin* and *GADD45α* genes independently from p53 to suppress mammary tumors. Oncogene 27:1045–1054

Maekawa T, Shinagawa T, Sano Y, Sakuma T, Nomura S, Nagasaki K, Miki Y, Saito-Ohara F, Inazawa J, Kohno T, Yokota J, Ishii S (2007) Reduced levels of ATF-2 predispose mice to mammary tumors. Mol Cell Biol 27:1730–1744

Matsuda S, Maekawa T, Ishii S (1991) Identification of the functional domains of the transcriptional regulator CRE-BP1. J Biol Chem 266:18188–18193

Mayer W, Niveleau A, Walter J, Fundele R, Haaf T (2000) Demethylation of the zygotic paternal genome. Nature 403:501–502

Mikula BC (1995) Environmental programming of heritable epigenetic changes in paramutant r-gene expression using temperature and light at a specific stage of early development in maize seedlings. Genetics 140:1379–1387

Molinier J, Ries G, Zipfel C, Hohn B (2006) Transgeneration memory of stress in plants. Nature 442:1046–1049

Montminy MR, Sevarino KA, Wagner JA, Mandel G, Goodman RH (1986) Identification of a cyclic-AMP-responsive element within the rat somatostatin gene. Proc Natl Acad Sci USA 83:6682–6686

Nagadoi A, Nakazawa K, Uda H, Okuno K, Maekawa T, Ishii S, Nishimura Y (1999) The solution structure of the transactivation domain of ATF-2 comprising a zinc finger-like subdomain and a flexible subdomain. J Mol Biol 287:593–607

Ng SF, Lin RC, Laybutt DR, Barres R, Owens JA, Morris MJ (2010) Chronic high-fat diet in fathers programs β-cell dysfunction in female rat offspring. Nature 467:963–966

Nomura N, Zu Y-L, Maekawa T, Tabata S, Akiyama T, Ishii S (1993) Isolation and characterization of a novel member of the gene family encoding the cAMP response element-binding protein CRE-BP1. J Biol Chem 268:4259–4266

Okamura T, Shimizu H, Nagao T, Ueda R, Ishii S (2007) ATF-2 regulates fat metabolism in *Drosophila*. Mol Biol Cell 18:1519–1529

Rassoulzadegan M, Grandjean V, Gounon P, Vincent S, Gillot I, Cuzin F (2006) RNA-mediated non-mendelian inheritance of an epigenetic change in the mouse. Nature 441:469–474

Reimold AM, Grusby MJ, Kosaras B, Fries JW, Mori R, Maniwa S, Clauss IM, Collins T, Sidman RL, Glimcher MJ, Glimcher LH (1996) Chondrodysplasia and neurological abnormalities in ATF-2-deficient mice. Nature 379:262–265

Rodgers RJ, Cole JC (1993) Influence of social isolation, gender, strain, and prior novelty on plus-maze behaviour in mice. Physiol Behav 54:729–736

Sano Y, Akimaru H, Okamura T, Nagao T, Okada M, Ishii S (2005) *Drosophila* activating transcription factor-2 is involved in stress response via activation by p38, but not c-Jun NH(2)-terminal kinase. Mol Biol Cell 16:2934–2946

Sano Y, Harada J, Tashiro S, Gotoh-Mandeville R, Maekawa T, Ishii S (1999) ATF-2 is a common nuclear target of Smad and TAK1 pathways in TGF-β signaling. J Biol Chem 274:8949–8957

Sano Y, Tokitou F, Dai P, Maekawa T, Yamamoto T, Ishii S (1998) CBP alleviates the intramolecular inhibition of ATF-2 function. J Biol Chem 273:29098–29105

Seong KH, Li D, Shimizu H, Nakamura R, Ishii S (2011) Inheritance of stress-induced, ATF-2-dependent epigenetic change. Cell 145:1049–1061

Shimizu H, Shimoda M, Yamaguchi T, Seong KH, Ishii S (2008) *Drosophila* ATF-2 regulates sleep and locomotor activity in pacemaker neurons. Mol Cell Biol 28:6278–6289

Shivers RP, Pagano DJ, Kooistra T, Richardson CE, Reddy KC, Whitney JK, Kamanzi O, Matsumoto K, Hisamoto N, Kim DH (2010) Phosphorylation of the conserved transcription factor ATF-7 by PMK-1 p38 MAPK regulates innate immunity in Caenorhabditis elegans. PLoS Genet 6:e1000892

van Dam H, Wilhelm D, Herr I, Steffen A, Herrlich P, Angel P (1997) ATF-2 is preferentially activated by stress-activated protein kinases to mediate c-*jun* induction in response to genotoxic agents. EMBO J 14:31798–31811
van der Heijden GW, Derijck AA, Ramos L, Giele M, van der Vlag J, de Boer P (2006) Transmission of modified nucleosomes from the mouse male germline to the zygote and subsequent remodeling of paternal chromatin. Dev Biol 298:458–469
Volpe TA, Kidner C, Hall IM, Teng G, Grewal SI, Martienssen RA (2002) Regulation of heterochromatic silencing and histone H3 lysine-9 methylation by RNAi. Science 297:1833–1837
Wang H, An W, Cao R, Xia L, Erdjument-Bromage H, Chatton B, Tempst P, Roeder RG, Zhang Y (2003) mAM facilitates conversion by ESET of dimethyl to trimethyl lysine 9 of histone H3 to cause transcriptional repression. Mol Cell 12:475–487
Waterland RA, Michels KB (2007) Epigenetic epidemiology of the developmental origins hypothesis. Annu Rev Nutr 27:363–388
Wu F, Caron C, De Robertis C, Khochbin S, Rousseaux S (2008) Testis-specific histone variants H2AL1/2 rapidly disappear from paternal heterochromatin after fertilization. J Reprod Dev 54:413–417

Chapter 7
Microsatellite Repeats: Canaries in the Coalmine

Nimrat Chatterjee, Beatriz A. Santillan, and John H. Wilson

Abstract Our genomes are much more dynamic than we ever imagined, creating variation that is good for evolution, but often bad for the individual. Microsatellite repeats constitute especially plastic genomic elements that can operate as rheostats to fine-tune gene function. Microsatellite repeats mutate at high frequency in predictable ways, and the mutations can confer incremental changes on the activities of gene regulatory elements and encoded proteins. The downside of microsatellite instability is highlighted by a collection of human diseases that are caused by expansion of microsatellite repeats. Studies with disease-associated trinucleotide repeats (TNRs) reveal that changes in DNA methylation, transcription, and the Hsp90 chaperone increase the mutation rates of TNRs. The observation that environmental stress can alter epigenetic marks, transcription rates, and Hsp90 activity raises the possibility that stress modifies the mutation rates of microsatellite repeats. Thus, microsatellite repeats may behave as genomic detectors of environmental stress.

Nimrat Chatterjee and Beatriz A. Santillan contributed equally.

N. Chatterjee
Verna and Marrs McLean Department of Biochemistry
and Molecular Biology, Baylor College of Medicine, Houston, TX 77030, USA

B.A. Santillan
Department of Molecular and Human Genetics, Baylor College of Medicine,
Houston, TX 77030, USA

J.H. Wilson (✉)
Verna and Marrs McLean Department of Biochemistry and Molecular Biology,
Department of Molecular and Human Genetics, Baylor College of Medicine,
Houston, TX 77030, USA
e-mail: jwilson@bcm.edu

D. Mittelman (ed.), *Stress-Induced Mutagenesis*, DOI 10.1007/978-1-4614-6280-4_7,

Introduction

Canaries once served as sentinels for poison gas in coalmines. Their extreme sensitivity to this hazard gave miners early warning—a chance to protect themselves and enhance their survival. Canaries-in-the-coalmine may provide an apt metaphor for the role of tandemly repeated sequences in the genome. Units of one to nine nucleotides repeated in tandem, which are termed microsatellite repeats, permeate the human genome, constituting roughly 3% of all genomic sequences (Ellegren 2004). One of their defining characteristics is a heightened instability in which a repeat tract gains or loses one or more units, typically at much higher rates than point mutations occur in unique sequences. It is clear that in some cases this extreme variability plays a key role in the evolution of phenotypic traits (Rando and Verstrepen 2007; Fondon and Garner 2004). Does this elevated instability provide the equivalent of an early warning system for the genome? Do repeats become more unstable in response to external stresses? Does repeat instability allow cells, or organisms, or populations, to improve their chances of survival in a stressful environment? In this chapter, we explore these questions in the context of research on trinucleotide repeats (TNRs).

All organisms engage various homeostatic mechanisms that buffer the individual from a range of perturbations to their normal operating environment. At the edges of this range and beyond, organisms experience stress—an abnormal condition that exceeds a threshold of normal tolerance. Not surprisingly, cells have adopted a variety of coping mechanisms to deal with stress, often tailored to the common stresses they encounter. Many microorganisms, for example, alternate expression of specific genes, termed contingency genes, which encode alternative forms of cell-surface proteins that allow organisms to adapt quickly to recurrent stressful environments or to evade a host immune response (Barry and McCulloch 2001; van der Woude and Baumler 2004). In other cases, similar variations in gene expression are controlled by epigenetic changes in DNA methylation or chromatin structure that switch genes on or off (Gottschling et al. 1990; Cubas et al. 1999; Hernday et al. 2002). In addition, as shown for *Drosophila* and *Arabidopsis*, the stress-responsive chaperone Hsp90 "hides" preexisting genetic variation by helping to stabilize mutant proteins, and then "releases" this selectable variation in times of stress, when Hsp90 becomes otherwise engaged (Queitsch et al. 2002; Rutherford and Lindquist 1998). As amazing as these examples are—and as illuminating for evolutionary biology—perhaps the most unexpected response to stress is the ability of organisms to change the rate at which they introduce mutations into their genomes (Rando and Verstrepen 2007; Fonville et al. 2011; Wright 2004).

In the past 25 years—against much initial resistance—it has become clear that mutation rates can vary in the face of environmental stress, as is evident now in bacteria, yeast, and mammalian cells (Hastings et al. 2004; Forche et al. 2011; Bjedov et al. 2003; Coyle and Kroll 2008; Mittelman et al. 2010). Remarkably, the varied mechanisms responsible for such stress-induced mutagenesis focus genomic change to times of stress and, in some cases, to genes that might make a difference

(Rando and Verstrepen 2007; Fonville et al. 2011; Wright 2004). In the context of stress-induced mutagenesis, microsatellite repeats deserve special consideration. They are prevalent in genes and highly mutable; they mutate in predictable ways and the mutations are readily reversible; and they can give rise to incremental, quantitative changes in gene function (Ellegren 2004; Rando and Verstrepen 2007; Fonville et al. 2011). Investigations over the past two decades have shown that repeat instability is stimulated by all aspects of DNA metabolism—replication, transcription, repair, and recombination—which are sensitive indicators of a cell's internal environment (Lopez Castel et al. 2011; Lin et al. 2009; Mirkin 2007). These same processes likely provide links to the external environment and key elements in the routes by which stresses produce changes to the genome. In this chapter we review the properties of repeated DNA sequences—focusing special attention on TNRs—and discuss ways in which the external environment of human cells might influence microsatellite instability via DNA methylation, transcription, and the Hsp90 chaperone.

Tandem Repeats and Microsatellite Instability

A high fraction of the human genome consists of repetitive DNA sequences, which are present in all genomes to a greater or lesser extent (Richard et al. 2008). One category of such sequences is tandem DNA repeats, which are repeated end-to-end in a strictly head-to-tail arrangement. Microsatellite repeats have unit sizes from one to nine nucleotides, while minisatellite repeats are composed of motifs of ten or more nucleotides (Gemayel et al. 2010). In addition to the repeat motif, a repeat tract is characterized by two other properties: the overall length of the tract, which can range from just a few to well over a thousand, and the purity of the tract, which is the percentage of nucleotides in the tract that match the repeat motif. The sequence of the repeat unit, along with the length of the repeat tract and its purity, is a major determinant in how sensitive an individual repeat tract is to mutation.

A defining property of microsatellite repeats, in particular, is their exceptionally high mutation rate, which typically ranges from 10^{-5} to 10^{-2} (Rando and Verstrepen 2007; Gemayel et al. 2010). Unlike unique sequences in the genome, which mutate most commonly by point mutations at a rate of 10^{-8} to 10^{-5}, tandem repeats mutate almost exclusively by altering the number of repeat units in the repeat tract, usually through the addition or loss of one or more units at a time. Thus, mutations at microsatellite repeats are predictable: they change the length of the repeat tract. Unlike point mutations, repeat mutations are readily reversible, allowing a variety of tract lengths to be tested for evolutionary fitness over a relatively short time (Kashi and King 2006; Nithianantharajah and Hannan 2007). Rates of mutation are largely determined by properties of the repeat tract itself (Legendre et al. 2007). The longer the repeat and the more pure it is, the more prone it will be to mutate. Mutation rates also depend on genomic context, with similar repeats at different locations in the genome mutating at different rates. These context effects may reflect the location of nearby origins of replication, promoters, chromatin structure, or epigenetic status, among other possibilities (Richard et al. 2008). If we understood these context effects better, we would likely gain deeper insight into the mechanisms of microsatellite instability.

Investigations in a variety of model organism, including bacteria, yeast, flies, mammalian cells, and mice have shown that the whole spectrum of DNA transactions—replication, recombination, DNA repair, and transcription—can contribute to the instability of microsatellite repeats (Lin et al. 2009; Mirkin 2007; Richard et al. 2008; Gemayel et al. 2010). Each of these processes has in common the exposure of single strands of DNA, which allows one strand to occasionally misalign with its complement in an array of repeat units. For example, during replication, slippage of the primer strand relative to the template strand could increase or decrease the length of the repeat tract (Fig. 7.1a) (Richard et al. 2008). Recombination, as well, can stimulate microsatellite repeat instability. In yeast, for example, long CAG tracts tend to be sites of chromosome breakage whose repair is dependent on annealing of CAG units on either side of the break: the so-called single-strand annealing pathway of recombination (Freudenreich et al. 1998) (Fig. 7.1b). Similarly in human cells, breaks in CAG tracts targeted by zinc-finger nucleases trigger repair that is also consistent with the single-strand annealing pathway (Mittelman et al. 2009).

Perhaps most surprising, transcription in bacteria, yeast, flies, and human cells has been found to destabilize microsatellite repeats, especially those with a propensity to form non-B-DNA structures such as G-quartets, hairpins, and slipped-strand duplexes (Lin et al. 2006a, 2009, 2010a; Kim and Jinks-Robertson 2011, 2012; Wierdl et al. 1996; Mochmann and Wells 2004; Nakamori et al. 2011; Jung and Bonini 2007). In human cells, the CAG repeat instability induced by transcription through the repeat tract is complex: it is stimulated by R-loop formation, requires the MutSβ recognition component of mismatch repair (MMR), and depends on the complete pathway of transcription-coupled nucleotide excision repair (TC-NER) (Fig. 7.1c) (Lin et al. 2010b; Lin and Wilson 2007). The strong dependence of microsatellite instability on the fundamental processes of DNA metabolism offers a multitude of potential links between repeat mutation and internal and external cell stressors.

Functional Significance of Microsatellite Instability

The biological consequences of microsatellite instability depend critically on the distribution of repeats in the genome: repeat mutations in genes are more likely to alter fitness than those in intergenic regions. Scans of microsatellite repeats in various genomes show that a substantial percentage of genes—17% in humans—contain repeats in their coding regions (Gemayel et al. 2010). The most common exonic microsatellites are trinucleotide and hexanucleotide repeats, whose mutation would not affect translation, unlike changes in most other microsatellite repeats,

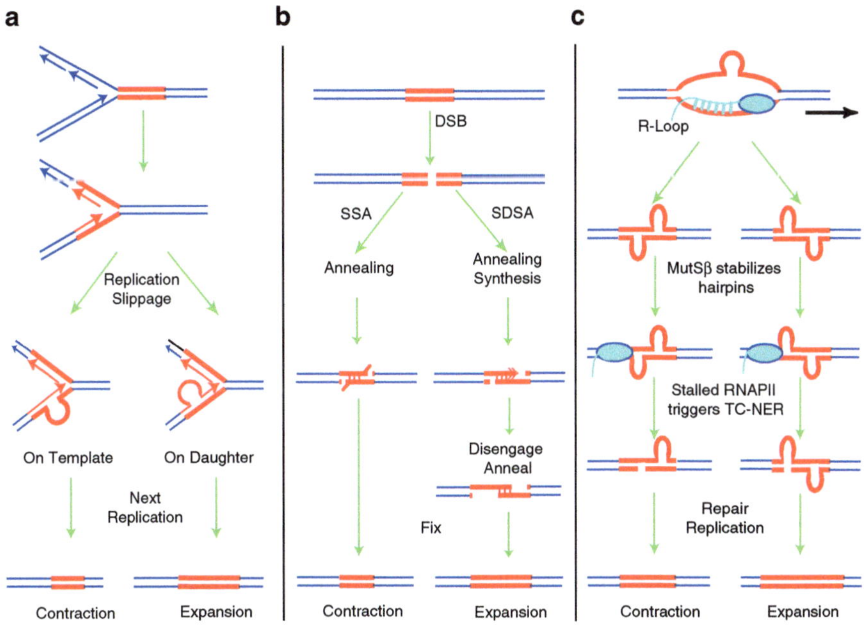

Fig. 7.1 Three mechanisms for promoting microsatellite repeat instability. (**a**) Contraction and expansion of microsatellite repeats by replication slippage. During replication, the newly synthesized strand can mispair with its template, looping out a stretch of repeat units on either the newly synthesized (daughter) strand or the template strand. Subsequent replication of a loop-out on the template strand will generate a contraction, while replication of a loop-out on the daughter strand will generate an expansion. (**b**) Contraction and expansion of repeats mediated by double-strand breaks (DSBs). Exposure of single-stranded regions on either side of the DSB allows strands to pair directly by single-strand annealing (SSA). Repair of the paired intermediate would generate a repeat tract with fewer repeat units, a contraction. Alternatively, one of the paired strands could be extended by DNA synthesis, using the repeats in the complement as a template. If the extended strand then disengaged and repaired, it could generate an expanded repeat. This pathway is a version of the synthesis-dependent strand-annealing (SDSA) pathway (Paques and Haber 1999). (**c**) Contraction and expansion induced by transcription through a repeat. Formation of an R-loop behind a transcribing RNA polymerase would enhance loop-out formation in the nontranscribed (nontemplate) strand. Once RNase H enzymes resolve an R-loop, the re-pairing of a loop-out-containing nontranscribed strand would entail the formation of a compensating loop-out in the template strand, generating a slipped-strand duplex. The binding of MutSβ to CAG and CTG loop-outs (Pearson et al. 1997; Owen et al. 2005) might stabilize them sufficiently to block RNA polymerase, which in cell free extracts has been shown to stall periodically at repeat loop-outs on both the transcribed and nontranscribed strands (Salinas-Rios et al. 2011). A stalled RNA polymerase is the primary signal to initiate TC-NER. Repair by removal of a loop-out would generate a contraction on the template strand, whereas repair by copying opposite a loop-out would generate an expansion on the template strand. Subsequent replication would generate duplex expansions and contractions

which would disrupt the reading frame. Intriguingly, trinucleotide and hexanucleotide repeats are overrepresented in exons relative to introns or intergenic regions, consistent with the idea that conserved exonic repeats serve a function that has been selected for during evolution (Richard et al. 2008; Kashi and King 2006; Subramanian et al. 2003; Verstrepen et al. 2005). This presumptive biological significance is reinforced by analysis of repeat-containing genes, which include a disproportionate number of genes for transcription factors and regulatory proteins (Legendre et al. 2007; Bacolla et al. 2008; Kozlowski et al. 2010; Mularoni et al. 2010).

These computational inferences are supported by a variety of observations in eukaryotes, the most robust of which involve skeletal morphology in dogs and regulation of the circadian clock in several species. A targeted study in dogs, which examined 17 developmental genes containing 36 trinucleotide or hexanucleotide repeats, revealed that most repeats (29) were more pure than their human homologs and displayed a higher level of tract-length polymorphism, consistent with the idea that repeat instability depends on tract purity (Fondon and Garner 2004). Tract-length polymorphism correlated in some cases with skull morphology and digit polydactyly. In one case, for example, a 51-nucleotide contraction of a hexanucleotide repeat (CCGCAG) encoding proline and glutamine in the *Alx4* gene correlated perfectly with the bilateral rear first digit polydactyly that is a defining characteristic of the Great Pyrenees breed (Fondon and Garner 2004). Moreover, the polydactyly observed in these dogs is similar to that seen in *Alx4* knockout mice (Qu et al. 1998). The high variability of these repeat loci in dogs may contribute to the ease with which so many morphologically distinct breeds of dogs have been generated in the relatively short time since their domestication (Fondon and Garner 2004).

A similar, striking correlation exists for circadian clock genes in a variety of species. In the fungus *Neurospora crassa*, a key transcription factor controlling circadian rhythm carries a CAG repeat that encodes a polyglutamine segment whose length correlates with the period of the clock (Lee et al. 2003; Michael et al. 2007). Similarly, in several species, the *Clock* gene, which encodes one component of a heterodimeric transcription factor, also contains a variable CAG repeat for polyglutamine in the C-terminal domain of the protein. In the nonmigratory blue tit bird, Northern European populations are enriched for alleles with longer repeats relative to more southern populations (Liedvogel et al. 2009; Johnsen et al. 2007). Similarly, long polyglutamine repeats in the migratory barn swallow have been found to be associated with late breeding behavior in females, which is timed to the arrival date in migration (Caprioli et al. 2012). In the Chinook salmon, a longer CAG repeat allele of one of their clock genes occurs at higher frequencies in fish that breed at higher latitudes, while the shorter allele is enriched in those that breed at lower latitudes (O'Malley and Banks 2008). Finally, in *Drosophila melanogaster* the *period* gene contains a hexanucleotide repeat (ACAGGA) that codes for a segment of poly-threonine-glycine, with the two most common alleles carrying 17 and 20 repeats. In warmer climates, the 17-repeat allele predominates over the longer allele, while at colder latitudes the reverse is true (Sawyer et al. 1997). These observations suggest that repeat sequences may function as tuning knobs for circadian clocks (Gemayel et al. 2010).

These examples and others support the notion that variable lengths of polymeric amino acid segments can modulate the function of proteins that contain them. But how might they do that? Repetitive amino acid stretches in proteins tend to consist of polar amino acids, suggesting that they are likely located on protein surfaces

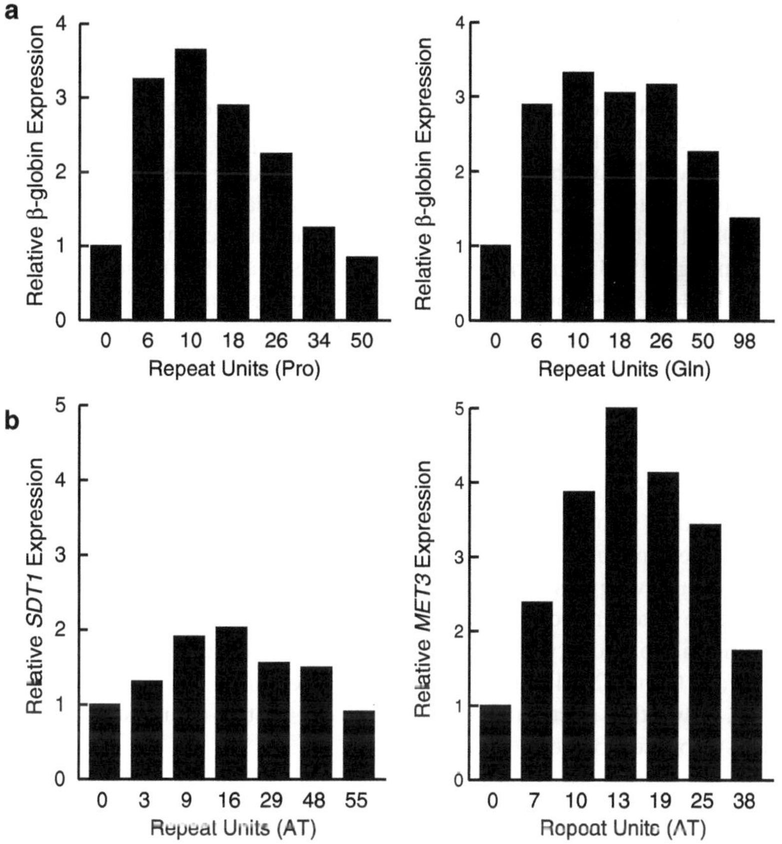

Fig. 7.2 Effects of microsatellite repeats on gene expression. (**a**) Gene expression stimulated by GAL4 transcriptional activators carrying various lengths of polyglutamine or polyproline fused to their C-termini. Transcription from a rabbit β-globin construct with five adjacent GAL4 binding sites was measured in human cells after co-transfection of plasmids expressing GAL4 genes with 3′ TNR sequences encoding C-terminal segments of polyglutamine or polyproline. Adapted from (Gerber et al. 1994). (**b**) Expression from promoters carrying various lengths of AT repeats. Promoters in the yeast *SDT1* and *MET3* genes were modified to contain various lengths of AT repeats and their relative transcription levels were measured. Adapted from (Vinces et al. 2009)

(Mularoni et al. 2010; Faux et al. 2005). These repeats are commonly found in disordered regions of the protein that can serve to separate protein domains or to modulate protein–protein interactions (Karlin and Burge 1996; Simon and Hancock 2009). Changes in repeat length might be able to subtly alter domain orientations relative to one another, affecting overall protein function, or to incrementally modify binding interfaces to adjust the strength of protein associations. Just such quantitative effects were demonstrated in a set of experiments in which a series of fusion genes encoding the yeast GAL4 transcriptional activator carrying a variable C-terminal segment with different lengths of polyglutamine or polyproline were transfected into mammalian cells (Fig. 7.2a) (Gerber et al. 1994).

Although we have focused above on the functional consequences of trinucleotide and hexanucleotide repeat variability in coding regions, there are many more examples of biologically relevant microsatellite variability in noncoding regions, where the size of the repeat unit is less constraining (Gemayel et al. 2010). Numerous studies suggest that the length of a microsatellite repeat can modulate promoter activity in several ways: by altering the number of binding sites for gene regulatory proteins (Contente et al. 2002; Johnson et al. 1988; Matsuura et al. 2000; Gangwal et al. 2008), by changing the spacing between functional elements in the promoter (van Ham et al. 1993; Sarkari et al. 1994; Willems et al. 1990), and by influencing local nucleosome positioning and chromatin structure (Yuan et al. 2005; Iyer and Struhl 1995; Vinces et al. 2009). The potential for tuning expression by changes in repeat lengths is shown in Fig. 7.2b, where the lengths of TA repeats were systematically varied in two yeast promoters (Vinces et al. 2009). Variable microsatellite repeats in introns and UTRs also have been linked to differences in gene expression or RNA splicing (Hui et al. 2003; Gebhardt et al. 1999; Albanese et al. 2001). Finally, it is clear that some pathogenic bacteria use coding-region repeats of 4, 5, or 8 nucleotides (whose variability would alter the reading frame) as on/off switches for key genes in phase variation, ensuring a diverse population with maximal chances of surviving host immune attacks. Collectively, these examples illustrate the immense potential benefit inherent in microsatellite instability, conferring on organisms an evolutionary agility that cannot be readily achieved by point mutation (Fondon and Garner 2004; Caburet et al. 2005). The central question for this chapter, however, is whether microsatellite instability can bestow a similar nimbleness on genomic responses to stress.

Microsatellite Repeats and Human Disease

The benefits of microsatellite instability come with a cost. At more than 30 loci in the human genome, expansions of microsatellite repeats cause devastating neurological, muscular, or developmental diseases (Orr and Zoghbi 2007; Brouwer et al. 2009; Pearson et al. 2005; Albrecht and Mundlos 2005; La Spada and Taylor 2010). These repeats are located in a variety of positions within their affected genes, as illustrated schematically in Fig. 7.3. The expanded repeats cause disease through both loss-of-function and gain-of-function effects, each of which illuminates key features of the underlying cell biology (Orr and Zoghbi 2007; Lopez Castel et al. 2010). For example, fragile X syndrome (FRAXA) and Friedreich ataxia (FRDA), which are caused by expansions of CGG in the 5′UTR and by GAA in an intron, respectively, both cause a recessive loss-of-function phenotype by shutting off gene expression. The intronic GAA interferes with transcription elongation (Bidichandani et al. 1998; Grabczyk and Usdin 2000; Soragni et al. 2008), whereas the CGG repeat becomes hypermethylated and inactivates the promoter (Pieretti et al. 1991). Interestingly, at an intermediate level of expansion, from 55 to 200 repeats, the CGG repeats at the fragile X locus cause a gain-of-function disease, fragile

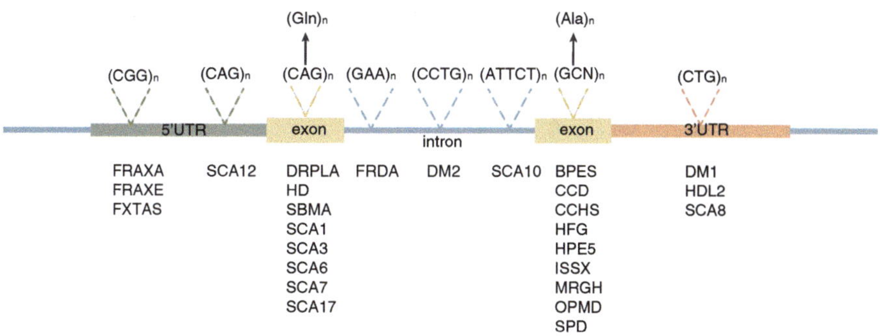

Fig. 7.3 Microsatellite repeats associated with human diseases. The units of the microsatellite repeats whose expansions lead to disease are indicated above the appropriate coding or noncoding region within a representative gene schematic. Associated diseases are shown below the repeat units. *BPES* blepharophimosis, ptosis, and epicanthus inversus, *CCD* cleidocranial dysplasia, *CCHS* congenital central hypoventilation syndrome, *DM* myotonic dystrophy types 1 and 2, *DRPLA* dentatorubral pallidoluysian atrophy, *FRAXA* fragile X syndrome, *FRAXE* fragile X mental retardation associated with FRAXE site, *FRDA* Friedreich ataxia, *FXTAS* fragile X tremor and ataxia syndrome, *HD* Huntington disease, *HDL2* Huntington-disease-like 2, *HFG* hand-food-genital syndrome, *HPE5* holoprosencephaly 5, *ISSX* X-linked infantile spasm syndrome, *MRGH* mental retardation with isolated growth hormone deficiency, *OPMD* oculopharyngeal muscular dystrophy, *SBMA* spinal and bulbar muscular atrophy, *SCA* spinocerebellar ataxia types 1, 3, 6, 7, 8, 10, 12, and 17, *SPD* synpolydactyly

X-associated tremor and ataxia (FXTAS), due to overexpression of the RNA, which is toxic to the cell (Li and Jin 2012; Willemsen et al. 2011).

The phenomenon of "RNA toxicity" was first identified at the myotonic dystrophy type I (DM1) locus (CTG repeat in the 3′UTR) and subsequently at the DM2 locus (intronic CCTG repeat) (Osborne and Thornton 2006; Lee and Cooper 2009). The expanded repeats in these DM RNAs bind the alternative splicing factor muscleblind, leading to the aberrant splicing of a number of other transcripts, which is the basis for the dominant phenotype of these diseases (Osborne and Thornton 2006). The dominant gain-of-function phenotype of a number of other diseases, including Huntington disease (HD) and several spinocerebellar ataxias (SCAs), is due to "protein toxicity" caused by exonic CAG repeats that encode polyglutamine (polyQ). In each case, the extended polyQ segment is thought to alter the properties of the mutated protein, making it toxic and leading to the development of disease symptoms (Orr and Zoghbi 2007). A recent observation that convergent transcription through CAG repeat tracts causes cell death raises the possibility that "DNA toxicity" may contribute to cell death in some of these diseases (Lin et al. 2010a; Lin and Wilson 2011), especially given the unexpectedly high frequency of antisense transcription in the human genome (Katayama et al. 2005; He et al. 2008).

All but a couple of microsatellite-associated diseases are due to expansions of TNRs (Fig. 7.3). These disease-associated repeats include only four of the ten possible families of TNRs: CAG, CGG, GAA, and GAC. (The family of CAG repeats, for example, includes the complementary CTG, as well as the permuted AGC and

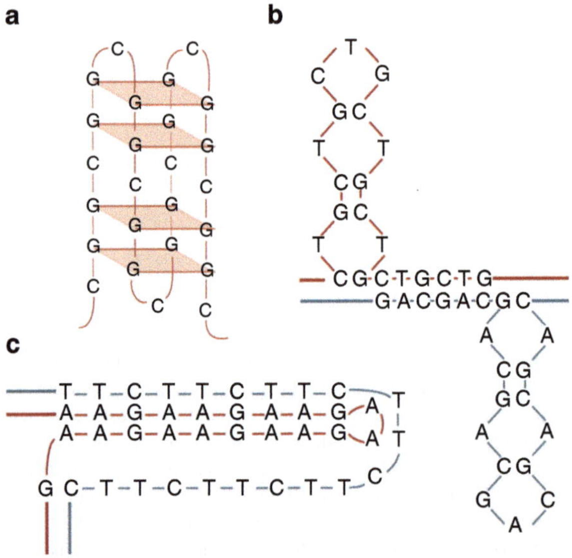

Fig. 7.4 Non-B-DNA structures formed by TNR repeats. (**a**) A G-quadruplex-like structure formed by a $(CGG)_n$ repeat tract. (**b**) A slipped duplex structure formed by out-of-register pairing of repeat units within the repeat tract. The looped-out sequences can form a CTG hairpin in one strand and a CAG hairpin in the complementary strand. (**c**) A triple helical H-DNA structure formed by a GAA repeat tract. Adapted from (Mirkin 2007)

GCA, along with their complements: GCT and TGC.) These disease-associated families of TNRs readily form non-B-DNA structures, including hairpins and slipped duplexes (CAG, CGG, and GAC), G-quartets (CGG), and triplexes (GAA) (Fig. 7.4) (Wells 2007; Sinden et al. 2002). The consensus view is that these TNRs tend to expand because they can form non-B-DNA secondary structures when exposed to the normal processes of DNA metabolism. The links between disease and the extreme instability of TNRs—approaching 100% in some cases (Yoon et al. 2003; Shelbourne et al. 2007)—have spurred numerous studies that have provided insights into how environmental stress might alter the stability of these repeats and, by extension, other microsatellite repeats as well.

Disease-associated TNRs display characteristically high rates of mutation—expansions and contractions—in the germ line, with a typical bias toward expansion in one or the other parent, depending on the disease. For example, patients with HD or any of several SCAs, which are all caused by long CAG repeats, typically show expansion bias in the paternal germ line. By contrast, patients with FRAXA (caused by a CGG repeat), FRDA (caused by a GAA repeat) or DM1 (caused by a CTG repeat, which is a CAG repeat family member) display a maternal bias for expansion (Pearson et al. 2005). This expansion bias underlies the increasing severity of disease symptoms commonly observed from one generation to the next, a phenomenon known as anticipation. It is important to note that TNR instability is not always biased toward expansion; for example, in patients with FRDA and FRAXA, there is a marked contraction bias in the paternal germ line (Pearson et al. 2005; Reyniers et al. 1999). The basis for these parent-of-origin effects is not known.

TNR instability is not limited to the germ line. As affected individuals age, expansion-biased TNR instability typically occurs in a variety of somatic tissues, including the pathologically important tissues of the brain, which may accelerate the onset of neuron dysfunction and hasten their death (Shelbourne et al. 2007; De Biase et al. 2007a; Swami et al. 2009). The TNR diseases caused by the CAG family of repeats display a similar pattern of instability (Lin et al. 2006b). Repeats in blood and heart are fairly stable, for example, while those in liver and kidney show intermediate instability. Similarly, in the brain, the striatum displays very high instability, the cerebral cortex and hippocampus, intermediate instability, and the cerebellum, very low instability (Lin et al. 2006b). By contrast, the GAA TNRs in FRDA patients show a high level of instability in a variety of tissues, including blood and cerebellum, which distinguishes its instability from that of CAG repeats (De Biase et al. 2007b). Moreover, GAA repeat instability is biased toward contractions in all tissues, but with a notably higher frequency of large expansions in the dorsal root ganglia, whose degeneration gives rise to the disease symptoms of FRDA (De Biase et al. 2007a, b). In FRAXA patients, it is clear that very long repeats that are methylated at their CpG sequences (their usual state) display minor instability (Reyniers et al. 1999; Tassone et al. 1999), while their rare, unmethylated counterparts are very unstable (Taylor et al. 1999; Wohrle et al. 1998).

The variation in repeat instability from disease to disease, from germ line to somatic tissues, and from one tissue to another remains one of the most puzzling features of TNR instability (Lin et al. 2006b). At a minimum, it suggests that TNR repeat instability probably involves multiple mechanisms—operating independently or together—that depend not only on the repeat sequence and its genomic context, but also on the type of tissue and its developmental status. TNRs in proliferating cells in the male germ line, for example, may expand via replication slippage (Mirkin 2007), but in terminally differentiated brain neurons, which do not replicate their DNA, expansion likely involves DNA repair processes potentially linked to transcription (Fig. 7.1) (Lin et al. 2009; Hubert et al. 2011a; Kaytor et al. 1997). Genetic experiments in mouse models of CAG repeat diseases reinforce the idea of multiple mechanisms. The Msh2 and Msh3 recognition components of mismatch repair, which form the MutSβ complex, affect instability in the male and female germ lines and in a variety of somatic tissues; the major maintenance DNA methyltransferase, Dnmt1, affects repeat instability in the germ line, but not in somatic tissues (Dion et al. 2008a); DNA ligase 1 alters repeat instability in the female germ line, but has no effect in the male germ line (Tome et al. 2011); the glycosylase Ogg1 selectively changes instability in somatic tissues (Kovtun et al. 2007, 2011); and the key nucleotide excision repair component Xpa selectively affects instability in neuronal tissues (Hubert et al. 2011a).

The diversity of mechanisms identified in these mouse studies pale in comparison to the multiplicity of factors and processes that have been shown to influence TNR instability in model organisms and in cell-free systems. (For an extensive list, see (Lopez Castel et al. 2010).) Just a sampling of the processes and factors affecting CAG TNRs is daunting: replicative polymerases (Schweitzer and Livingston 1999), flap structure-specific endonuclease (Freudenreich et al. 1998), replication factors (Schweitzer and Livingston 1999; Rosche et al. 1996; Lopez Castel et al. 2009; Hou et al. 2009; Pelletier et al. 2003; Razidlo and Lahue 2008), translesion synthesis (Collins et al. 2007), supercoiling and topoisomerases (Napierala et al. 2005; Hubert et al. 2011b), helicases (Bhattacharyya and Lahue 2004; Dhar and Lahue 2008; Kerrest et al. 2009), mismatch repair proteins (Lin et al. 2006a; Pearson et al. 1997; Manley et al. 1999; Schweitzer and Livingston 1997; Jaworski et al. 1995; Gomes-Pereira et al. 2004), components of base-excision repair (Kovtun et al. 2007; Kovtun et al. 2011; Jarem et al. 2009), nucleotide excision repair (Jung and Bonini 2007; Lin and Wilson 2007; Hubert et al. 2011a; Parniewski et al. 1999), single-strand break repair (Hubert et al. 2011b), double-strand break repair and homologous recombination (Mittelman et al. 2010; Freudenreich et al. 1998; Mittelman et al. 2009; Richard et al. 1999, 2000; Hebert et al. 2004), transcription (Lin et al. 2006a; Jung and Bonini 2007; Bowater et al. 1997), convergent transcription (Lin et al. 2010a; Nakamori et al. 2011), R-loops (Lin et al. 2010b; Reddy et al. 2011), E3 ubiquitin ligases and the proteasome (Lin and Wilson 2007), CpG methylation (Dion et al. 2008a; Gorbunova et al. 2004), histone deacetylases (Debacker et al. 2012), Hsp90 chaperone (Mittelman et al. 2010), and DNA-damage checkpoint pathways (Lin et al. 2010a; Razidlo and Lahue 2008; Lahiri et al. 2004). Understanding the tissue specificity of these processes and factors and how they are linked into pathways and interconnected into the networks that control TNR instability will be a challenge, to say the least. Nevertheless, the complexity itself demonstrates that TNRs, and maybe other microsatellite repeats as well, are exquisitely sensitive to a large number of processes that affect genome integrity and cell physiology. Thus, they are situated to serve the role of canaries in the coalmine. In the next section, we examine in more detail the potential links between environmental stress and the instability of microsatellite repeats mediated by CpG methylation, transcription, and the Hsp90 chaperone.

Environmental Stress and TNR Instability

Environmental stress comes in a variety of forms, including those encountered as part of a person's occupation or as a set of lifestyle choices such as smoking or overeating. These external stresses translate to a set of cellular stresses that include genotoxic stress, oxidative stress, nitrosative stress, hyper- and hypo-osmotic stress, hypoxia and hyperoxia, heat and cold, starvation, infection, and inflammation. In a few instances, environmental stress has been directly linked to TNR instability. Not surprisingly, direct exposure of cells to genotoxic chemicals such as mitomycin C alters repeat instability (Pineiro et al. 2003); however, even exposure to such benign drugs as caffeine and aspirin can modulate TNR repeat instability in cultured cells (Gomes-Pereira and Monckton 2004).

In one case, the effects of oxidative stress on TNR instability has been tested in a mouse model carrying a tract of 130 CGG repeats by giving the mice drinking water containing potassium bromate ($KBrO_3$) for two generations (Entezam et al. 2010). Bromate, which is sometimes found in municipal drinking water, certain foods, and hair permanent wave kits, is a powerful oxidizing agent that generates 8-oxoG modified bases, along with other damage to cells. Exposed mice were about twofold more likely to contribute an altered CGG repeat tract, usually expanded, to their progeny (Entezam et al. 2010). In related work, it was shown that deficiency of Ogg1, the glycosylase responsible for removing 8-oxoG lesions from DNA, is required for the instability of long CAG repeat tracts in the mouse brain, which, like the human brain, is a highly oxidative environment (Kovtun et al. 2007). These examples demonstrate that TNR instability can be linked to environmental stress via the avenue of DNA damage and repair. Here, we explore potential connections that involve epigenetics, transcription, and the Hsp90 chaperone.

CpG Methylation

Studies of human patients with expanded repeats and mouse models of TNR diseases have pointed to two periods in development—early embryogenesis and germ-line development—at which TNRs seem to be especially unstable (MacDonald et al. 1993; Helderman-van den Enden et al. 1999; Wohrle et al. 1993; Devys et al. 1992; Savouret et al. 2003). These two periods of TNR instability overlap the two major cycles of epigenetic reprogramming that occur during mammalian development (Reik et al. 2001). Immediately after fertilization and again in the differentiating germ line, genomic patterns of CpG methylation are largely erased genome-wide and then reestablished in a tissue-specific manner (Reik 2007; Li and Sasaki 2011). In human somatic cells, 70–80% of all CpG dinucleotides are methylated, with most of the remainder clustered in so-called CpG islands—unmethylated GC-rich regions with a high frequency of CpG—which are overrepresented at the 5′ ends of genes (Deaton and Bird 2011). Most CpG islands remain unmethylated, but a small percentage are methylated during development and the associated gene is stably silenced (Deaton and Bird 2011). CpG islands in adult tissues can also become methylated, as occurs in certain tissues with age and in many cancer cells (Issa 2000, 2004; Jones and Baylin 2007). The variability inherent in the epigenome may contribute to differences between monozygotic twins, including susceptibility to disease (Coolen et al. 2011; Fraga et al. 2005; Oates et al. 2006).

Epigenetic differences relative to the normal alleles are evident in the chromatin environment of expanded TNRs, many of which are associated with CpG islands (Brock et al. 1999; Gourdon et al. 1997). TNRs tend to differ from the normal locus in their patterns of DNA methylation, histone modification, and chromatin structure (Dion and Wilson 2009). For example, DNA methylation at the highly expanded CGG repeats from FRAXA patients is associated with high levels of modified histones such as histone 3 methylated at Lys9 (H3K9me) typical of heterochromatin (Coffee et al. 2002). Similarly, the 3′UTR around the expanded CTG repeats in DM1 patient cells is heterochromatic, with highly methylated DNA that is enriched for H3K9me (Cho et al. 2005; Steinbach et al. 1998; Filippova et al. 2001). Likewise, the region just upstream of the intronic GAA repeats from FRDA patients is more highly methylated at CpG sites with increased levels of H3K9me, also suggesting a more heterochromatin-like structure (Soragni et al. 2008; Al-Mahdawi et al. 2008; Greene et al. 2007). Finally, in a SCA1 mouse model, high mouse-to-mouse variability was detected in the epigenetic marks adjacent to the long CAG repeat, suggesting that the TNR at this locus creates a metastable epiallele, an allele whose epigenetic pattern is highly variable (Dion et al. 2008a). Such alleles are susceptible to environmental influences (Waterland et al. 2010; Waterland and Jirtle 2003).

The effects of CpG methylation on repeat instability was first noted in studies in cultured cells from FRAXA patients, which showed that while methylated CGG repeats were stable, unmethylated ones were unstable (Taylor et al. 1999; Glaser et al. 1999). The potential link between epigenetic reprogramming and repeat instability was strengthened by observations in bacteria, where methylation of plasmids was shown to stabilize several different microsatellite repeats, even when the methylated CpG dinucleotides were outside the repeat tract (Nichol and Pearson 2002). Studies using a selectable system for CAG repeat contractions in mammalian cells showed that treatment with 5-aza-deoxycytidine and hydralazine, which inhibit DNA methyltransferase activity by different mechanisms, and thereby decrease CpG methylation, destabilize CAG repeats by 100- to 1,000-fold (Gorbunova et al. 2004). Similar treatments in DM1 patient cell lines also destabilized repeats with a distinct bias toward expansion (Gorbunova et al. 2004). The implied link between TNR instability and CpG methylation is additionally supported by studies in a mouse model of SCA1 (CAG repeat), which showed that deficiency of the major maintenance DNA methyltransferase Dnmt1 led to significant increases in the frequency of expansions in both the male and female germ lines (Dion et al. 2008a). These studies bolster the idea that the normal cycle of CpG demethylation during germ-line development contributes to TNRs instability (Dion and Wilson 2009). The dynamic nature of the epigenome and its link to TNR instability raise the possibility that environmental stress could influence TNR instability via epigenetic changes.

There is a growing body of evidence suggesting that diet can modulate epigenetic marks (Link et al. 2010; Davis and Uthus 2004; Wolff et al. 1998). In mice, where it has been best studied, metastable alleles of genes affecting fur color and tail development vary in their expression in a way that can be influenced by maternal nutrition (Waterland and Jirtle 2003; Waterland et al. 2006; Dolinoy et al. 2007). For example, in females fed high methylation diets—extra folic acid, vitamin B_{12}, betaine, and choline—these epialleles were more highly methylated, which was reflected in changes in coat color and tail morphology (Waterland and Jirtle 2003; Waterland et al. 2006). Similar metastable alleles have been identified in humans, where they have been shown to vary with the season of conception (Waterland et al. 2010). Other genes also show methylation patterns that respond to maternal diet. In rats, a protein-restricted diet decreased CpG methylation and increased expression of the peroxisome proliferator-activated receptor and the glucocorticoid receptor: effects that were reversed by folate supplementation (Lillycrop et al. 2005, 2008).

Epigenetic profiling of the livers of offspring revealed many modest changes in CpG methylation that depend on paternal diet (Carone et al. 2010).

In addition to diet, a variety of environmental stresses can also influence DNA methylation. Lifestyle-associated stresses such as ethanol consumption can interfere with folate absorption and lead to methylation changes (Hamid et al. 2009; Ouko et al. 2009). In mice, radiation-induced transgenerational genome instability correlates with epigenetic changes in the offspring (Koturbash et al. 2006). In *Arabidopsis*, UVC irradiation stimulates homologous recombination in the whole population of plants in a way that persists in subsequent generations, consistent with epigenetic changes (Molinier et al. 2006). Salt, cold, heat, and flood stresses in *Arabidopsis* also induce similar transgenerational adaptation, which is associated with changes in DNA methylation (Boyko et al. 2010). Collectively, these studies on diet and stress indicate clearly that the environment can lead to an alteration in genomic DNA methylation.

The observations listed above indicate a potential connection between the environment and TNR instability, but they do not prove it. Even with the established link between CpG methylation and TNR instability, we do not know the mechanism of instability, which could be due to more efficient recruitment of repair enzymes, enhanced recognition of DNA secondary structure, or enhanced transcription, among other possible mechanisms (Dion et al. 2008a). In the specific instance of induced CpG demethylation in cultured cells, the mechanism apparently does not involve homologous recombination (Dion et al. 2008b). Critically, we do not know whether epigenetic changes are linked to alterations in other microsatellite repeats or to shorter repeat tracts. At a gross level, however, we know that epigenetic changes and microsatellite instability are both common in cancer cells (Jones and Baylin 2007; Thibodeau et al. 1993).

Transcription

Viewed simply as a mechanism for copying a DNA template, transcription seems an unlikely source of genetic variation. Nevertheless, as was shown first in bacteria and later in yeast, transcription through a target gene increases its rate of spontaneous mutagenesis in a way that correlates with the rate of transcription (Kim and Jinks-Robertson 2011; Reimers et al. 2004; Kim et al. 2007). These transcription-induced mutations arise predominantly on the nontranscribed strand, a pattern that is also evident in evolutionary comparisons of mammalian genomes and in the genomic mutations that arise in rapidly evolving tumor cells (Rubin and Green 2009; Majewski 2003; Green et al. 2003). Damage to the exposed nontranscribed strand, unprotected by its complement in the duplex or by pairing with the nascent RNA transcript (Lindahl 1993), accounts for much of the observed strand bias (Kim and Jinks-Robertson 2011; Datta and Jinks-Robertson 1995). The effects of transcription on microsatellite repeat instability may also stem from initial events on the nontranscribed strand, which when freed from its partner, can explore alternative base-pairing arrangements. It has been shown, for example, that TTAGGG repeats can form G-quartet structures on the nontranscribed strand (Duquette et al. 2004). The general ability of microsatellite repeats to mispair after strand separation likely contributes to transcription-induced repeat instability, as illustrated in Fig. 7.1c (Lin et al. 2009, 2010b; Kim and Jinks-Robertson 2011).

Transcription was initially proposed as a possible source for TNR instability based on patterns of instability of a CAG repeat tract in transgenic mice; in three lines that transcribed the transgene the repeat was unstable, whereas in the one line with a silent transgene the repeat was stable (Mangiarini et al. 1997). This correlation between transcription and microsatellite instability was confirmed and extended in studies in bacteria (Mochmann and Wells 2004; Bowater et al. 1997; Schumacher et al. 2001), yeast (Wierdl et al. 1996), flies (Jung and Bonini 2007), and human cells (Nakamori et al. 2011; Lin and Wilson 2007; Lin et al. 2006b; Ditch et al. 2009). In bacteria and yeast, transcription-induced instability was interpreted in terms of collisions between RNA and DNA polymerase complexes (Wierdl et al. 1996; Parniewski et al. 1999; Bowater et al. 1997). In human cells, however, transcription-induced instability of a CAG repeat tract was shown to be independent of DNA replication (Lin et al. 2006a, 2010a). Thus, transcription offers a potential mechanism to explain TNR instability in both proliferating and nonproliferating cells, including terminally differentiated neurons, which often display high levels of TNR instability (Lin et al. 2009). As expected for a transcription-based mechanism, TNR-associated disease genes appear to be widely transcribed in germ-line and somatic tissues (Dixon et al. 2004; Banfi et al. 1994; Doyu et al. 1994; Nagafuchi et al. 1994; Li et al. 1993; Strong et al. 1993).

In human cells, siRNA knockdowns have revealed that transcription-induced CAG repeat instability is a surprisingly complex process. It involves transcription-coupled nucleotide excision repair (TC-NER)—a major DNA repair pathway—triggered by stalled RNA polymerase II complexes, as well as multiple other components that include topoisomerase 1, RNase H, the mismatch recognition element Msh2/Msh3 (MutSβ), transcription factor TFIIS, the Brca1/Bard1 E3 ubiquitin ligase, and the proteasome (Lin et al. 2006a, 2009, 2010b; Lin and Wilson 2007, 2009; Hubert et al. 2011b), as described in Fig. 7.1c. The complexity of the pathway for transcription-induced repeat instability serves to illustrate some of the many ways that environmental stress might impact the stability of microsatellite repeats (Lin et al. 2009; Lin and Wilson 2007).

Sense transcription, however, is not the whole story. Analyses of the human transcriptome in cells suggest that RNA polymerase transcribes a majority of the human genome (Mattick and Makunin 2006; Kapranov et al. 2007; Lindberg and Lundeberg 2010). More surprisingly, some 30% of human genes are transcribed in both the sense and antisense directions (Katayama et al. 2005; Chen et al. 2004; Yelin et al. 2003). Antisense transcription has been reported for eight TNR disease genes (Hubert et al. 2011a; Cho et al. 2005; Moseley et al. 2006; Ladd et al. 2007; De Biase et al. 2009; Chen et al. 2008; Chung et al. 2011; Wilburn et al. 2011; Sopher et al. 2011) and many other TNR disease genes are represented in the antisense transcriptome of several human cells lines (He et al. 2008). In a selection assay for

CAG repeat contractions in human cells, antisense transcription was found to promote repeat instability with an efficiency similar to sense transcription (Lin et al. 2006a, 2010a). Unexpectedly, induction of both sense and antisense transcription—convergent transcription—was shown to destabilize repeats to a significantly higher frequency that just the sum of the sense- and antisense-induced frequencies (Lin et al. 2010a). The synergistic destabilizing effects of convergent transcription have been confirmed and extended in a cell line with a very large CAG tract consisting of 800 units, where it was shown that convergent transcription induced about equal frequencies of repeat contractions and expansions (Nakamori et al. 2011). Although the functions of antisense transcription are unclear, the ratios of sense and antisense transcripts appear to be tissue specific (Klostermeier et al. 2012) and they can change in cancer cells (Maruyama et al. 2012). In addition, antisense transcription is regulated by CTCF (CCCTC binding factor) at both the DM1 and SCA7 loci (Filippova et al. 2001; Sopher et al. 2011). Moreover, the activity of CTCF is altered under hypoxic and oxidative stress, providing another potential connection between transcription and stress (Wang et al. 2012).

Classically, animals respond to stress in ways that are regulated by hormones, as, for example, when an animal encounters a predator, which triggers the "flight-or-fight" response. But a host of other external conditions can threaten the internal environment of the cell, including temperature changes, oxidative stress, osmotic shocks, and nutrient depletion, to name a few (Kultz 2005). In yeast, experiments testing a diverse range of such environmental challenges have identified a large, overlapping set of genes (about 15% of the genome) whose transcription is altered in response to almost all of the environmental insults that were tested (Gasch et al. 2000). These results suggest that cells respond in a similar manner to a wide variety of environmental insults (Gasch et al. 2000). This cellular stress response is highly conserved across eukaryotes and is triggered by damage to membranes, proteins, and DNA and by changes to the redox state of the cell (Kultz 2005). Endogenous sensors, which are not fully defined, stimulate a variety of signaling pathways that generally terminate in the activation of a set of transcription factors (Kultz 2005; Nebert 2002; Boutahar et al. 2011). Thus, environmental stress is linked to transcription, and transcription renders microsatellite repeats more mutable. Nevertheless, no connection has yet been forged between a non-genotoxic environmental stress and microsatellite instability. It is worth noting that since environmental stress activates transcription in a subset of genes, and about 17% of those are likely to have microsatellite repeats (Gemayel et al. 2010), the mutagenic effect of transcription on repeat instability would be targeted to the very genes that mediate the response, perhaps allowing the response itself to be refined and fine-tuned (Wright 2004).

Hsp90 Chaperone

Hsp90 chaperone is a highly conserved, ubiquitously expressed member of the heat shock family of proteins, which guard against the consequences of environmental and cellular stress (Csermely et al. 1998; Lindquist 1986; Pearl and Prodromou 2000). Hsp90 family members, which are found in all cells with the exception of archaea, perform several essential functions: they help fold newly synthesized proteins into their active conformations; they aid in the assembly of macromolecular complexes; and they target misfolded proteins for ubiquitination and proteasomal degradation (Taipale et al. 2010). Aided by a set of some 20 co-chaperones, Hsp90 interacts with hundreds of client proteins that are involved in signal transduction, protein trafficking, receptor maturation, and genome stability, among others (Taipale et al. 2010; Zhao and Houry 2007; Oda et al. 2007; Pozo et al. 2011; Sekimoto et al. 2010; Noguchi et al. 2006; Arlander et al. 2003; Dote et al. 2006; Yao et al. 2007). Although Hsp90 is highly abundant, constituting 1–2% of cytoplasmic proteins in eukaryotic cells, it is induced to even higher levels in response to environmental stress (Borkovich et al. 1989). When cells are heavily stressed, however, Hsp90 is titrated away from its natural clients to deal with the increased load of improperly folded proteins, potentially compromising the pathways in which Hsp90 clients normally function.

The critical role of Hsp90 in controlling genetic variation has been illuminated by several studies that examine the consequences of deficient Hsp90 function. In both *Drosophila* and *Arabidopsis*, impairment of Hsp90 function exposes cryptic genetic variation that was hidden by Hsp90 (Queitsch et al. 2002; Rutherford and Lindquist 1998; Sangster et al. 2007, 2008). Apparently, Hsp90 helps stabilize mutant client proteins that do not fold adequately on their own. When stress overwhelms Hsp90's buffering capacity, however, this previously hidden genetic variation is revealed (Rutherford and Lindquist 1998). In addition, Hsp90 can modulate the expression of genetic variation in flies by epigenetic mechanisms in ways that suggest that Hsp90 plays a role in the maintenance of active chromatin (Tariq et al. 2009; Sollars et al. 2003; Ruden et al. 2003; Gangaraju et al. 2011). Beyond revealing hidden variation and inducing epigenetic changes, Hsp90 also regulates transposon-mediated mutagenesis in the fly germ line by interfering with transposon mobilization (Specchia et al. 2010). When Hsp90 activity is diverted or inhibited, the normal suppression of transposable elements by piRNA and Piwi protein is compromised, allowing transposons to hop to new places in the genome, causing mutations as they go (Gangaraju et al. 2011; Specchia et al. 2010).

In addition to these effects on genetic variation, Hsp90 is required to maintain TNR stability in human cells (Mittelman et al. 2010; Mittelman and Wilson 2010). In a selection assay for CAG repeat contractions, treatment with an Hsp90 inhibitor (the geldanamycin derivative 17-AAG (Schulte and Neckers 1998)), which mimics cell stress, was shown to increase CAG contractions tenfold. The Rad51 recombinase, an Hsp90 client whose cellular levels drop substantially in the presence of 17-AAG, likely mediates the increase in contractions, since its direct knockdown by siRNA also stimulated contractions tenfold (Mittelman et al. 2010). Rad51 is the primary recombinase involved in the homology-dependent repair of double-strand breaks (DSBs), which in mammalian cells are repaired either by strand invasion, which requires Rad51, or by single-strand annealing (SSA), which does not (Paques and Haber 1999). The ability of Hsp90 to modulate Rad51 suggests that it regulates the choice between these two pathways. Under stress, when Hsp90 is diverted from its normal function, the decreased levels of Rad51 will enhance the SSA pathway, leading to changes in the length of the repeat tract (Fig. 7.1b) (Mittelman et al. 2009, 2010; Richard et al. 1999). Thus, above a certain threshold of stress, Hsp90 acts as a rheostat to increase the activity of the error-prone SSA pathway and promote microsatellite instability (Fonville et al. 2011).

Hsp90 chaperone clearly occupies a central position in the cellular response to external stresses (Taipale et al. 2010). Nearly 20% of the yeast proteome and about 1,600 human proteins are regulated by Hsp90 chaperone within complex interaction networks (Taipale et al. 2010; Zhao and Houry 2007). It is equally clear that one consequence of perturbing Hsp90 function is the induction of genetic variation, including TNR instability. Together, these observations establish a strong inferential link between environmental stress and TNR instability; however, it has not yet been shown experimentally that a particular stress can induce a change in TNR repeat length. It remains an open question whether microsatellite repeats, in general, respond to stress.

Summary and Perspectives

The genome is much more dynamic than we once thought. In addition to point mutations, genomes suffer small deletions and insertions, large rearrangements, copy number variations, and transposition events. Microsatellite repeats are the most mutationally dynamic elements in the genome. They are also special because they mutate at a high frequency in predictable ways and the mutations can confer incremental changes on the activities of gene regulatory elements and encoded proteins. The evolutionary importance of microsatellite mutation for fine-tuning gene function is reasonably clear (Gemayel et al. 2010). What is less clear is whether microsatellite instability is completely independent of the evolutionary forces acting on the organism. Or, as framed here, can environmental stress modulate the mutation rates of microsatellite repeats? Do microsatellite repeats behave as genomic detectors of environmental stress, like canaries sensing poison gas in a coalmine?

Studies of long CAG repeats in human cells, which mutate at a high rate already, have revealed three potential connections between the environment and TNR instability: CpG methylation, transcription, and the Hsp90 chaperone (Fig. 7.5) (Lin et al. 2009; Dion and Wilson 2009; Mittelman and Wilson 2010). Environmental stress can alter each of these processes, and changes in these processes have been shown to modulate TNR instability. It will be critical, however, to show that particular environmental stresses can alter TNRs and to define the molecular steps that lead to TNR instability. It will be equally important to determine whether such links are specific for long TNRs or represent a common property of microsatellite repeats. Experiments in *Arabidopsis* suggest that microsatellite instability may be a universal response to stress: exposure to heat, cold, drought, flood, and UVC irradiation

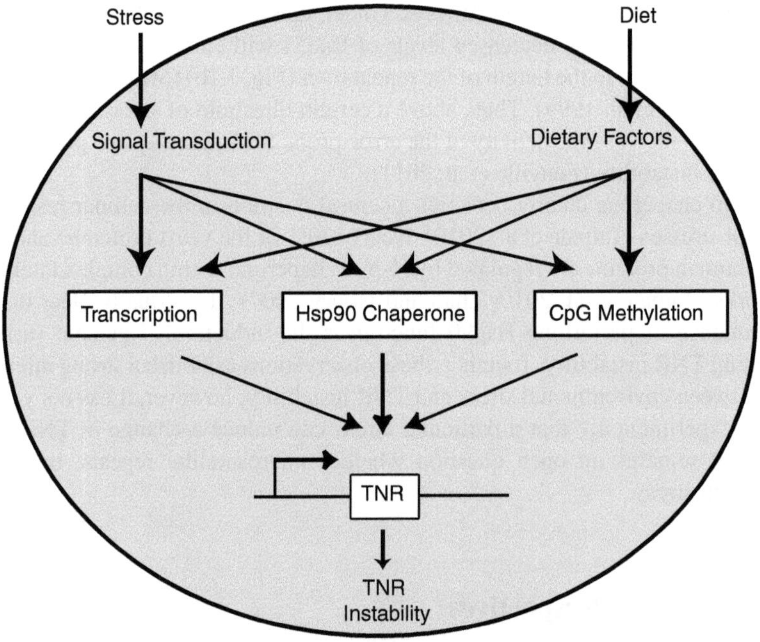

Fig. 7.5 Potential connections between the environment and TNR instability. Stress, via signal transduction pathways, and components of the diet can modulate transcription, CpG methylation, and the Hsp90 chaperone, which, in turn, can modulate TNR instability

stimulated instability in a run of 16 Gs in the somatic tissues of treated plants and in their progeny (Yao and Kovalchuk 2011).

Does microsatellite instability improve survival in a stressful environment? Stress-induced mutagenesis in bacteria and fungi, which elevates the rates of mutation during times of stress such as aging, starvation, onset of stationary phase, and exposure to drugs, is clearly beneficial to the organism (Bjedov et al. 2003; Galhardo et al. 2007; Hall 1992). The overrepresentation of TNRs in exons, coupled with the incremental functional differences associated with various repeat-tract lengths, suggests that microsatellite instability is also advantageous. Examining the stress genes themselves illustrates the potential usefulness of TNRs. While less than 5% of human genes (878) contain TNRs in their coding regions, more than 50% of the genes (111 out of 204) involved in the cellular stress response contain TNRs (www.geneontology.org/) (Kozlowski et al. 2010). This remarkable overrepresentation suggests that microsatellite instability has shaped the cellular stress response and raises the interesting possibility that stress itself might fine-tune the cellular response.

While is it easy to see how microsatellite instability might play a role in the adaptation of microorganisms to stress, it is not so clear how instability might benefit multicellular organisms. All the same, cells in organisms are exposed to stressful environments: liver cells detoxify chemicals from the diet; kidney cells are exposed to hyperosmotic, low oxygen conditions; and the brain is highly oxidative. Do these stresses induce higher levels of microsatellite instability in the cells in these organs? Do the potential differences in microsatellite repeat lengths from cell to cell in these organs cause a problem, or might they be beneficial? These questions are crucial, but difficult to address in the normal setting.

In cancer cells, however, the roles of stress and microsatellite instability may be more accessible. In the evolving cancer cell, stress may drive mutagenesis, which is critical to the process of oncogenesis. Cancer cells typically arise in a hypoxic environment with low pH and semi-starved for nutrients (Chiosis and Neckers 2006). Cancer cells have severely altered transcription, extensive epigenetic modification of the genome, and increased synthesis of Hsp90 chaperone; that is, the common conditions that lead to TNR instability in normal cells (Weis and Cheresh 2011; Johanning et al. 2002; Whitesell and Lindquist 2005; Travers et al. 2012). Interestingly, 90% of the stress response genes that contain TNRs (101/111) have been implicated in cancer (Chatterjee, unpublished). Whether environmental stress and the stress response contribute to the genesis of cancer cells via their effects on microsatellite instability remains to be determined.

Acknowledgments This work was supported by an RO1 grant from the NIH (GM38219) to J.H.W. and a Baylor Research Advocates for Student Scientists (BRASS) scholarship to B.A.S.

References

Albanese V, Biguet NF, Kiefer H, Bayard E, Mallet J, Meloni R (2001) Quantitative effects on gene silencing by allelic variation at a tetranucleotide microsatellite. Hum Mol Genet 10: 1785–1792

Albrecht A, Mundlos S (2005) The other trinucleotide repeat: polyalanine expansion disorders. Curr Opin Genet Dev 15:285–293

Al-Mahdawi S, Pinto RM, Ismail O, Varshney D, Lymperi S, Sandi C, Trabzuni D, Pook M (2008) The Friedreich ataxia GAA repeat expansion mutation induces comparable epigenetic changes in human and transgenic mouse brain and heart tissues. Hum Mol Genet 17:735–746

Arlander SJ, Eapen AK, Vroman BT, McDonald RJ, Toft DO, Karnitz LM (2003) Hsp90 inhibition depletes Chk1 and sensitizes tumor cells to replication stress. J Biol Chem 278: 52572–52577

Bacolla A, Larson JE, Collins JR, Li J, Milosavljevic A, Stenson PD, Cooper DN, Wells RD (2008) Abundance and length of simple repeats in vertebrate genomes are determined by their structural properties. Genome Res 18:1545–1553

Banfi S, Servadio A, Chung MY, Kwiatkowski TJ Jr, McCall AE, Duvick LA, Shen Y, Roth EJ, Orr HT, Zoghbi HY (1994) Identification and characterization of the gene causing type 1 spinocerebellar ataxia. Nat Genet 7:513–520

Barry JD, McCulloch R (2001) Antigenic variation in trypanosomes: enhanced phenotypic variation in a eukaryotic parasite. Adv Parasitol 49:1–70

Bhattacharyya S, Lahue RS (2004) *Saccharomyces cerevisiae* Srs2 DNA helicase selectively blocks expansions of trinucleotide repeats. Mol Cell Biol 24:7324–7330

Bidichandani SI, Ashizawa T, Patel PI (1998) The GAA triplet-repeat expansion in Friedreich ataxia interferes with transcription and may be associated with an unusual DNA structure. Am J Hum Genet 62:111–121

Bjedov I, Tenaillon O, Gerard B, Souza V, Denamur E, Radman M, Taddei F, Matic I (2003) Stress-induced mutagenesis in bacteria. Science 300:1404–1409

Borkovich KA, Farrelly FW, Finkelstein DB, Taulien J, Lindquist S (1989) hsp82 is an essential protein that is required in higher concentrations for growth of cells at higher temperatures. Mol Cell Biol 9:3919–3930

Boutahar N, Wierinckx A, Camdessanche JP, Antoine JC, Reynaud E, Lassabliere F, Lachuer J, Borg J (2011) Differential effect of oxidative or excitotoxic stress on the transcriptional profile of amyotrophic lateral sclerosis-linked mutant SOD1 cultured neurons. J Neurosci Res 89:1439–1450

Bowater RP, Jaworski A, Larson JE, Parniewski P, Wells RD (1997) Transcription increases the deletion frequency of long CTG.CAG triplet repeats from plasmids in *Escherichia coli*. Nucleic Acids Res 25:2861–2868

Boyko A, Blevins T, Yao Y, Golubov A, Bilichak A, Ilnytskyy Y, Hollunder J, Meins F Jr, Kovalchuk I (2010) Transgenerational adaptation of Arabidopsis to stress requires DNA methylation and the function of Dicer-like proteins. PLoS One 5:e9514

Brock GJ, Anderson NH, Monckton DG (1999) Cis-acting modifiers of expanded CAG/CTG triplet repeat expandability: associations with flanking GC content and proximity to CpG islands. Hum Mol Genet 8:1061–1067

Brouwer JR, Willemsen R, Oostra BA (2009) Microsatellite repeat instability and neurological disease. Bioessays 31:71–83

Caburet S, Cocquet J, Vaiman D, Veitia RA (2005) Coding repeats and evolutionary "agility". Bioessays 27:581–587

Caprioli M, Ambrosini R, Boncoraglio G, Gatti E, Romano A, Romano M, Rubolini D, Gianfranceschi L, Saino N (2012) Clock gene variation is associated with breeding phenology and maybe under directional selection in the migratory barn swallow. PLoS One 7:e35140

Carone BR, Fauquier L, Habib N, Shea JM, Hart CE, Li R, Bock C, Li C, Gu H, Zamore PD et al (2010) Paternally induced transgenerational environmental reprogramming of metabolic gene expression in mammals. Cell 143:1084–1096

Chen J, Sun M, Kent WJ, Huang X, Xie H, Wang W, Zhou G, Shi RZ, Rowley JD (2004) Over 20% of human transcripts might form sense-antisense pairs. Nucleic Acids Res 32:4812–4820

Chen WL, Lin JW, Huang HJ, Wang SM, Su MT, Lee-Chen GJ, Chen CM, Hsieh-Li HM (2008) SCA8 mRNA expression suggests an antisense regulation of KLHL1 and correlates to SCA8 pathology. Brain Res 1233:176–184

Chiosis G, Neckers L (2006) Tumor selectivity of Hsp90 inhibitors: the explanation remains elusive. ACS Chem Biol 1:279–284

Cho DH, Thienes CP, Mahoney SE, Analau E, Filippova GN, Tapscott SJ (2005) Antisense transcription and heterochromatin at the DM1 CTG repeats are constrained by CTCF. Mol Cell 20:483–489

Chung DW, Rudnicki DD, Yu L, Margolis RL (2011) A natural antisense transcript at the Huntington's disease repeat locus regulates HTT expression. Hum Mol Genet 20:3467–3477

Coffee B, Zhang F, Ceman S, Warren ST, Reines D (2002) Histone modifications depict an aberrantly heterochromatinized FMR1 gene in fragile x syndrome. Am J Hum Genet 71:923–932

Collins NS, Bhattacharyya S, Lahue RS (2007) Rev1 enhances CAG.CTG repeat stability in *Saccharomyces cerevisiae*. DNA Repair (Amst) 6:38–44

Contente A, Dittmer A, Koch MC, Roth J, Dobbelstein M (2002) A polymorphic microsatellite that mediates induction of PIG3 by p53. Nat Genet 30:315–320

Coolen MW, Statham AL, Qu W, Campbell MJ, Henders AK, Montgomery GW, Martin NG, Clark SJ (2011) Impact of the genome on the epigenome is manifested in DNA methylation patterns of imprinted regions in monozygotic and dizygotic twins. PLoS One 6:e25590

Coyle S, Kroll E (2008) Starvation induces genomic rearrangements and starvation-resilient phenotypes in yeast. Mol Biol Evol 25:310–318

Csermely P, Schnaider T, Soti C, Prohaszka Z, Nardai G (1998) The 90-kDa molecular chaperone family: structure, function, and clinical applications. A comprehensive review. Pharmacol Ther 79:129–168

Cubas P, Vincent C, Coen E (1999) An epigenetic mutation responsible for natural variation in floral symmetry. Nature 401:157–161

Datta A, Jinks-Robertson S (1995) Association of increased spontaneous mutation rates with high levels of transcription in yeast. Science 268:1616–1619

Davis CD, Uthus EO (2004) DNA methylation, cancer susceptibility, and nutrient interactions. Exp Biol Med (Maywood) 229:988–995

De Biase I, Rasmussen A, Endres D, Al-Mahdawi S, Monticelli A, Cocozza S, Pook M, Bidichandani SI (2007a) Progressive GAA expansions in dorsal root ganglia of Friedreich's ataxia patients. Ann Neurol 61:55–60

De Biase I, Rasmussen A, Monticelli A, Al-Mahdawi S, Pook M, Cocozza S, Bidichandani SI (2007b) Somatic instability of the expanded GAA triplet-repeat sequence in Friedreich ataxia progresses throughout life. Genomics 90:1–5

De Biase I, Chutake YK, Rindler PM, Bidichandani SI (2009) Epigenetic silencing in Friedreich ataxia is associated with depletion of CTCF (CCCTC-binding factor) and antisense transcription. PLoS One 4:e7914

Deaton AM, Bird A (2011) CpG islands and the regulation of transcription. Genes Dev 25: 1010–1022

Debacker K, Frizzell A, Gleeson O, Kirkham-McCarthy L, Mertz T, Lahue RS (2012) Histone deacetylase complexes promote trinucleotide repeat expansions. PLoS Biol 10:e1001257

Devys D, Biancalana V, Rousseau F, Boue J, Mandel JL, Oberle I (1992) Analysis of full fragile X mutations in fetal tissues and monozygotic twins indicate that abnormal methylation and somatic heterogeneity are established early in development. Am J Med Genet 43:208–216

Dhar A, Lahue RS (2008) Rapid unwinding of triplet repeat hairpins by Srs2 helicase of *Saccharomyces cerevisiae*. Nucleic Acids Res 36:3366–3373

Dion V, Wilson JH (2009) Instability and chromatin structure of expanded trinucleotide repeats. Trends Genet 25:288–297

Dion V, Lin Y, Hubert L Jr, Waterland RA, Wilson JH (2008a) Dnmt1 deficiency promotes CAG repeat expansion in the mouse germline. Hum Mol Genet 17:1306–1317

Dion V, Lin Y, Price BA, Fyffe SL, Seluanov A, Gorbunova V, Wilson JH (2008b) Genome-wide demethylation promotes triplet repeat instability independently of homologous recombination. DNA Repair (Amst) 7:313–320

Ditch S, Sammarco MC, Banerjee A, Grabczyk E (2009) Progressive GAA.TTC repeat expansion in human cell lines. PLoS Genet 5:e1000704

Dixon KT, Cearley JA, Hunter JM, Detloff PJ (2004) Mouse Huntington's disease homolog mRNA levels: variation and allele effects. Gene Expr 11:221–231

Dolinoy DC, Huang D, Jirtle RL (2007) Maternal nutrient supplementation counteracts bisphenol A-induced DNA hypomethylation in early development. Proc Natl Acad Sci USA 104:13056–13061

Dote H, Burgan WE, Camphausen K, Tofilon PJ (2006) Inhibition of hsp90 compromises the DNA damage response to radiation. Cancer Res 66:9211–9220

Doyu M, Sobue G, Kimata K, Yamamoto K, Mitsuma T (1994) Androgen receptor mRNA with increased size of tandem CAG repeat is widely expressed in the neural and nonneural tissues of X-linked recessive bulbospinal neuronopathy. J Neurol Sci 127:43–47

Duquette ML, Handa P, Vincent JA, Taylor AF, Maizels N (2004) Intracellular transcription of G-rich DNAs induces formation of G-loops, novel structures containing G4 DNA. Genes Dev 18:1618–1629

Ellegren H (2004) Microsatellites: simple sequences with complex evolution. Nat Rev Genet 5:435–445

Entezam A, Lokanga AR, Le W, Hoffman G, Usdin K (2010) Potassium bromate, a potent DNA oxidizing agent, exacerbates germline repeat expansion in a fragile X premutation mouse model. Hum Mutat 31:611–616

Faux NG, Bottomley SP, Lesk AM, Irving JA, Morrison JR, de la Banda MG, Whisstock JC (2005) Functional insights from the distribution and role of homopeptide repeat-containing proteins. Genome Res 15:537–551

Filippova GN, Thienes CP, Penn BH, Cho DH, Hu YJ, Moore JM, Klesert TR, Lobanenkov VV, Tapscott SJ (2001) CTCF-binding sites flank CTG/CAG repeats and form a methylation-sensitive insulator at the DM1 locus. Nat Genet 28:335–343

Fondon JW 3rd, Garner HR (2004) Molecular origins of rapid and continuous morphological evolution. Proc Natl Acad Sci USA 101:18058–18063

Fonville NC, Ward RM, Mittelman D (2011) Stress-induced modulators of repeat instability and genome evolution. J Mol Microbiol Biotechnol 21:36–44

Forche A, Abbey D, Pisithkul T, Weinzierl MA, Ringstrom T, Bruck D, Petersen K, Berman J (2011) Stress alters rates and types of loss of heterozygosity in Candida albicans. MBio 2(4):e00129–11

Fraga MF, Ballestar E, Paz MF, Ropero S, Setien F, Ballestar ML, Heine-Suner D, Cigudosa JC, Urioste M, Benitez J et al (2005) Epigenetic differences arise during the lifetime of monozygotic twins. Proc Natl Acad Sci USA 102:10604–10609

Freudenreich CH, Kantrow SM, Zakian VA (1998) Expansion and length-dependent fragility of CTG repeats in yeast. Science 279:853–856

Galhardo RS, Hastings PJ, Rosenberg SM (2007) Mutation as a stress response and the regulation of evolvability. Crit Rev Biochem Mol Biol 42:399–435

Gangaraju VK, Yin H, Weiner MM, Wang J, Huang XA, Lin H (2011) Drosophila Piwi functions in Hsp90-mediated suppression of phenotypic variation. Nat Genet 43:153–158

Gangwal K, Sankar S, Hollenhorst PC, Kinsey M, Haroldsen SC, Shah AA, Boucher KM, Watkins WS, Jorde LB, Graves BJ et al (2008) Microsatellites as EWS/FLI response elements in Ewing's sarcoma. Proc Natl Acad Sci USA 105:10149–10154

Gasch AP, Spellman PT, Kao CM, Carmel-Harel O, Eisen MB, Storz G, Botstein D, Brown PO (2000) Genomic expression programs in the response of yeast cells to environmental changes. Mol Biol Cell 11:4241–4257

Gebhardt F, Zanker KS, Brandt B (1999) Modulation of epidermal growth factor receptor gene transcription by a polymorphic dinucleotide repeat in intron 1. J Biol Chem 274:13176–13180

Gemayel R, Vinces MD, Legendre M, Verstrepen KJ (2010) Variable tandem repeats accelerate evolution of coding and regulatory sequences. Annu Rev Genet 44:445–477

Gerber HP, Seipel K, Georgiev O, Hofferer M, Hug M, Rusconi S, Schaffner W (1994) Transcriptional activation modulated by homopolymeric glutamine and proline stretches. Science 263:808–811

Glaser D, Wohrle D, Salat U, Vogel W, Steinbach P (1999) Mitotic behavior of expanded CGG repeats studied on cultured cells: further evidence for methylation-mediated triplet repeat stability in fragile X syndrome. Am J Med Genet 84:226–228

Gomes-Pereira M, Monckton DG (2004) Chemically induced increases and decreases in the rate of expansion of a CAG*CTG triplet repeat. Nucleic Acids Res 32:2865–2872

Gomes-Pereira M, Fortune MT, Ingram L, McAbney JP, Monckton DG (2004) Pms2 is a genetic enhancer of trinucleotide CAG{middle dot}CTG repeat somatic mosaicism: implications for the mechanism of triplet repeat expansion. Hum Mol Genet 13:1815–1825

Gorbunova V, Seluanov A, Mittelman D, Wilson JH (2004) Genome-wide demethylation destabilizes CTG.CAG trinucleotide repeats in mammalian cells. Hum Mol Genet 13:2979–2989

Gottschling DE, Aparicio OM, Billington BL, Zakian VA (1990) Position effect at *S. cerevisiae* telomeres: reversible repression of Pol II transcription. Cell 63:751–762

Gourdon G, Dessen P, Lia AS, Junien C, Hofmann-Radvanyi H (1997) Intriguing association between disease associated unstable trinucleotide repeat and CpG island. Ann Genet 40:73–77

Grabczyk E, Usdin K (2000) The GAA*TTC triplet repeat expanded in Friedreich's ataxia impedes transcription elongation by T7 RNA polymerase in a length and supercoil dependent manner. Nucleic Acids Res 28:2815–2822

Green P, Ewing B, Miller W, Thomas PJ, Green ED (2003) Transcription-associated mutational asymmetry in mammalian evolution. Nat Genet 33:514–517

Greene E, Mahishi L, Entezam A, Kumari D, Usdin K (2007) Repeat-induced epigenetic changes in intron 1 of the frataxin gene and its consequences in Friedreich ataxia. Nucleic Acids Res 35:3383–3390

Hall BG (1992) Selection-induced mutations occur in yeast. Proc Natl Acad Sci USA 89:4300–4303
Hamid A, Wani NA, Kaur J (2009) New perspectives on folate transport in relation to alcoholism-induced folate malabsorption–association with epigenome stability and cancer development. FEBS J 276:2175–2191
Hastings PJ, Slack A, Petrosino JF, Rosenberg SM (2004) Adaptive amplification and point mutation are independent mechanisms: evidence for various stress-inducible mutation mechanisms. PLoS Biol 2:e399
He Y, Vogelstein B, Velculescu VE, Papadopoulos N, Kinzler KW (2008) The antisense transcriptomes of human cells. Science 322:1855–1857
Hebert ML, Spitz LA, Wells RD (2004) DNA double-strand breaks induce deletion of CTG.CAG repeats in an orientation-dependent manner in *Escherichia coli*. J Mol Biol 336:655–672
Helderman-van den Enden AT, Maaswinkel-Mooij PD, Hoogendoorn E, Willemsen R, Maat-Kievit JA, Losekoot M, Oostra BA (1999) Monozygotic twin brothers with the fragile X syndrome: different CGG repeats and different mental capacities. J Med Genet 36:253–257
Hernday A, Krabbe M, Braaten B, Low D (2002) Self-perpetuating epigenetic pili switches in bacteria. Proc Natl Acad Sci USA 99(Suppl 4):16470–16476
Hou C, Chan NL, Gu L, Li GM (2009) Incision-dependent and error-free repair of (CAG)(n)/(CTG)(n) hairpins in human cell extracts. Nat Struct Mol Biol 16:869–875
Hubert L Jr, Lin Y, Dion V, Wilson JH (2011a) Xpa deficiency reduces CAG trinucleotide repeat instability in neuronal tissues in a mouse model of SCA1. Hum Mol Genet 20:4822–4830
Hubert L Jr, Lin Y, Dion V, Wilson JH (2011b) Topoisomerase 1 and single-strand break repair modulate transcription-induced CAG repeat contraction in human cells. Mol Cell Biol 31:3105–3112
Hui J, Stangl K, Lane WS, Bindereif A (2003) HnRNP L stimulates splicing of the eNOS gene by binding to variable-length CA repeats. Nat Struct Biol 10:33–37
Issa JP (2000) CpG-island methylation in aging and cancer. Curr Top Microbiol Immunol 249:101–118
Issa JP (2004) CpG island methylator phenotype in cancer. Nat Rev Cancer 4:988–993
Iyer V, Struhl K (1995) Poly(dA:dT), a ubiquitous promoter element that stimulates transcription via its intrinsic DNA structure. EMBO J 14:2570–2579
Jarem DA, Wilson NR, Delaney S (2009) Structure-dependent DNA damage and repair in a trinucleotide repeat sequence. Biochemistry 48:6655–6663
Jaworski A, Rosche WA, Gellibolian R, Kang S, Shimizu M, Bowater RP, Sinden RR, Wells RD (1995) Mismatch repair in *Escherichia coli* enhances instability of (CTG)n triplet repeats from human hereditary diseases. Proc Natl Acad Sci USA 92:11019–11023
Johanning GL, Heimburger DC, Piyathilake CJ (2002) DNA methylation and diet in cancer. J Nutr 132:3814S–3818S
Johnsen A, Fidler AE, Kuhn S, Carter KL, Hoffmann A, Barr IR, Biard C, Charmantier A, Eens M, Korsten P et al (2007) Avian Clock gene polymorphism: evidence for a latitudinal cline in allele frequencies. Mol Ecol 16:4867–4880
Johnson AC, Jinno Y, Merlino GT (1988) Modulation of epidermal growth factor receptor proto-oncogene transcription by a promoter site sensitive to S1 nuclease. Mol Cell Biol 8:4174–4184
Jones PA, Baylin SB (2007) The epigenomics of cancer. Cell 128:683–692
Jung J, Bonini N (2007) CREB-binding protein modulates repeat instability in a Drosophila model for polyQ disease. Science 315:1857–1859
Kapranov P, Willingham AT, Gingeras TR (2007) Genome-wide transcription and the implications for genomic organization. Nat Rev Genet 8:413–423
Karlin S, Burge C (1996) Trinucleotide repeats and long homopeptides in genes and proteins associated with nervous system disease and development. Proc Natl Acad Sci USA 93:1560–1565
Kashi Y, King DG (2006) Simple sequence repeats as advantageous mutators in evolution. Trends Genet 22:253–259
Katayama S, Tomaru Y, Kasukawa T, Waki K, Nakanishi M, Nakamura M, Nishida H, Yap CC, Suzuki M, Kawai J et al (2005) Antisense transcription in the mammalian transcriptome. Science 309:1564–1566

Kaytor MD, Burright EN, Duvick LA, Zoghbi HY, Orr HT (1997) Increased trinucleotide repeat instability with advanced maternal age. Hum Mol Genet 6:2135–2139

Kerrest A, Anand RP, Sundararajan R, Bermejo R, Liberi G, Dujon B, Freudenreich CH, Richard GF (2009) SRS2 and SGS1 prevent chromosomal breaks and stabilize triplet repeats by restraining recombination. Nat Struct Mol Biol 16:159–167

Kim N, Jinks-Robertson S (2011) Transcription as a source of genome instability. Nat Rev Genet 13:204–214

Kim N, Jinks-Robertson S (2012) Guanine repeat-containing sequences confer transcription-dependent instability in an orientation-specific manner in yeast. DNA Repair (Amst) 10:953–960

Kim N, Abdulovic AL, Gealy R, Lippert MJ, Jinks-Robertson S (2007) Transcription-associated mutagenesis in yeast is directly proportional to the level of gene expression and influenced by the direction of DNA replication. DNA Repair (Amst) 6:1285–1296

Klostermeier UC, Barann M, Wittig M, Hasler R, Franke A, Gavrilova O, Kreck B, Sina C, Schilhabel MB, Schreiber S et al (2012) A tissue-specific landscape of sense/antisense transcription in the mouse intestine. BMC Genomics 12:305

Koturbash I, Baker M, Loree J, Kutanzi K, Hudson D, Pogribny I, Sedelnikova O, Bonner W, Kovalchuk O (2006) Epigenetic dysregulation underlies radiation-induced transgenerational genome instability in vivo. Int J Radiat Oncol Biol Phys 66:327–330

Kovtun IV, Liu Y, Bjoras M, Klungland A, Wilson SH, McMurray CT (2007) OGG1 initiates age-dependent CAG trinucleotide expansion in somatic cells. Nature 447:447–452

Kovtun IV, Johnson KO, McMurray CT (2011) Cockayne Syndrome B protein antagonizes OGG1 in modulating CAG repeat length in vivo. Aging (Albany NY) 3:509–514

Kozlowski P, de Mezer M, Krzyzosiak WJ (2010) Trinucleotide repeats in human genome and exome. Nucleic Acids Res 38:4027–4039

Kultz D (2005) Molecular and evolutionary basis of the cellular stress response. Annu Rev Physiol 67:225–257

La Spada AR, Taylor JP (2010) Repeat expansion disease: progress and puzzles in disease pathogenesis. Nat Rev Genet 11:247–258

Ladd PD, Smith LE, Rabaia NA, Moore JM, Georges SA, Hansen RS, Hagerman RJ, Tassone F, Tapscott SJ, Filippova GN (2007) An antisense transcript spanning the CGG repeat region of FMR1 is upregulated in premutation carriers but silenced in full mutation individuals. Hum Mol Genet 16:3174–3187

Lahiri M, Gustafson TL, Majors ER, Freudenreich CH (2004) Expanded CAG repeats activate the DNA damage checkpoint pathway. Mol Cell 15:287–293

Lee JE, Cooper TA (2009) Pathogenic mechanisms of myotonic dystrophy. Biochem Soc Trans 37:1281–1286

Lee K, Dunlap JC, Loros JJ (2003) Roles for WHITE COLLAR-1 in circadian and general photoperception in Neurospora crassa. Genetics 163:103–114

Legendre M, Pochet N, Pak T, Verstrepen KJ (2007) Sequence-based estimation of minisatellite and microsatellite repeat variability. Genome Res 17:1787–1796

Li Y, Jin P (2012) RNA-mediated neurodegeneration in fragile X-associated tremor/ataxia syndrome. Brain Res 1462:112–117

Li Y, Sasaki H (2011) Genomic imprinting in mammals: its life cycle, molecular mechanisms and reprogramming. Cell Res 21:466–473

Li SH, Schilling G, Young WS 3rd, Li XJ, Margolis RL, Stine OC, Wagster MV, Abbott MH, Franz ML, Ranen NG et al (1993) Huntington's disease gene (IT15) is widely expressed in human and rat tissues. Neuron 11:985–993

Liedvogel M, Szulkin M, Knowles SC, Wood MJ, Sheldon BC (2009) Phenotypic correlates of Clock gene variation in a wild blue tit population: evidence for a role in seasonal timing of reproduction. Mol Ecol 18:2444–2456

Lillycrop KA, Phillips ES, Jackson AA, Hanson MA, Burdge GC (2005) Dietary protein restriction of pregnant rats induces and folic acid supplementation prevents epigenetic modification of hepatic gene expression in the offspring. J Nutr 135:1382–1386

Lillycrop KA, Phillips ES, Torrens C, Hanson MA, Jackson AA, Burdge GC (2008) Feeding pregnant rats a protein-restricted diet persistently alters the methylation of specific cytosines in the hepatic PPAR alpha promoter of the offspring. Br J Nutr 100:278–282
Lin Y, Wilson JH (2007) Transcription-induced CAG repeat contraction in human cells is mediated in part by transcription-coupled nucleotide excision repair. Mol Cell Biol 27:6209–6217
Lin Y, Wilson JH (2009) Diverse effects of individual mismatch repair components on transcription-induced CAG repeat instability in human cells. DNA Repair (Amst) 8:878–885
Lin Y, Wilson JH (2011) Transcription-induced DNA toxicity at trinucleotide repeats: double bubble is trouble. Cell Cycle 10:611–618
Lin Y, Dion V, Wilson JH (2006a) Transcription promotes contraction of CAG repeat tracts in human cells. Nat Struct Mol Biol 13:179–180
Lin Y, Dion V, Wilson JH (2006b) Transcription and triplet repeat instability. In: Wells RD, Ashizawa T (eds) Genetic Instabilities and Neurological Diseases, 2nd edn. Academic, San Diego, CA, pp 691–704
Lin Y, Hubert L Jr, Wilson JH (2009) Transcription destabilizes triplet repeats. Mol Carcinog 48:350–361
Lin Y, Leng M, Wan M, Wilson JH (2010a) Convergent transcription through a long CAG tract destabilizes repeats and induces apoptosis. Mol Cell Biol 30:4435–4451
Lin Y, Dent SY, Wilson JH, Wells RD, Napierala M (2010b) R loops stimulate genetic instability of CTG.CAG repeats. Proc Natl Acad Sci USA 107:692–697
Lindahl T (1993) Instability and decay of the primary structure of DNA. Nature 362:709–715
Lindberg J, Lundeberg J (2010) The plasticity of the mammalian transcriptome. Genomics 95:1–6
Lindquist S (1986) The heat-shock response. Annu Rev Biochem 55:1151–1191
Link A, Balaguer F, Goel A (2010) Cancer chemoprevention by dietary polyphenols: promising role for epigenetics. Biochem Pharmacol 80:1771–1792
Lopez Castel A, Tomkinson AE, Pearson CE (2009) CTG/CAG repeat instability is modulated by the levels of human DNA ligase I and its interaction with proliferating cell nuclear antigen: a distinction between replication and slipped-DNA repair. J Biol Chem 284:26631–26645
Lopez Castel A, Cleary JD, Pearson CE (2010) Repeat instability as the basis for human diseases and as a potential target for therapy. Nat Rev Mol Cell Biol 11:165–170
Lopez Castel A, Nakamori M, Tomo S, Chitayat D, Gourdon G, Thornton CA, Pearson CE (2011) Expanded CTG repeat demarcates a boundary for abnormal CpG methylation in myotonic dystrophy patient tissues. Hum Mol Genet 20:1–15
MacDonald ME, Barnes G, Srinidhi J, Duyao MP, Ambrose CM, Myers RH, Gray J, Conneally PM, Young A, Penney J et al (1993) Gametic but not somatic instability of CAG repeat length in Huntington's disease. J Med Genet 30:982–986
Majewski J (2003) Dependence of mutational asymmetry on gene-expression levels in the human genome. Am J Hum Genet 73:688–692
Mangiarini L, Sathasivam K, Mahal A, Mott R, Seller M, Bates GP (1997) Instability of highly expanded CAG repeats in mice transgenic for the Huntington's disease mutation. Nat Genet 15:197–200
Manley K, Shirley TL, Flaherty L, Messer A (1999) Msh2 deficiency prevents in vivo somatic instability of the CAG repeat in Huntington disease transgenic mice. Nat Genet 23:471–473
Maruyama R, Shipitsin M, Choudhury S, Wu Z, Protopopov A, Yao J, Lo PK, Bessarabova M, Ishkin A, Nikolsky Y et al (2012) Altered antisense-to-sense transcript ratios in breast cancer. Proc Natl Acad Sci USA 109:2820–2824
Matsuura T, Yamagata T, Burgess DL, Rasmussen A, Grewal RP, Watase K, Khajavi M, McCall AE, Davis CF, Zu L et al (2000) Large expansion of the ATTCT pentanucleotide repeat in spinocerebellar ataxia type 10. Nat Genet 26:191–194
Mattick JS, Makunin IV (2006) Non-coding RNA. Hum Mol Genet *15 Spec No 1*, R17–29.
Michael TP, Park S, Kim TS, Booth J, Byer A, Sun Q, Chory J, Lee K (2007) Simple sequence repeats provide a substrate for phenotypic variation in the Neurospora crassa circadian clock. PLoS One 2:e795

Mirkin SM (2007) Expandable DNA repeats and human disease. Nature 447:932–940
Mittelman D, Wilson JH (2010) Stress, genomes, and evolution. Cell Stress Chaperones 15:463–466
Mittelman D, Moye C, Morton J, Sykoudis K, Lin Y, Carroll D, Wilson JH (2009) Zinc-finger directed double-strand breaks within CAG repeat tracts promote repeat instability in human cells. Proc Natl Acad Sci USA 106:9607–9612
Mittelman D, Sykoudis K, Hersh M, Lin Y, Wilson JH (2010) Hsp90 modulates CAG repeat instability in human cells. Cell Stress Chaperones 15:753–759
Mochmann LH, Wells RD (2004) Transcription influences the types of deletion and expansion products in an orientation-dependent manner from GAC*GTC repeats. Nucleic Acids Res 32:4469–4479
Molinier J, Ries G, Zipfel C, Hohn B (2006) Transgeneration memory of stress in plants. Nature 442:1046–1049
Moseley ML, Zu T, Ikeda Y, Gao W, Mosemiller AK, Daughters RS, Chen G, Weatherspoon MR, Clark HB, Ebner TJ et al (2006) Bidirectional expression of CUG and CAG expansion transcripts and intranuclear polyglutamine inclusions in spinocerebellar ataxia type 8. Nat Genet 38:758–769
Mularoni L, Ledda A, Toll-Riera M, Alba MM (2010) Natural selection drives the accumulation of amino acid tandem repeats in human proteins. Genome Res 20:745–754
Nagafuchi S, Yanagisawa H, Ohsaki E, Shirayama T, Tadokoro K, Inoue T, Yamada M (1994) Structure and expression of the gene responsible for the triplet repeat disorder, dentatorubral and pallidoluysian atrophy (DRPLA). Nat Genet 8:177–182
Nakamori M, Pearson CE, Thornton CA (2011) Bidirectional transcription stimulates expansion and contraction of expanded (CTG)*(CAG) repeats. Hum Mol Genet 20:580–588
Napierala M, Bacolla A, Wells RD (2005) Increased negative superhelical density in vivo enhances the genetic instability of triplet repeat sequences. J Biol Chem 280:37366–37376
Nebert DW (2002) Transcription factors and cancer: an overview. Toxicology 181–182:131–141
Nichol K, Pearson CE (2002) CpG methylation modifies the genetic stability of cloned repeat sequences. Genome Res 12:1246–1256
Nithianantharajah J, Hannan AJ (2007) Dynamic mutations as digital genetic modulators of brain development, function and dysfunction. Bioessays 29:525–535
Noguchi M, Yu D, Hirayama R, Ninomiya Y, Sekine E, Kubota N, Ando K, Okayasu R (2006) Inhibition of homologous recombination repair in irradiated tumor cells pretreated with Hsp90 inhibitor 17-allylamino-17-demethoxygeldanamycin. Biochem Biophys Res Commun 351:658–663
O'Malley KG, Banks MA (2008) A latitudinal cline in the Chinook salmon (Oncorhynchus tshawytscha) Clock gene: evidence for selection on PolyQ length variants. Proc Biol Sci 275:2813–2821
Oates NA, van Vliet J, Duffy DL, Kroes HY, Martin NG, Boomsma DI, Campbell M, Coulthard MG, Whitelaw E, Chong S (2006) Increased DNA methylation at the AXIN1 gene in a monozygotic twin from a pair discordant for a caudal duplication anomaly. Am J Hum Genet 79:155–162
Oda T, Hayano T, Miyaso H, Takahashi N, Yamashita T (2007) Hsp90 regulates the Fanconi anemia DNA damage response pathway. Blood 109:5016–5026
Orr HT, Zoghbi HY (2007) Trinucleotide repeat disorders. Annu Rev Neurosci 30:575–621
Osborne RJ, Thornton CA (2006) RNA-dominant diseases. Hum Mol Genet *15 Spec No 2*, R162–169.
Ouko LA, Shantikumar K, Knezovich J, Haycock P, Schnugh DJ, Ramsay M (2009) Effect of alcohol consumption on CpG methylation in the differentially methylated regions of H19 and IG-DMR in male gametes: implications for fetal alcohol spectrum disorders. Alcohol Clin Exp Res 33:1615–1627
Owen BA, Yang Z, Lai M, Gajec M, Badger JD 2nd, Hayes JJ, Edelmann W, Kucherlapati R, Wilson TM, McMurray CT (2005) (CAG)(n)-hairpin DNA binds to Msh2-Msh3 and changes properties of mismatch recognition. Nat Struct Mol Biol 12:663–670

Paques F, Haber JE (1999) Multiple pathways of recombination induced by double-strand breaks in *Saccharomyces cerevisiae*. Microbiol Mol Biol Rev 63:349–404

Parniewski P, Bacolla A, Jaworski A, Wells RD (1999) Nucleotide excision repair affects the stability of long transcribed (CTG*CAG) tracts in an orientation-dependent manner in *Escherichia coli*. Nucleic Acids Res 27:616–623

Pearl LH, Prodromou C (2000) Structure and in vivo function of Hsp90. Curr Opin Struct Biol 10:46–51

Pearson CE, Ewel A, Acharya S, Fishel RA, Sinden RR (1997) Human MSH2 binds to trinucleotide repeat DNA structures associated with neurodegenerative diseases. Hum Mol Genet 6:1117–1123

Pearson CE, Edamura KN, Cleary JD (2005) Repeat instability: mechanisms of dynamic mutations. Nat Rev Genet 6:729–742

Pelletier R, Krasilnikova MM, Samadashwily GM, Lahue R, Mirkin SM (2003) Replication and expansion of trinucleotide repeats in yeast. Mol Cell Biol 23:1349–1357

Pieretti M, Zhang FP, Fu YH, Warren ST, Oostra BA, Caskey CT, Nelson DL (1991) Absence of expression of the FMR-1 gene in fragile X syndrome. Cell 66:817–822

Pineiro E, Fernandez-Lopez L, Gamez J, Marcos R, Surralles J, Velazquez A (2003) Mutagenic stress modulates the dynamics of CTG repeat instability associated with myotonic dystrophy type 1. Nucleic Acids Res 31:6733–6740

Pozo FM, Oda T, Sekimoto T, Murakumo Y, Masutani C, Hanaoka F, Yamashita T (2011) Molecular chaperone Hsp90 regulates REV1-mediated mutagenesis. Mol Cell Biol 31:3396–3409

Qu S, Tucker SC, Ehrlich JS, Levorse JM, Flaherty LA, Wisdom R, Vogt TF (1998) Mutations in mouse Aristaless-like4 cause Strong's luxoid polydactyly. Development 125:2711–2721

Queitsch C, Sangster TA, Lindquist S (2002) Hsp90 as a capacitor of phenotypic variation. Nature 417:618–624

Rando OJ, Verstrepen KJ (2007) Timescales of genetic and epigenetic inheritance. Cell 128:655–668

Razidlo DF, Lahue RS (2008) Mrc1, Tof1 and Csm3 inhibit CAG.CTG repeat instability by at least two mechanisms. DNA Repair (Amst) 7:633–640

Reddy K, Tam M, Bowater RP, Barber M, Tomlinson M, Nichol Edamura K, Wang YH, Pearson CE (2011) Determinants of R-loop formation at convergent bidirectionally transcribed trinucleotide repeats. Nucleic Acids Res 39:1749–1762

Reik W (2007) Stability and flexibility of epigenetic gene regulation in mammalian development. Nature 447:425–432

Reik W, Dean W, Walter J (2001) Epigenetic reprogramming in mammalian development. Science 293:1089–1093

Reimers JM, Schmidt KH, Longacre A, Reschke DK, Wright BE (2004) Increased transcription rates correlate with increased reversion rates in leuB and argH *Escherichia coli* auxotrophs. Microbiology 150:1457–1466

Reyniers E, Martin JJ, Cras P, Van Marck E, Handig I, Jorens HZ, Oostra BA, Kooy RF, Willems PJ (1999) Postmortem examination of two fragile X brothers with an FMR1 full mutation. Am J Med Genet 84:245–249

Richard GF, Dujon B, Haber JE (1999) Double-strand break repair can lead to high frequencies of deletions within short CAG/CTG trinucleotide repeats. Mol Gen Genet 261:871–882

Richard GF, Goellner GM, McMurray CT, Haber JE (2000) Recombination-induced CAG trinucleotide repeat expansions in yeast involve the MRE11-RAD50-XRS2 complex. EMBO J 19:2381–2390

Richard GF, Kerrest A, Dujon B (2008) Comparative genomics and molecular dynamics of DNA repeats in eukaryotes. Microbiol Mol Biol Rev 72:686–727

Rosche W, Jaworski A, Kang S, Kramer S, Larson J, Geidroc D, Wells R, Sinden R (1996) Single-stranded DNA-binding protein enhances the stability of CTG triplet repeats in *Escherichia coli* [published erratum appears in J Bacteriol 1996 Dec;178(23):7024]. J Bacteriol 178:5042–5044

Rubin AF, Green P (2009) Mutation patterns in cancer genomes. Proc Natl Acad Sci USA 106:21766–21770

Ruden DM, Garfinkel MD, Sollars VE, Lu X (2003) Waddington's widget: Hsp90 and the inheritance of acquired characters. Semin Cell Dev Biol 14:301–310
Rutherford SL, Lindquist S (1998) Hsp90 as a capacitor for morphological evolution. Nature 396:336–342
Salinas-Rios V, Belotserkovskii BP, Hanawalt PC (2011) DNA slip-outs cause RNA polymerase II arrest in vitro: potential implications for genetic instability. Nucleic Acids Res 39(17):7444–7454
Sangster TA, Bahrami A, Wilczek A, Watanabe E, Schellenberg K, McLellan C, Kelley A, Kong SW, Queitsch C, Lindquist S (2007) Phenotypic diversity and altered environmental plasticity in *Arabidopsis thaliana* with reduced Hsp90 levels. PLoS One 2:e648
Sangster TA, Salathia N, Undurraga S, Milo R, Schellenberg K, Lindquist S, Queitsch C (2008) HSP90 affects the expression of genetic variation and developmental stability in quantitative traits. Proc Natl Acad Sci USA 105:2963–2968
Sarkari J, Pandit N, Moxon ER, Achtman M (1994) Variable expression of the Opc outer membrane protein in Neisseria meningitidis is caused by size variation of a promoter containing poly-cytidine. Mol Microbiol 13:207–217
Savouret C, Brisson E, Essers J, Kanaar R, Pastink A, te Riele H, Junien C, Gourdon G (2003) CTG repeat instability and size variation timing in DNA repair-deficient mice. EMBO J 22:2264–2273
Sawyer LA, Hennessy JM, Peixoto AA, Rosato E, Parkinson H, Costa R, Kyriacou CP (1997) Natural variation in a Drosophila clock gene and temperature compensation. Science 278:2117–2120
Schulte TW, Neckers LM (1998) The benzoquinone ansamycin 17-allylamino-17-demethoxygeldanamycin binds to HSP90 and shares important biologic activities with geldanamycin. Cancer Chemother Pharmacol 42:273–279
Schumacher S, Pinet I, Bichara M (2001) Modulation of transcription reveals a new mechanism of triplet repeat instability in *Escherichia coli*. J Mol Biol 307:39–49
Schweitzer JK, Livingston DM (1997) Destabilization of CAG trinucleotide repeat tracts by mismatch repair mutations in yeast. Hum Mol Genet 6:349–355
Schweitzer JK, Livingston DM (1999) The effect of DNA replication mutations on CAG tract stability in yeast. Genetics 152:953–963
Sekimoto T, Oda T, Pozo FM, Murakumo Y, Masutani C, Hanaoka F, Yamashita T (2010) The molecular chaperone Hsp90 regulates accumulation of DNA polymerase eta at replication stalling sites in UV-irradiated cells. Mol Cell 37:79–89
Shelbourne PF, Keller-McGandy C, Bi WL, Yoon SR, Dubeau L, Veitch NJ, Vonsattel JP, Wexler NS, Arnheim N, Augood SJ (2007) Triplet repeat mutation length gains correlate with cell-type specific vulnerability in Huntington disease brain. Hum Mol Genet 16:1133–1142
Simon M, Hancock JM (2009) Tandem and cryptic amino acid repeats accumulate in disordered regions of proteins. Genome Biol 10:R59
Sinden RR, Potaman VN, Oussatcheva EA, Pearson CE, Lyubchenko YL, Shlyakhtenko LS (2002) Triplet repeat DNA structures and human genetic disease: dynamic mutations from dynamic DNA. J Biosci 27:53–65
Sollars V, Lu X, Xiao L, Wang X, Garfinkel MD, Ruden DM (2003) Evidence for an epigenetic mechanism by which Hsp90 acts as a capacitor for morphological evolution. Nat Genet 33:70–74
Sopher BL, Ladd PD, Pineda VV, Libby RT, Sunkin SM, Hurley JB, Thienes CP, Gaasterland T, Filippova GN, La Spada AR (2011) CTCF regulates ataxin-7 expression through promotion of a convergently transcribed, antisense noncoding RNA. Neuron 70:1071–1084
Soragni E, Herman D, Dent SY, Gottesfeld JM, Wells RD, Napierala M (2008) Long intronic GAA*TTC repeats induce epigenetic changes and reporter gene silencing in a molecular model of Friedreich ataxia. Nucleic Acids Res 36:6056–6065
Specchia V, Piacentini L, Tritto P, Fanti L, D'Alessandro R, Palumbo G, Pimpinelli S, Bozzetti MP (2010) Hsp90 prevents phenotypic variation by suppressing the mutagenic activity of transposons. Nature 463:662–665

Steinbach P, Glaser D, Vogel W, Wolf M, Schwemmle S (1998) The DMPK gene of severely affected myotonic dystrophy patients is hypermethylated proximal to the largely expanded CTG repeat. Am J Hum Genet 62:278–285

Strong TV, Tagle DA, Valdes JM, Elmer LW, Boehm K, Swaroop M, Kaatz KW, Collins FS, Albin RL (1993) Widespread expression of the human and rat Huntington's disease gene in brain and nonneural tissues. Nat Genet 5:259–265

Subramanian S, Mishra RK, Singh L (2003) Genome-wide analysis of microsatellite repeats in humans: their abundance and density in specific genomic regions. Genome Biol 4:R13

Swami M, Hendricks AE, Gillis T, Massood T, Mysore J, Myers RH, Wheeler VC (2009) Somatic expansion of the Huntington's disease CAG repeat in the brain is associated with an earlier age of disease onset. Hum Mol Genet 18:3039–3047

Taipale M, Jarosz DF, Lindquist S (2010) HSP90 at the hub of protein homeostasis: emerging mechanistic insights. Nat Rev Mol Cell Biol 11:515–528

Tariq M, Nussbaumer U, Chen Y, Beisel C, Paro R (2009) Trithorax requires Hsp90 for maintenance of active chromatin at sites of gene expression. Proc Natl Acad Sci USA 106:1157–1162

Tassone F, Hagerman RJ, Gane LW, Taylor AK (1999) Strong similarities of the FMR1 mutation in multiple tissues: postmortem studies of a male with a full mutation and a male carrier of a premutation. Am J Med Genet 84:240–244

Taylor AK, Tassone F, Dyer PN, Hersch SM, Harris JB, Greenough WT, Hagerman RJ (1999) Tissue heterogeneity of the FMR1 mutation in a high-functioning male with fragile X syndrome. Am J Med Genet 84:233–239

Thibodeau SN, Bren G, Schaid D (1993) Microsatellite instability in cancer of the proximal colon. Science 260:816–819

Tome S, Panigrahi GB, Lopez Castel A, Foiry L, Melton DW, Gourdon G, Pearson CE (2011) Maternal germline-specific effect of DNA ligase I on CTG/CAG instability. Hum Mol Genet 20:2131–2143

Travers J, Sharp S, Workman P (2012) HSP90 inhibition: two-pronged exploitation of cancer dependencies. Drug Discov Today 17:242–252

van der Woude MW, Baumler AJ (2004) Phase and antigenic variation in bacteria. Clin Microbiol Rev 17:581–611, table of contents

van Ham SM, van Alphen L, Mooi FR, van Putten JP (1993) Phase variation of H. influenzae fimbriae: transcriptional control of two divergent genes through a variable combined promoter region. Cell 73:1187–1196

Verstrepen KJ, Jansen A, Lewitter F, Fink GR (2005) Intragenic tandem repeats generate functional variability. Nat Genet 37:986–990

Vinces MD, Legendre M, Caldara M, Hagihara M, Verstrepen KJ (2009) Unstable tandem repeats in promoters confer transcriptional evolvability. Science 324:1213–1216

Wang J, Wang Y, Lu L (2012) De-SUMOylation of CCCTC binding factor (CTCF) in hypoxic stress-induced human corneal epithelial cells. J Biol Chem 287:12469–12479

Waterland RA, Jirtle RL (2003) Transposable elements: targets for early nutritional effects on epigenetic gene regulation. Mol Cell Biol 23:5293–5300

Waterland RA, Dolinoy DC, Lin JR, Smith CA, Shi X, Tahiliani KG (2006) Maternal methyl supplements increase offspring DNA methylation at Axin Fused. Genesis 44:401–406

Waterland RA, Kellermayer R, Laritsky E, Rayco-Solon P, Harris RA, Travisano M, Zhang W, Torskaya MS, Zhang J, Shen L et al (2010) Season of conception in rural gambia affects DNA methylation at putative human metastable epialleles. PLoS Genet 6:e1001252

Weis SM, Cheresh DA (2011) Tumor angiogenesis: molecular pathways and therapeutic targets. Nat Med 17:1359–1370

Wells RD (2007) Non-B DNA conformations, mutagenesis and disease. Trends Biochem Sci 32:271–278

Whitesell L, Lindquist SL (2005) HSP90 and the chaperoning of cancer. Nat Rev Cancer 5:761–772

Wierdl M, Greene CN, Datta A, Jinks-Robertson S, Petes TD (1996) Destabilization of simple repetitive DNA sequences by transcription in yeast. Genetics 143:713–721

Wilburn B, Rudnicki DD, Zhao J, Weitz TM, Cheng Y, Gu X, Greiner E, Park CS, Wang N, Sopher BL et al (2011) An antisense CAG repeat transcript at JPH3 locus mediates expanded polyglutamine protein toxicity in Huntington's disease-like 2 mice. Neuron 70:427–440

Willems R, Paul A, van der Heide HG, ter Avest AR, Mooi FR (1990) Fimbrial phase variation in Bordetella pertussis: a novel mechanism for transcriptional regulation. EMBO J 9:2803–2809

Willemsen R, Levenga J, Oostra BA (2011) CGG repeat in the FMR1 gene: size matters. Clin Genet 80:214–225

Wohrle D, Hennig I, Vogel W, Steinbach P (1993) Mitotic stability of fragile X mutations in differentiated cells indicates early post-conceptional trinucleotide repeat expansion. Nat Genet 4:140–142

Wohrle D, Salat U, Glaser D, Mucke J, Meisel-Stosiek M, Schindler D, Vogel W, Steinbach P (1998) Unusual mutations in high functioning fragile X males: apparent instability of expanded unmethylated CGG repeats. J Med Genet 35:103–111

Wolff GL, Kodell RL, Moore SR, Cooney CA (1998) Maternal epigenetics and methyl supplements affect agouti gene expression in Avy/a mice. FASEB J 12:949–957

Wright BE (2004) Stress-directed adaptive mutations and evolution. Mol Microbiol 52:643–650

Yao Y, Kovalchuk I (2011) Abiotic stress leads to somatic and heritable changes in homologous recombination frequency, point mutation frequency and microsatellite stability in Arabidopsis plants. Mutat Res 707:61–66

Yao Q, Weigel B, Kersey J (2007) Synergism between etoposide and 17-AAG in leukemia cells: critical roles for Hsp90, FLT3, topoisomerase II, Chk1, and Rad51. Clin Cancer Res 13:1591–1600

Yelin R, Dahary D, Sorek R, Levanon EY, Goldstein O, Shoshan A, Diber A, Biton S, Tamir Y, Khosravi R et al (2003) Widespread occurrence of antisense transcription in the human genome. Nat Biotechnol 21:379–386

Yoon SR, Dubeau L, de Young M, Wexler NS, Arnheim N (2003) Huntington disease expansion mutations in humans can occur before meiosis is completed. Proc Natl Acad Sci USA 100:8834–8838

Yuan GC, Liu YJ, Dion MF, Slack MD, Wu LF, Altschuler SJ, Rando OJ (2005) Genome-scale identification of nucleosome positions in *S. cerevisiae*. Science 309:626–630

Zhao R, Houry WA (2007) Molecular interaction network of the Hsp90 chaperone system. Adv Exp Med Biol 594:27–36

Chapter 8
Genetic Instability Induced by Hypoxic Stress

Susan E. Scanlon and Peter M. Glazer

Abstract Lack of sufficient oxygen, known as hypoxia, constitutes a significant physiological stress, particularly relevant to cancer biology. In similarity to many other stresses, hypoxia induces an integrated response that both adapts cellular machinery to the stress and generates genomic instability, which can promote accelerated genetic evolution. The cellular response to hypoxia, mediated by hypoxia-inducible factor (HIF) signaling and HIF-independent signaling pathways, drives changes in metabolism, oxygen supply, and cell growth. In concert with these adaptations, genetic instability occurs through the down-regulation of DNA repair pathways—particularly nucleotide excision repair, mismatch repair, and homologous recombination repair—and increased DNA damage rates, especially during reoxygenation. To prevent mutagenesis, cells respond to DNA damage with the activation of cell cycle checkpoints and apoptotic pathways, which thus interconnect closely with the hypoxic stress response. As a pervasive feature of solid tumors, hypoxia may promote cancer progression through its impact on genomic integrity.

Introduction

When exposed to a wide variety of exogenous stressors, lower organisms, including bacteria and yeast, often exhibit enhanced mutagenesis as part of their cellular stress response (Rosenberg 2001; Galhardo et al. 2007). This phenomenon of stress-induced mutagenesis promotes more rapid genetic change, allowing a population of cells to better adapt to the stress as those with favorable mutations preferentially survive.

This work was supported by NIH MSTP TG T32GM07205 and by NIH grants R01ES005775, R01CA148996, and P01CA129186 to P.M.G.

S.E. Scanlon • P.M. Glazer (✉)
Department of Therapeutic Radiology,
Yale University School of Medicine, New Haven, CT, USA
e-mail: peter.glazer@yale.edu

D. Mittelman (ed.), *Stress-Induced Mutagenesis*, DOI 10.1007/978-1-4614-6280-4_8

A possible evolutionary vestige of this process occurs in mammalian biology in the form of cancer. Tumor cells often exist under adverse conditions, including low oxygen and nutrient availability and decreased pH, commonly referred to as the "tumor microenvironment" (Yuan and Glazer 1998). Concurrently, many malignant cells exhibit a mutator phenotype, generating many more mutations than can be accounted for by the basal mutation rate (Loeb 2001). This genomic instability is now recognized as an "enabling characteristic" of cancer that allows for the selection of cells proficient for growth, invasion, and metastasis (Hanahan and Weinberg 2011). Though defects in genes responsible for caretaking of the genome may sometimes be present, the stresses of the tumor microenvironment, particularly hypoxia, have also been implicated in promoting genomic instability (Yuan and Glazer 1998; Vaupel et al. 2004; Bindra and Glazer 2005; Huang et al. 2007). Knowledge of the ways in which hypoxic stress induces genetic instability is critical to fully understanding cancer progression, and cellular stress responses in general.

This chapter begins by presenting the observations that tumors nearly universally contain hypoxic areas and that hypoxia is associated with genomic instability. After reviewing the general cellular response to hypoxia, the mechanisms by which hypoxia induces genetic instability are discussed in detail. Finally, the regulation of hypoxia-induced mutagenesis in the context of DNA damage checkpoint signaling and apoptotic pathways is considered.

Observations of Tumor Hypoxia and Genomic Instability

The first suggestions that tumors may contain regions of hypoxia were made in the 1930s based on the observations that tumors contain cells resistant to gamma radiation and that malignant cells cultured in anaerobic conditions were similarly insensitive to radium (Crabtree and Cramer 1933). Radioresistant cells were moreover found at the center of tumors and in tumors with scanty stroma, suggesting that diminished blood oxygen supply was responsible for their phenotype (Mottram 1936). Histological observations of tumors later revealed a pattern of central necrosis at the center of tumor cords and the absence of intact tumor cells further than 180 μm from a blood vessel (Thomlinson and Gray 1955). Based on calculations that a complete absence of oxygen, or anoxia, would occur approximately 150 μm from a vessel, it was hypothesized that hypoxia develops in tumors as they grow and outstrip their blood supply. The resistance of hypoxic cells to radiation was used to demonstrate that most transplanted mouse tumors and human xenografted tumors contain significant hypoxic fractions (Rockwell and Moulder 1990; Moulder and Rockwell 1987). Modern techniques of oxygen measurement, such as invasive polarographic or luminescence-based microelectrodes, various magnetic resonance imaging procedures, and immunohistochemistry of endogenous or exogenous hypoxic markers, have confirmed that while normal tissues have an oxygen tension of 24–66 mmHg, most tumors exhibit values of less than 20 mmHg (Vaupel et al. 1989, 2004; Vaupel 2004).

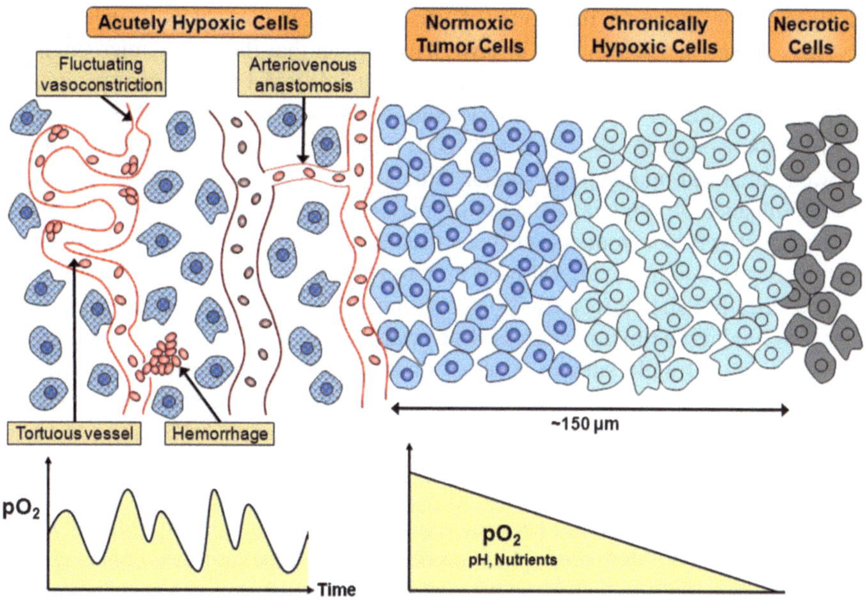

Fig. 8.1 The hypoxic tumor microenvironment

In addition to the chronic hypoxia due to abnormal blood vessel density, it also became apparent that some tumor cells undergo hypoxic stress on a shorter timescale. In the late 1970s, observations of radioresistant cells at the outer well-vascularized rim of tumors that had not previously been hypoxic led researchers to propose that "acutely hypoxic cells are produced by intermittent opening and closing of tumor blood vessels" (Brown 1979). Experiments using microspheres with two different fluorescent agents injected into mice 20 min apart demonstrated that, indeed, tumor vessels have fluctuating patency (Chaplin et al. 1987). Intravascular imaging techniques of hemoglobin saturation have revealed further complexity as microvascular changes, including tortuous vessels, arteriovenous anastomoses, and regions of hemorrhage, occur very early in tumor development and have functional consequences on tumor oxygenation (Huang et al. 2007). Thus, tumor hypoxia comprises both diffusion-limited chronic hypoxia and perfusion-limited acute hypoxia with the possibility of hypoxia–reoxygenation cycles (Fig. 8.1).

The tumor microenvironment, as well as artificially imposed hypoxia, has been implicated in a variety of DNA lesions, including gene amplification, point mutation, deletion, fragile site induction, and microsatellite instability (MSI), both in mouse models and in vitro [reviewed in (Yuan and Glazer 1998; Bindra and Glazer 2005)]. The first reports of hypoxia-induced genetic instability demonstrated that the exposure of Chinese hamster ovary cells to hypoxia, followed by return to normoxia, led to DNA overreplication, amplification of the dihydrofolate reductase (DHFR) gene, and resultant methotrexate resistance (Rice et al. 1986). Several experiments comparing the mutation frequency of various reporter constructs in cells grown in vitro, versus in mice tumors in vivo, have shown that the tumor microenvironment can enhance overall mutation frequency by 3.4- to 5-fold and can specifically increase the level of deletions, transversion mutations, and MSI (Reynolds et al. 1996; Wilkinson et al. 1995; Li et al. 2001). In one case, cells grown in hypoxia showed enhanced mutagenesis similar to those grown in tumors, pointing towards the significance of hypoxia in the tumor microenvironment (Reynolds et al. 1996). More recently, flow cytometry has been used to show increased deletion mutations in cells exposed to hypoxia, followed by reoxygenation (Keysar et al. 2010), while cytogenetics has shown increased sister chromatid exchange abnormalities in cells that have undergone hypoxic stress (Lee et al. 2010). Finally, hypoxia can induce DNA breaks at fragile sites to generate double minute chromosomes (DMs) and other DNA rearrangements (inversions, ring chromosomes, and translocations), can promote the fusion and amplification of DMs to form larger DMs, and can bring about the chromosomal integration of DMs at other fragile sites (Coquelle et al. 1998, 2002; Fischer et al. 2008).

In support of a connection between oxygen deprivation, genetic instability, and cancer progression, DNA overreplication occurring in hypoxia-treated fibrosarcoma and melanoma cell lines enhanced the potential of the cells to form metastatic colonies in mice (Young et al. 1988). Cycles of acute hypoxia imposed exogenously upon mice also increased metastasis from developing KHT tumors (Cairns et al. 2001). Finally, human tumor hypoxia may lead to unfavorable disease outcomes as more than ten different clinical studies have found the level of tumor hypoxia to be an independent prognostic factor for decreased survival or increased metastasis in cervical cancer, head and neck carcinomas, and soft tissue sarcomas [reviewed in (Vaupel and Mayer 2007)]. Thus, over the past 80 years, it has become clear that hypoxia is prevalent in tumors, promotes genomic instability, and may contribute to cancer progression.

Cellular Response to Hypoxia

When cells experience hypoxic stress, they activate a variety of responsive mechanisms including angiogenesis, cell cycle checkpoints, and changes in cellular metabolism and RNA and protein synthesis [reviewed in (Majmundar et al. 2010; Nakayama 2009; Kaluz et al. 2008)]. Severe and prolonged hypoxia leads to cell death by apoptosis or necrosis. Many of these cellular responses are modulated by hypoxia-inducible factors (HIFs) though important HIF-independent pathways respond as well (Fig. 8.2). Chromatin modifications and microRNAs (miRNAs), either HIF-dependent or HIF-independent, also play important roles in the cellular response to hypoxia (Fig. 8.3).

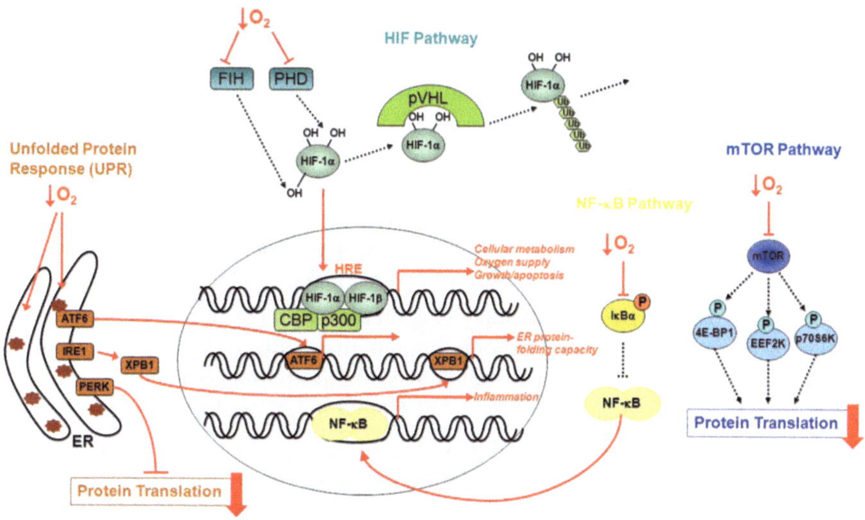

Fig. 8.2 HIF and HIF-independent signaling pathways activated in hypoxia

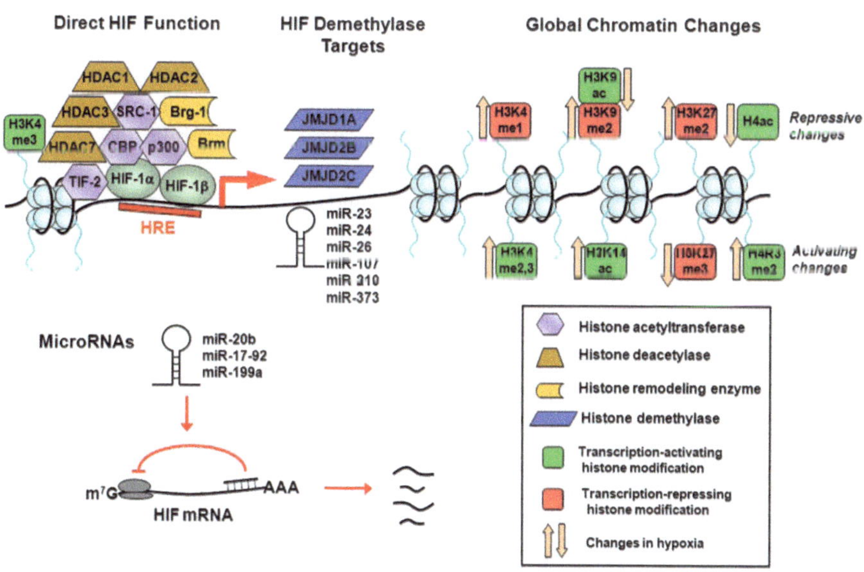

Fig. 8.3 Roles of epigenetics and microRNAs in hypoxia

Hypoxia-Inducible Factor Signaling

HIFs, heterodimers composed of α and β subunits, were first identified as factors that stimulate transcription of the erythropoietin (EPO) gene by binding to its enhancer region in hypoxic conditions (Wang and Semenza 1993). The 120 kDa α-subunit has three isoforms, HIF-1α, HIF-2α, and HIF-3α, while the 91–94 kDa β-subunit is one of three paralogs, HIF-1β/ARNT1, ARNT2, and ARNT3, which are equivalent to the aryl hydrocarbon nuclear translocator proteins that also dimerize with the aryl hydrocarbon receptor (AhR) (Loboda et al. 2010). HIF-1 (consisting of HIF-1α and HIF-1β subunits) is expressed in most tissues and cell types while the other four subunits have only limited, tissue-specific expression (Loboda et al. 2010). Both HIF-1 subunits contain basic helix-loop-helix (bHLH) and PER/ARNT/SIM (PAS) domains responsible for DNA-binding and dimerization, respectively, and C-terminal and N-terminal activation domains (CAD, NAD) which mediate transcriptional activation (Loboda et al. 2010).

Oxygen tension-mediated regulation of HIF activity proceeds through the HIF-α subunit while the HIF-β subunit is constitutively active (Huang et al. 1996). Under normoxic conditions, HIF-α is targeted for degradation by a proyl-4-hydroxylase domain-containing enzyme (PHD), which hydroxylates two conserved proline residues in an oxygen-dependent degradation (ODD) domain of HIF-α (Jaakkola et al. 2001; Huang et al. 1998). HIF-α proline hydroxylation promotes its association with the von Hippel-Lindau tumor suppressor, pVHL, a component of an E3 ubiquitin ligase complex (Maxwell et al. 1999), and ubiquitination of HIF-α in turn leads to its degradation by the proteosome (Cockman et al. 2000). The PHDs that hydroxylate HIF-α require Fe(II) as a cofactor and oxygen as a substrate and, consequently, are inhibited by hypoxia or iron chelators, which mimic the effects of hypoxia (Jaakkola et al. 2001). Therefore, in hypoxia, HIF-α is stabilized, dimerizes with HIF-β, translocates to the nucleus, binds to hypoxia response elements (HREs) in the promoter regions of certain genes, recruits the CBP/p300 coactivators via its CAD domain, and up-regulates the target gene transcription (Kallio et al. 1998). Several additional factors further modulate HIF activation, including an oxygen-dependent asparaginyl hydroxylase (FIH), which modifies HIF's CAD preventing its interaction with CBP/p300; multiple kinases, which phosphorylate HIF; sirtuins, which deacetylate HIF; and many HIF-binding proteins, which affect its stabilization (Majmundar et al. 2010; Wenger et al. 2005). Although hypoxia is the primary stimulus for HIF activation, these additional mechanisms may lead to cell type and situation-specific activity.

There are about 2.5 million HRE consensus sequences in the human genome, yet chromatin immunoprecipitation microarray analysis (ChIP-chip) with HIF-1 antibody identified only several hundred HIF-1 binding sites (Xia et al. 2009), suggesting further regulation by non-conserved flanking sequences or by chromatin structure. Approximately 70 confirmed genes are direct HIF targets, many of which fall into the categories of cellular metabolism, oxygen supply, and cell growth and apoptosis (Wenger et al. 2005). In order to enact a shift towards non-oxidative metabolic pathways, HIF-1 increases the expression of glucose transporters and glycolytic enzymes, while HIF-2 inhibits fatty acid oxidation and promotes fat storage (Semenza et al. 1994; Rankin et al. 2009). HIF-1 induces the expression of many genes to promote angiogenesis and erythropoiesis, such as VEGF and its receptor, EPO; transferrin; and heme oxygenase (Wenger et al. 2005). The role of HIF-1 in growth and apoptosis is less distinct, as it up-regulates both pro- and anti-growth genes, and pro- and anti-apoptotic genes (Wenger et al. 2005). In addition to its direct coactivation of genes containing HREs, HIF-1 modulates the expression of many more by up-regulating other transcription factors such as DEC1 and DEC2 (Miyazaki et al. 2002). Lastly, independent of its DNA-binding and transcriptional activity, HIF-1 can interact with and antagonize the function of Myc, and for example, leads to cell cycle arrest by displacing the repressive Myc from the promoter of the cyclin-dependent kinase inhibitor $p21^{Cip1}$ (Koshiji et al. 2004).

HIF-Independent Signaling Pathways

Hypoxia impacts several HIF-independent pathways, three of which are the unfolded protein response (UPR), the mTOR pathway, and the NF-κB pathway, which also control gene transcription and protein translation [reviewed in (Taylor and Cummins 2009; Wouters and Koritzinsky 2008)]. Severe hypoxia induces the UPR as oxygen depletion is hypothesized to inhibit certain endoplasmic reticulum (ER) functions such as oligosaccharide modifications, disulfide bond formation, and protein isomerization and cause misfolded proteins to accumulate in the ER (Wouters and Koritzinsky 2008). Briefly, UPR involves the dissociation of three ER stress sensors, PERK, IRE1, and ATF6, from the chaperone BiP or their direct binding to unfolded proteins with their downstream signaling cascades leading to translation attenuation and transcriptional activation of genes that serve to improve ER protein folding capacity. In energy rich conditions, the mTOR pathway promotes cell growth via phosphorylation of eukaryotic initiation factor 4E binding protein 1 (4E BP1), eukaryotic elongation factor 2 kinase (EEF2K), and ribosomal protein S6 kinase (p70S6K), all of which enhance protein translation. Under hypoxic stress, several different oxygen-sensing pathways converge to inhibit mTOR, resulting in slowed protein translation and cell growth (Wouters and Koritzinsky 2008). Finally, hypoxia moderately up-regulates the NF-κB pathway via phosphorylation of inhibitor of NF-κB α (IκBα), allowing NF-κB to translocate to the nucleus and up-regulate transcription of inflammatory genes (Koong et al. 1994; Culver et al. 2010). The precise oxygen-sensing mechanism, however, remains unclear.

Transcription and Chromatin Alterations

The combination of HIF and HIF-independent signaling in hypoxic stress leads to significant global changes in transcription. Under hypoxic conditions of 0.2% oxygen, overall RNA synthesis was found to decrease 42 and 65% after 24 and 48 h respectively (Johnson et al. 2008). Transcriptome profiling by microarray analysis has shown that after 24 h hypoxia in vitro, about 5% of genes are up-regulated greater than twofold, and about 10% are down-regulated greater than twofold, while about 85% are not altered compared to normoxic cells (Bindra et al. 2004). Laser-capture microdissection of gliomas to study gene expression of hypoxic tumor cells in vivo demonstrated that even more genes are altered in vivo than in vitro (Marotta et al. 2011). These studies confirmed that many of the HIF-induced genes are up-regulated and identified certain DNA repair genes that are down-regulated (discussed later).

Epigenetics plays an important role in the hypoxic stress response through several different mechanisms: direct HIF-1 function, HIF-1 demethylase targets, and global chromatin changes (Watson et al. 2010). First, in transactivating its target genes, HIF-1 recruits many different chromatin-modifying enzymes, in addition to its strong association with CBP/p300. Histone acetyltransferases (CBP/p300, SRC-1, TIF-2), histone deacetylases (HDAC1, 2, 3, 7), and histone remodeling enzymes (Brm, Brg-1) can all contribute to the expression of HIF target genes (Wenger et al. 2005). In addition, the initial structure of chromatin has shown to be a factor in determining to which of the approximately 2.5 million HRE consensus sites HIF-1 binds (Xia and Kung 2009). Under acute hypoxia, HIF-1 binds preferentially to transcriptionally active genes marked by histone H3 lysine 4 trimethylation (H3K4me3), possibly contributing to cell type specific hypoxia-induced gene expression (Xia and Kung 2009).

The second manifestation of epigenetic regulation in hypoxia occurs through histone H3 lysine 9 (H3K9) demethylases of the Jumonji family (Watson et al. 2010). Hypoxia causes HRE- and HIF-dependent increases in the mRNA and protein levels of three Jumonji genes, JMJD1A, JMJD2B, and JMJD2C (Wellmann et al. 2008; Pollard et al. 2008; Beyer et al. 2008), and several hypoxia-inducible genes require JMJD1A-dependent decreased H3K9 methylation for their expression (Krieg et al. 2010). However, Jumonji demethylases, like PHDs, require oxygen and Fe(II) for their catalytic activity, and thus are functionally inhibited by hypoxia (Tausendschon et al. 2011). Increased H3K9 methylation has also been reported in hypoxia and found to repress certain genes (Tausendschon et al. 2011; Chen et al. 2010). A possible explanation for these apparent contradictions is that Jumonji demethylase activity may be dynamically regulated based on the severity and duration of hypoxia, with mild hypoxia up-regulating their activity and the expression of specific hypoxia-responsive genes and severe or extended hypoxia down-regulating their activity and overall transcription.

Finally, hypoxic stress results in a wide range of global chromatin alterations. Histone modifications typically associated with transcriptional activation that have been reported in hypoxia include increased H3K4me2, H3K4me3, H3K14 acetylation (H3K14ac), and H4R3me2 and decreased H3K27me3, while those generally coupled to transcriptional repression include increased H3K4me1, H3K9me2, and H3K27me2 and decreased H3K9ac and H4ac (Johnson et al. 2008; Zhou et al. 2010; Li and Costa 2009; Chen et al. 2006). In hypoxia, some of these alterations correlate specifically with increased or decreased transcription, whereas others are present regardless of the transcriptional status. For example, increased H3K4me3 and decreased H3K27me3 are found at the promoters of genes that are both up- and down-regulated in hypoxia (Johnson et al. 2008). These gene-activating alterations, together with global repressive modifications, might serve to "poise hypoxia-responsive chromatin and facilitate downstream flexibility in response to induced or inhibited trans-acting regulators" (Johnson et al. 2008). In most cases, the mechanisms underlying these hypoxia-induced histone modifications are unknown; however, increased H3K9me2 is associated with heightened G9a methyltransferase activity and potentially diminished Jumonji family demethylase activity (Chen et al. 2006), and increased H3K4me3 is associated with decreased activity of the demethylase JARID1A, without changes in its mRNA or protein level (Zhou et al. 2010). Finally, hypoxia also causes alterations in global DNA methylation. Interestingly, after 24 h of severe hypoxia, colorectal carcinoma cell lines exhibited DNA hypomethylation (Shahrzad et al. 2005), whereas chronically and mildly hypoxic benign prostate cells had DNA hypermethylation which correlated with up-regulation of the DNA methyltransferase DNMT3b (Watson et al. 2009). Thus, the severity and duration of hypoxic stress should be expected to affect chromatin changes as well. The significance of many of these hypoxia-induced epigenetic changes remains to be elucidated.

MicroRNAs

MicroRNAs (miRNAs), which offer the advantages of rapid, reversible, and localized protein expression regulation, have recently been recognized as important contributors to the hypoxic stress response (Pocock 2011). Studies have found in excess of 90 miRNAs regulated by hypoxia, though only a few of these have been studied in detail (Pocock 2011). HIF directly up-regulates at least six miRNAs (miR-23, 24, 26, 107, 210, 373) and is itself regulated during hypoxia by additional miRNAs (miR-199a, 17-92, 20b) (Huang et al. 2010). Of all of these, miR-210 is most strongly induced by HIF and has been ascribed the most diverse functions in hypoxia [reviewed in (Devlin et al. 2011)]. MiR-210 plays a role in the switch to anaerobic glycolysis by decreasing Krebs cycle and electron transport chain enzyme activity by down-regulating the iron sulfur scaffold protein (ISCU) required for assembly of iron–sulfur clusters (Chan et al. 2009; Favaro et al. 2010). It also contributes to angiogenesis by stimulating endothelial cell migration via down-regulation of the receptor tyrosine kinase ligand Ephrin-A3 (Fasanaro et al. 2008). In addition, miR-210 and miR-373 both target DNA repair proteins (discussed later). Notably, increased miR-210 levels in tumors in vivo correlate with a worse prognosis in breast, head, and neck cancers (Camps et al. 2008; Gee et al. 2010), highlighting the importance of understanding the significance of the "miRNA signature" (Kulshreshtha et al. 2007) in the hypoxic stress response.

Mechanisms of Genetic Instability: Decreased DNA Repair

Genetic instability arises when DNA damage exceeds the capacity of DNA repair mechanisms. Under hypoxic stress, evidence exists for both increased DNA damage and an impairment of three of the major DNA repair pathways: mismatch repair (MMR), nucleotide excision repair (NER), and DNA double-strand break (DSB) repair. Moreover, the genetic instability observed under hypoxic stress appears to be an integrated part of the cellular response to hypoxia. This section and the next discuss how the hypoxic stress response can lead to DNA repair deficiencies and an increase in DNA damage during hypoxia (Fig. 8.4).

Mismatch Repair

MMR, a DNA repair pathway responsible for correcting DNA base mismatches and small insertions and deletions, is mediated primarily by hMutSα and hMutLα, composed of MSH2-MSH6 and MLH1-PMS1 heterodimers respectively [reviewed in (Li 2008)]. hMutSα recognizes the DNA mismatches while hMutLα contains an endonuclease activity that can generate an incision 5′ to the mismatch. The exonuclease ExoI, single-stranded DNA binding protein RPA, DNA polymerase Pol δ, and DNA ligase I are involved in the downstream steps of mismatch excision, DNA resynthesis, and nick ligation. Since MMR repair is critical for repairing replication errors, inadequate MMR leads to an increase in spontaneous point mutations and frameshifts, particularly in simple sequence repeats known as microsatellites. Inherited mutations in MSH2, MLH1, and PMS2 cause hereditary nonpolyposis colon cancer (HNPCC) while inactivation of MLH1 by promoter hypermethylation is often found in sporadic colon cancer. Thus, decreased MMR function can promote genetic instability and enable cancer progression.

It was first discovered that hypoxia could impact the MMR pathway when mouse and human cells subjected to chronic hypoxia in vitro, or to the iron chelator desferoxamine, exhibited decreased MLH1 mRNA and protein levels (Mihaylova et al. 2003). PMS2 protein levels were also decreased, presumably due to its destabilization in the absence of its binding partner MLH1. Subsequently, MSH2 and MSH6 mRNA and protein levels have also been found to be down-regulated by hypoxic stress (Shahrzad et al. 2005; Koshiji et al. 2005; Bindra and Glazer 2007a; Rodriguez-Jimenez et al. 2008). Decreased MSH2 protein levels were also found in mouse xenografted tumors relative to cells grown in normoxic cell culture, and MSH2 immunostaining inversely correlated with hypoxyprobe staining (Shahrzad et al. 2005). Finally, in a mouse model of spontaneous colitis, MMR protein levels are initially normal, but MLH1 and PMS2 decrease after the onset of hypoxic colitis (Edwards et al. 2009).

The hypoxic down-regulation of MLH1 was first reported to be dependent on HDAC activity, and independent of HIF-1 (Mihaylova et al. 2003). However,

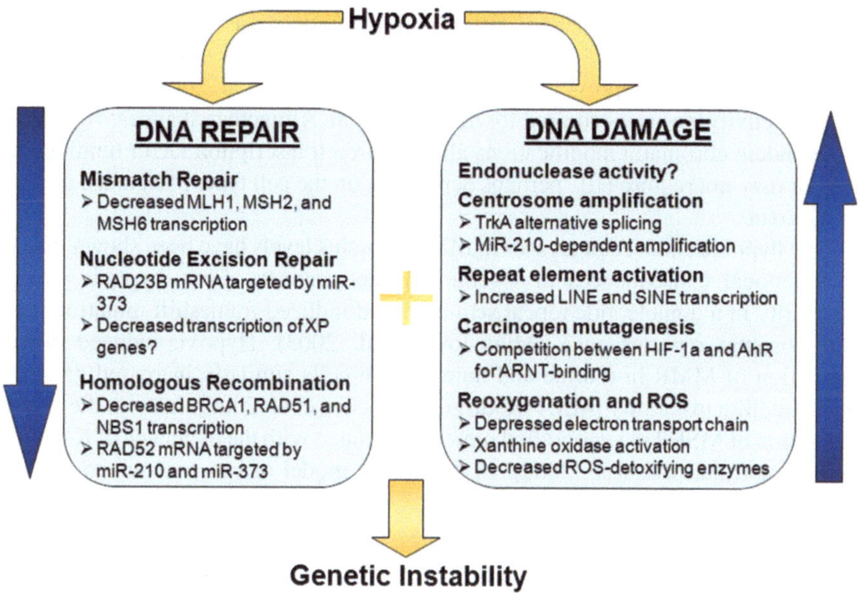

Fig. 8.4 Mechanisms of genetic instability induced by hypoxia

additional studies have suggested that there may be multiple HIF-dependent and HIF-independent pathways of MMR down-regulation in hypoxia. In addition to HDAC activity, MLH1 and MSH2 are regulated (HIF-independently) by changes in the occupancy of their proximal promoter regions by members of the Myc/Max network (Bindra and Glazer 2007a). Under normoxic conditions, the activating c-Myc/Max complex predominantly occupies the promoters, while under hypoxic conditions, a decrease in c-Myc levels induces a shift towards greater binding of the repressive Mad1/Max and Mnt/Max complexes, leading to decreased MLH1 and MSH2 transcription. It is not known how the required HDAC activity and the Myc/Max network are connected. However, a separate study found that in hypoxic stem cells, histone H3 hypoacetylation and hypermethylation at the MLH1 and MSH6 promoters were required to decrease binding of the transcription factor Sp1, and to down-regulate MLH1 and MSH6 transcription (Rodriguez-Jimenez et al. 2008). Therefore, chromatin modifications and altered transcription factor binding likely cooperate in hypoxia-induced MMR gene repression.

Although MLH1, MSH2, and MSH6 down-regulation in hypoxia can occur independently of HIF (Mihaylova et al. 2003; Bindra and Glazer 2007a; Rodriguez-Jimenez et al. 2008), and although constitutively active HIF-1α is not sufficient for MMR depression (Lehtonen et al. 2007), HIF nevertheless has been shown to modulate the down-regulation of all three genes. HIF-1α was found to down-regulate MutSα expression by antagonizing Myc in a p53-dependent manner (Koshiji et al. 2005). Specifically, HIF-1α bound to the transcription factor Sp1 at the MSH2 and MSH6 promoters, displacing the Myc activator and leading to decreased transcription.

In a completely different manner, HIF was found to decrease MLH1 expression via up-regulation of DEC1 and DEC2, which bind to an E-box-like motif in the MLH1 promoter region and repress its transcription (Nakamura et al. 2008). Interestingly, HDAC activity was also required for this repression. Altogether, it appears that HIF-independent chromatin modifications allow altered transcription factor binding that may or may not require HIF, perhaps depending on the cell type, p53 status, or level of hypoxia.

The hypoxia-induced decreases in MMR proteins levels have been shown to lead to functional consequences in several different manners. First, hypoxia caused instability in a dinucleotide repeat sequence and induced frameshift mutations in a chromosomal reporter gene (Mihaylova et al. 2003). Hypoxia-induced down-regulation of MMR in murine and human stem cells similarly increased genomic microsatellite instability (MSI) (Rodriguez-Jimenez et al. 2008). In vivo, the down-regulation of MSH2 in xenografted tumors correlated with the incidence of K-ras^{G13D} mutations (Shahrzad et al. 2005). In the mouse model of spontaneous colitis, the colitic crypts with decreased MMR protein expression frequently progress to colon carcinoma (Edwards et al. 2009). Thus, the down-regulation of MMR induced by hypoxic stress may contribute to genetic instability and cancer progression.

Nucleotide Excision Repair

NER functions to repair a variety of bulky lesions that cause DNA helix distortion, including chemical base adducts, interstrand cross-links, the UV-induced cyclobutane pyrimidine dimers (CPDs) and (6-4) pyrimidine-pyrimidine photoproducts (6-4PPs), and the oxidative species-induced cyclopurines [reviewed in (Nouspikel 2009)]. Inherited mutations in NER genes cause a number of photosensitivity diseases including xeroderma pigmentosum (XP) and Cockayne syndrome (CS). While XP individuals have extremely high risks of developing skin cancer, genetic instability also causes growth failure and neurodevelopmental abnormalities in CS. NER can occur by two main pathways: global genomic repair (GGR) which removes damage from throughout the genome and transcription-coupled repair (TCR) which removes damage from the transcribed strand of active genes. In GGR, the XPC-RAD23B complex recognizes a damaged site and recruits TFIIH, which uses its ATP-dependent helicase activity to unwind the DNA. Next, XPA and RPA bind, followed by XPG and ERCC1-XPF which make incisions on the 3′ and 5′ sides of the damaged site. Finally, DNA polymerase δ or ε synthesizes DNA across the gap and Ligase III seals the nick. In TCR, stalled RNA Pol II recruits CSA, CSB, and XAB2, which in turn recruit TFIIH, bypassing the XPC-RAD23B complex.

Like MMR, the NER capacity of cells decreases under hypoxic stress, but the underlying mechanism is far less understood. To demonstrate diminished NER in hypoxia, a reporter plasmid was pre-irradiated to generate CPDs and 6-4PPs, and the ability of mouse fibroblasts to reactivate the plasmid was measured, thereby isolating NER capacity from potential DNA damage induced by hypoxia (Yuan et al. 2000).

When cells were placed under hypoxic and acidic conditions, their ability to repair and reactivate the reporter plasmid decreased to approximately 50% of that of cells under normal oxygen tension and pH. In addition, cells placed under hypoxic and acidic conditions following a dose of UV irradiation accumulated approximately twice as many mutations, all of which were point mutations and adhered to the UV mutation spectrum. Though these findings demonstrate significantly decreased NER in hypoxia, corresponding changes in the levels of XPA and XPD proteins were not observed (Yuan et al. 2000). More recently, it was found that miR-373 targets RAD23B, suggesting that this HIF-induced miRNA might be the connection, or perhaps just one of multiple connections, between hypoxia and impaired NER (Crosby et al. 2009). Finally, HIF-1α can regulate the expression of XPC in keratinocytes following UVB exposure, both directly by binding an HRE and indirectly through competition with Sp1 (Rezvani et al. 2010). Potential HREs are also located in the XPB, XPD, XPG, CSA, and CSB genes (Rezvani et al. 2010), but how these findings might translate under hypoxia remains to be elucidated.

DNA Double-Strand Break Repair

DNA DSBs, considered the most deleterious form of DNA damage, can be generated by both endogenous and exogenous sources, including ionizing radiation (IR), certain chemotherapeutic drugs, reactive oxygen species, nuclear enzymes, or collapsed replication forks (Khanna and Jackson 2001). The failure to correctly repair DSBs can lead to chromosomal rearrangements, amplifications, deletions, and translocations, as well as to the onset of apoptosis (Khanna and Jackson 2001). Two primary pathways serve to repair DSBs: homologous recombination (HR), which utilizes a homologous sister chromatid present during S or G2 phases to complete error-free repair, and non-homologous end joining (NHEJ), which can occur throughout the cell cycle but is usually error-prone (Khanna and Jackson 2001). HR, a highly regulated process requiring the coordinated actions of many protein factors, involves recognition and resection of the DSB ends, leaving a 3′ single-strand DNA overhang; the formation of a RAD51-ssDNA filament, which mediates invasion of a homologous duplex; homology-directed DNA synthesis; and resolution of the resultant Holliday junctions (San Filippo et al. 2008). In NHEJ, a Ku70/Ku80 heterodimer binds each of the DSB ends and recruits nuclease, polymerase, and ligase enzymes to process and rejoin the ends (Lieber 2010). Cells with defects in many of the DSB repair pathway proteins exhibit hypersensitivity to IR and spontaneous chromosomal aberrations, and, in humans, inherited HR and NHEJ mutations can result in vastly increased predisposition to cancers, including breast, ovarian, and hematologic malignancies (Khanna and Jackson 2001).

Hypoxic stress was first reported to depress cellular HR capacity via the downregulation of RAD51 (Bindra et al. 2004, 2005a). RAD51 mRNA levels were found by microarray analysis to be decreased 4.3-fold under hypoxia, and it was subsequently found in several different human cancer cell lines that RAD51 protein levels were diminished after 24 and 48 h of moderate (0.5% oxygen) or severe (0.01% oxygen) hypoxia and did not return to baseline until 48 h after reoxygenation. Importantly, because HR proteins are known to be physiologically up-regulated during S phase, the decrease in RAD51 was proven to be independent of the cell cycle phase (Bindra et al. 2004, 2005a). Hypoxia-induced down-regulation of RAD51 has subsequently been confirmed in several other studies (Bindra and Glazer 2007b; Meng et al. 2005; Chan et al. 2008). In vivo, RAD51 immunostaining is decreased in hypoxic regions of mouse cervical and prostate cancer xenografts (Bindra et al. 2004), and cells isolated from hypoxic regions of gliomas by laser-capture microdissection have decreased RAD51 mRNA levels (Marotta et al. 2011).

Following the discovery of hypoxia-induced RAD51 depression, a second essential HR factor, BRCA1 was also found to be down-regulated at the mRNA level in hypoxia, independent of the cell cycle (Bindra et al. 2005b). The mechanism underlying BRCA1 hypoxic down-regulation was elucidated and later found to apply to RAD51 down-regulation as well. In log phase cells under normoxic conditions, the activating E2F1 and the repressive E2F4 and p130/p107 transcription factors bind simultaneously to the BRCA1 promoter, resulting in a basal level of repression (Bindra and Glazer 2006). During hypoxia, a shift occurs in which the repressive E2Fs replace the activating the E2Fs at the BRCA1 promoter (Bindra et al. 2005b). RAD51 contains an E2F-binding site that is identical to the one in the BRCA1 promoter, and hypoxia similarly induces greater E2F4/p130 and less E2F1 binding, promoted by the dephosphorylation and nuclear accumulation of p130 (Bindra and Glazer 2007b). RAD51 and BRCA1 hypoxic down-regulation occurs in a HIF-independent manner (Bindra et al. 2004, 2005b; Bindra and Glazer 2007b). Instead, hypoxia-induced histone modifications at the RAD51 and BRCA1 promoters, specifically decreased H3K4 methylation, increased H3K9 methylation, and decreased H3K9 acetylation, promote their silencing (Lu et al. 2011). The E2F4/p130 complexes were neither necessary nor sufficient for the histone modifications, which persisted stably until cells were treated with an HDAC inhibitor. Thus, as seen with MMR, chromatin modifications and altered transcription factor binding play important roles in hypoxia-induced HR protein depression, perhaps with transcription factors generating immediate down-regulation, and chromatin modifications accounting for long-term silencing (Lu et al. 2011).

Importantly, hypoxic down-regulation of RAD51 and BRCA1 correlates with functional decreases in cellular HR capacity. Using a recombination assay with a mutated shuttle vector plasmid and a wild-type donor DNA fragment, recombination frequencies were shown to decrease threefold in hypoxia and fivefold during reoxygenation (Bindra et al. 2004, 2005a). Similarly, an intrachromosomal DSB assay, measuring repair of an I-*Sce*I endonuclease-induced DSB within a reporter gene, revealed a significant reduction in HR during hypoxic stress (Bindra et al. 2005b). Finally, hypoxic cells are more sensitive than normoxic cells to chemotherapeutics that generate lesions subject to repair by HR, such as cisplatin and mitomycin C. Although acute hypoxia decreases IR sensitivity due to the lack of oxidation of DNA free radicals, chronic hypoxia reduces this radioresistance, likely because cells are less able to repair DSBs that do form (Chan et al. 2008). Thus, hypoxia-induced RAD51 and BRCA1 depression is thought to contribute to genomic instability created by unrepaired DNA DSBs.

In addition to RAD51 and BRCA1, several other HR proteins may undergo hypoxic regulation by alternative mechanisms. First, RAD52 is down-regulated in hypoxia by miR-210 and miR-373, which are up-regulated directly by HIF-1 (Crosby et al. 2009). NBS1, a component of the MRN complex involved in sensing DSBs, was also found to decrease in hypoxia in a HIF-1α-dependent manner (To et al. 2006). As seen with MSH2, HIF-1α binds Sp1 and displaces Myc from the NBS1 promoter, resulting in decreased NBS1 mRNA transcription. Interestingly, the PAS-B domain of HIF-1α is necessary and sufficient for this function, and over-expression of this domain increased the number of nuclear γH2AX/53BP1 foci which are thought to mark DSBs. Finally, the RAD51 paralogs, RAD51B, RAD51C, RAD51D, and XRCC3, as well as RAD54 and BRCA2, have inconsistently been reported to be down-regulated at the mRNA and/or protein levels in hypoxia (Meng et al. 2005; Chan et al. 2008). Thus, HR depression in hypoxia appears to be a multifaceted process that can involve multiple pathways with the potential for cell type or condition-determined specificity.

In contrast to HR, the NHEJ pathway of DSB repair is generally not depressed during hypoxic stress. Though one study reported decreased mRNA levels of the NHEJ proteins RAD50, MRE11, DNA-PKcs, Ku70, XRCC4, and DNA Lig-IV, no corresponding decrease in their protein levels occurs (Meng et al. 2005; Chan et al. 2008; Bindra et al. 2005b). Moreover, the functional ability of cells to repair DSBs by NHEJ remains intact during hypoxia (Bindra et al. 2005b). In acute hypoxia (2–8 h) the mRNA and protein levels of DNA-PKcs and Ku70/Ku80 actually increase in an HIF-1α-dependent manner, while cycles of hypoxia and reoxygenation also lead to a gradual increase in DNA-PKcs and Ku70/Ku80 (Kang et al. 2008; Um et al. 2004). Hypoxia also leads to DNA-PK activation by autophosphorylation and nuclear accumulation, possibly triggered by global genomic changes in histone modifications, particularly increased acetylated H3K14 (Bouquet et al. 2011). Of note, the increase in activated DNA-PK is required for the full stabilization of HIF-1, though the mechanism remains unclear (Um et al. 2004; Bouquet et al. 2011). Since NHEJ usually repairs DNA imprecisely, the activation and maintenance of NHEJ in hypoxia might contribute to genetic instability, particularly in the context of diminished HR capacity.

Mechanisms of Genetic Instability: Increased DNA Damage

In addition to a decreased ability to repair damaged DNA by the MMR, NER, and HR pathways, an increased rate of DNA damage could also contribute to genetic instability generated by hypoxic stress. Indeed, hypoxia has been found to induce endonuclease activity, centrosome amplification, repeat element activation, and decreased carcinogen detoxification, though the significance of these phenomena has not been studied as intently as the depression of DNA repair. In addition, reoxygenation results in the generation of ROS that can overwhelm cellular antioxidant mechanisms and cause oxidative DNA damage. Increased DNA damage can thus act synergistically with decreased DNA repair functionality to induce genetic instability.

Endonuclease Activity

An endonuclease induced by hypoxia has been suggested to contribute to genetic instability (Russo et al. 1995; Stoler et al. 1992). Specifically, after exposure of cultured fibroblasts to at least 16 h of hypoxia, increased endonuclease activity was detected as DNA in cell lysates was digested into nucleosomal fragments, and an exogenous plasmid could be cleaved. The cells nevertheless remained viable and their DNA grossly intact up to the point of cell lysis, indicating that the cells were not undergoing apoptosis, and that a sequestered endonuclease was released during membrane digestion (Stoler et al. 1992). Flow cytometry revealed a threefold increase in DSBs, which paralleled the expression of an endonuclease appearing as 29 and 31 kDa peptides, potentially due to different posttranslational modifications. This endonuclease was also found expressed in four of ten colorectal carcinomas but not in eight normal colorectal tissue samples (Russo et al. 1995). However, as the identity of this endonuclease has not been discovered, its significance in inducing genetic instability remains unclear.

Centrosome Amplification

In cells and tumors originating from the neural crest, hypoxia has been shown to induce centrosome amplification via the action of the neurotrophin receptor tyrosine kinase TrkA (Tacconelli et al. 2004, 2005; Farina et al. 2009). Centrosome amplification wreaks havoc during mitosis, as it leads to abnormal mitotic spindle formation and chromosome mis-segregation. In hypoxia, an alternative splice variant of TrkA called TrkAIII is increased in expression relative to the normal TrkAI and TrkAII variants in a variety of cells and tumors originating from the neural crest (Tacconelli et al. 2004, 2005). TrkAIII, lacking introns 6, 7, and 9, and thus an IG-C1 domain with important *N*-glycosylation sites, fails to traffic to the plasma membrane and instead localizes to intracellular membranes of the endoplasmic reticulum and Golgi network (Farina et al. 2009). The cytoplasmic tyrosine kinase domain is activated independent of its ligand during interphase, undergoes autophosphorylation, associates with γ-tubulin, and is thereby recruited to the centrosome where it phosphorylates several components of the centrosome. These phosphorylation events induce increased association of polo kinase 4 (Plk4) and decreased association of separase with the centrosome, which in turn leads to centrosome amplification. Overexpression of TrkA in neural crest-derived cells promoted centrosome amplification and genetic instability manifested as polyploidy, aneuploidy, sister chromatid exchange, anaphase DNA bridging, and multinuclear cells (Farina et al. 2009). The aggressive phenotype of neural crest cancer cells in hypoxia may thus derive in part from TrkAIII-mediated centrosome amplification.

Recently, a mechanism of centrosome amplification dependent upon miR-210, which is induced directly by HIF-1 in hypoxia, has been reported (Nakada et al. 2011). Overexpression of miR-210 in renal carcinoma cells and in an immortalized kidney cell line promoted centrosome amplification, aneuploidy, and multinucleation. The mechanism depended in part on miR-210 down-regulation of the E2F3 transcription factor. If hypoxia-induced miR-210 is able to promote a similar process, then centrosome amplification could represent a more general mechanism of hypoxia-induced DNA damage.

Repeat Element Activation

DNA repeat elements, such as LINES (long interspersed nuclear elements) and SINES (short interspersed nuclear elements), can contribute to genetic instability through insertion mutagenesis or unbalanced recombination (Watanabe and Maekawa 2010). DNA hypomethylation of repeated sequences can stimulate recombination, translocations, or transpositions, and is associated with many types of cancer, including breast, ovarian, hepatocellular, renal cell, and urothelial carcinomas (Watanabe and Mackawa 2010). In cultured glial tumor and osteosarcoma cell lines, 6 weeks of hypoxia resulted in 9.6- and 6.3-fold increases of SINE expression and 19.5- and 5.9-fold increases in LINE 1 expression, which were associated with hypomethylation, particularly at the SINE and LINE foci (Pal et al. 2010). Furthermore, inter-Alu PCR demonstrated altered band sizes and intensity, revealing the occurrence of Alu mobilization (Pal et al. 2010). Chronic hypoxia, through its impact on global DNA methylation, may therefore increase the mutagenic potential of repeat elements.

Carcinogen Detoxification

The HIF-1β subunit of HIF-1 is identical to the aryl hydrocarbon nuclear translocator (ARNT), which also dimerizes with the aryl hydrocarbon receptor (AhR) and functions as a sensor of toxic halogenated aromatic carcinogens (Beischlag et al. 2008). Upon binding a toxin, the ARNT/AhR complex regulates mRNA transcription of genes involved in detoxification such as those of the cytochrome P450 (CYP) family (Beischlag et al. 2008). During hypoxic stress, HIF-1α competes with AhR for ARNT binding (Schults et al. 2010). Demonstrating that this competition results in a diminished response to carcinogens, HIF-1 stabilization with the hypoxia mimetic

$CoCl_2$ reduced benzo[a]pyrene (BaP)-induced CYP1A1 and CYP1A2 mRNA. Furthermore, when HIF was stabilized, the BaP mutation frequency in an HPRT reporter gene increased several fold and γH2AX foci marking DSBs increased significantly as well (Schults et al. 2010). Hypoxia may thus render cells more susceptible to mutagenesis by environmental carcinogens.

Reoxygenation and ROS

Although one might logically expect oxygen deprivation to result in decreased production of reactive oxygen species (ROS), real-time measurements with novel fluorescent ROS probes have established that the generation of ROS actually increases during hypoxia (Hamanaka and Chandel 2009; Guzy and Schumacker 2006). Complex III, the only mitochondrial complex capable of producing ROS in the intermembrane space, is the source of hypoxia-induced ROS; they are generated at the Complex III Qo site as electrons are transferred from ubisemiquinone to molecular oxygen (Bell et al. 2007). Interestingly, hypoxia-induced ROS act as second messenger signaling molecules and contribute to HIF-induced gene up-regulation (Ziel et al. 2005). Hypoxia-induced ROS induce "targeted" oxidative base modifications in or near the HREs of certain hypoxia-inducible genes, such as VEGF. These lesions, which are not found in unrelated genes, correlate with mRNA transcription and are repaired over the course of 6–24 h. The oxidized base, if converted to an abasic site, can be bound by the transcription coactivator Ref-1/Ape-1, leading to increased HIF-1 recruitment. Apart from their function as second messengers, it has been suggested, but not proven, that hypoxia-induced ROS could generate genomic instability (Ziel et al. 2005).

In contrast to the questionable role of hypoxia-induced ROS in causing genetic instability, reoxygenation following a period of hypoxic stress is widely known to generate a vast quantity of ROS which account for most of the cell damage and death observed in ischemia-reperfusion injuries (Li and Jackson 2002). Reoxygenation in tumors, as discussed previously, results from perfusion-limited fluctuations in blood flow due to the aberrant vasculature formed during tumor angiogenesis (Chaplin et al. 1987; Hardee et al. 2009). The first evidence that cells have increased capacity for ROS production during reoxygenation arose from reports that cultured human cells, while resistant to IR during hypoxia, actually have increased radiosensitivity in the post-hypoxic period (Hall et al. 1966). The maximum radiosensitivity occurs approximately 5 min after reoxygenation and returns to baseline after 12 h of normoxia (Kwok and Sutherland 1989), suggesting a correlation to rapid ROS production and delayed compensatory protein synthesis. Bursts of ROS during tissue reoxygenation have since been confirmed by electron spin resonance spectroscopy and ROS-sensitive fluorescent dyes (Kuznetsov et al. 2011), while treatment with exogenous antioxidants has been shown to decrease tissue injury, confirming the pathological significance of the ROS surge (Li and Jackson 2002). Furthermore, DNA damage detectable by the comet assay occurs in cultured cells subjected to hypoxia and reoxygenation (Hammond et al. 2003).

ROS levels increase during reoxygenation because hypoxia-induced cellular adaptations create an imbalance between ROS-generating and ROS-quenching mechanisms (Li and Jackson 2002). Though there are many contributing mechanisms, just three significant ones are mentioned here. First, as hypoxia induces a shift to glycolytic metabolism, it down-regulates the mitochondrial transport chain enzymes (Murphy et al. 1984). During reoxygenation, sufficient levels of cytochrome oxidase (complex IV) may not be available to complete the final electron transfer to oxygen, driving upstream complexes to transfer electrons to oxygen prematurely and to thereby generate superoxide radicals (Chen et al. 2008). Second, xanthine dehydrogenase/oxidase, an important generator of superoxide and hydrogen peroxide, is phosphorylated and activated during hypoxia (Kayyali et al. 2001). Finally, hypoxia leads to down-regulation of ROS-detoxifying enzymes, including manganese superoxide dismutase, glutathione peroxidase, glutathione reductase, and catalase (Plateel et al. 1995; Kirshenbaum and Singal 1992). As hypoxia thus renders cells exceptionally sensitive to ROS accumulation, subsequent reoxygenation constitutes one of the greatest threats to genomic integrity associated with hypoxic stress.

Regulation of Genetic Instability

Cells have evolved elaborate mechanisms to sense and respond to DNA damage and to avoid mutagenesis by preventing further cell division until the damage is resolved. Namely, DNA damage activates checkpoint signaling pathways, which can lead to cell cycle arrest, or, in the case of severe or irreparable damage, cell death by apoptosis. This final section discusses how the hypoxic stress response intertwines with the checkpoint signaling and apoptotic pathways (Fig. 8.5).

DNA Damage Checkpoint Signaling

DNA damage activates checkpoint signaling pathways mediated predominantly by the ataxia telangiectasia mutated (ATM) and ATM and Rad3-related (ATR) phosphatidylinositol-3-kinase (PI3K) family members, as well as their downstream targets checkpoint kinase 2 (CHK2) and checkpoint kinase 1 (CHK1) [reviewed in (Smith et al. 2010)]. The ATM-CHK2 and ATR-CHK1 pathways primarily sense and respond to DSBs and single-stranded DNA, respectively. Upon the formation of DSBs, ATM homodimers undergo autophosphorylation, dissociate into monomers, and are recruited to DSBs by the MRE11/RAD50/NBS1 (MRN) complex where they phosphorylate CHK2 and other targets such as histone H2AX, MRN, and cohesion SMC1. CHK2 phosphorylates a number of target proteins, including p53, BRCA1, and Cdc25 family phosphatases. In contrast, single-stranded DNA becomes coated by RPA and recruits ATR, with its binding partner ATRIP, whereupon ATR

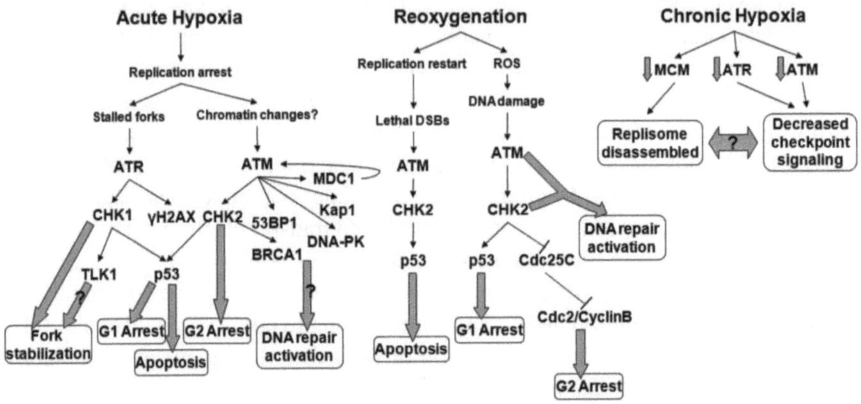

Fig. 8.5 DNA damage checkpoint signaling in acute and chronic hypoxia and during reoxygenation

phosphorylates CHK1 and other targets such as BRCA1, MCM proteins, and RPA. CHK1 downstream phosphorylation targets include p53, RAD51, BRCA2, Wee1 kinase, and Cdc25 phosphatases. The ultimate outcomes of the ATM-CHK2 and ATR-CHK1 pathways overlap significantly, and can include cell cycle arrest at the G1, intra-S, and G2 checkpoints, DNA repair activation, or the onset of apoptosis.

Hypoxia and reoxygenation activate components of the ATM-CHK2 and ATR-CHK1 signaling pathways, though with certain deviances from the classical pathways described above (Hammond et al. 2006). Acute severe hypoxia (less than 0.1% oxygen), but not milder hypoxia, activates both ATR-CHK1 and ATM-CHK2 signaling, and is associated with G1 and G2 checkpoint activation and replication arrest (Hammond et al. 2002; Gibson et al. 2005; Bencokova et al. 2009). This replication arrest is not a true cell cycle checkpoint as it occurs independently of ATR and ATM, as well as p53, p21, and HIF-1α (Hammond and Giaccia 2004). Instead, replication arrest appears to arise due to the depletion of ribonucleotides, as ribonucleotide reductase requires a tyrosal free radical in its active site, and is thus inhibited by severe, but not milder, hypoxia (Pires et al. 2010a). DNA fiber analysis has revealed that hypoxia both inhibits the initiation of new replication origins and induces elongating replication fork stalling (Pires et al. 2010a). Single-stranded DNA accumulates at the site of stalled forks, is coated with RPA, and recruits ATR/ATRIP, which form nuclear foci in cells under hypoxic stress (Hammond et al. 2002, 2004). Histone H2AX is phosphorylated in an ATR-dependent manner, and forms foci that are found in the same cells with RPA foci (Hammond et al. 2003). Activated ATR and CHK1 appear to be necessary for stabilizing the stalled replication forks during hypoxia, as their absence results in DNA damage accumulation in cells in S phase and reduced survival upon reoxygenation (Hammond et al. 2004). ATR and CHK1 are also required for p53 accumulation during hypoxia via phosphorylation of p53 on serines 15 and 37 (Hammond et al. 2002). p53 mediates

G1 arrest and is involved in regulating the apoptotic decision (discussed later). Finally, CHK1 phosphorylates and inactivates TLK1, a kinase involved in processing newly replicated DNA to form mature chromatin, and in regulating S phase progression (Pires et al. 2010b), perhaps suggesting a link to the role of ATR-CHK1 signaling in protecting and stabilizing stalled replication forks.

Acute, severe hypoxia also leads to activation of the ATM-CHK2 signaling pathway, though in an unconventional manner that does not require the MRN complex (Bencokova et al. 2009). Again independently of HIF-1, ATM undergoes autophosphorylation on serine 1981, localizes diffusely in the nucleus (not forming distinct foci as it does in the response to DSBs), and activates CHK2 by phosphorylation on threonine 68 (Bencokova et al. 2009). The stimulus for ATM-CHK2 activation in hypoxia is not well understood but may be due to chromatin changes induced by stalled replication forks. Hypoxia-induced CHK2 is required for G2 arrest and contributes to p53 accumulation by phosphorylating residue serine 20 (Gibson et al. 2005). In addition, CHK2 phosphorylates BRCA1 on serine 988 in hypoxia, though its role in the context of depressed HR capacity is unclear, and it may instead serve to regulate NHEJ (Gibson et al. 2006). Hypoxia-induced ATM is also required for DNA-PKcs, Kap1, and 53BP1 phosphorylation though the latter does not form foci as it does in the response to DSBs (Bencokova et al. 2009). MDC1, however, does form foci in hypoxia, and is required for full amplification of ATM signaling (Bencokova et al. 2009). In summary, acute hypoxia induces a classical ATR-CHK1 pathway that is required for stabilization of stalled replication forks and an unusual ATM-CHK2 pathway, which differs in its upstream activation and is directed more towards G2 cell cycle arrest rather than DNA repair initiation. Both pathways are important for genetic stability and cell survival.

During reoxygenation following acute hypoxia, the replication block disappears. However, cells, which resume replication ultimately, undergo p53-dependent apoptosis as unregulated DNA replication generates lethal chromosomal damage (Pires et al. 2010a, b). The number of cells that restart replication increases in the absence of CHK1, indicating its role in suppressing inappropriate replication (Pires et al. 2010b). Cells that do not undergo replication restart and apoptosis nevertheless accumulate DNA damage detectable by the comet assay, and activate an ATM-CHK2 DNA damage checkpoint pathway (Hammond et al. 2003). Activated CHK2 leads to G2 arrest via phosphorylation and inactivation of Cdc25C causing Cdc2 to remain phosphorylated and unable to interact with cyclin B (Freiberg et al. 2006a; Kim et al. 2007). Loss of CHK2 abrogates the G2 arrest and increases apoptosis (Freiberg et al. 2006a). ATM-CHK2 also leads to G1 arrest by maintaining the phosphorylation of p53 serine 15, though loss of this checkpoint in p53 deficient cells does not reduce cell survival (Hammond et al. 2003; Freiberg et al. 2006b). Overall, ATM-CHK2 signaling during reoxygenation prevents cell cycle progression to avoid genetic instability or p53-dependent apoptosis.

Chronic hypoxia offers yet another variation of checkpoint signaling. After approximately 16 h of severe hypoxia, replisomes disassemble in a process that involves decreased transcription of MCM proteins, which ensures against replication restart (Pires et al. 2010a). Indeed, DNA fiber analysis demonstrated that after

12–24 h of hypoxia, reoxygenation no longer results in replication restart (Pires et al. 2010b). In conjunction with replisome disassembly, ATR and ATM signaling decrease during chronic hypoxia, disappearing after approximately 18 h (Pires et al. 2010a), though how the two processes are connected remains to be determined.

Apoptotic Signaling and Resistance

While short-term hypoxia leads to cell cycle arrest, prolonged, severe hypoxia (0–0.5% oxygen) results in cell death (Brunelle and Chandel 2002). Under simultaneous glucose deprivation, energy depletion results in necrosis, but if energy is available, apoptosis can occur through the mitochondrial intrinsic pathway. Briefly, a pro-apoptotic Bcl-2 family member, Bax or Bak, stimulates mitochondrial release of cytochrome c, activating in turn Apaf-1, initiator caspase-9, and executioner caspases-3 and -7. The apoptotic decision is controlled primarily by the p53 and HIF-1 signaling pathways, which are themselves intricately connected [reviewed in (Sermeus and Michiels 2011)]. First, HIF-1 is required—though not sufficient for—p53 accumulation during hypoxia. HIF-1α down-regulates MDM2, resulting in decreased p53 ubiquitination and degradation. As described previously, ATM and ATR signaling pathways also contribute to p53 stabilization by phosphorylation. Conversely, p53 decreases HIF-1 activity by various condition-dependent mechanisms. Hypoxia-induced p53 acts through gene transrepression rather than its classical method of gene transactivation. Though p53 binds to its target genes in hypoxia, it fails to recruit CBP/p300, possibly due to competition with HIF-1 or hypoxia-specific posttranslational modifications; instead it recruits the co-repressor mSin3A and down-regulates the expression of certain anti-apoptotic genes such as the caspase inhibitor HIAP/BIRC3. HIF-1 itself also up-regulates the transcription of both pro- and anti-apoptotic genes. The combination of synergistic and antagonistic actions of p53 and HIF-1 on the apoptotic decision leads to different outcomes depending upon the severity and duration of hypoxia. Generally, under mild hypoxia high HIF-1 and low p53 levels favor cell survival, whereas under severe hypoxia, high levels of both HIF-1 and p53 promote apoptosis.

In addition to inducing apoptosis, hypoxia has been shown consequently to select for apoptotic-resistant clones and allow for genetic instability to accumulate. First, when a small number of p53-deficient cells were co-cultured with p53-positive cells, the percentage of p53-deficient cells increased about 2.4-fold following each successive treatment of hypoxia, and eventually overtook the p53-positive cells (Graeber et al. 1996). In another model, repeated treatments of hypoxia led to accelerated formation and selection of HPV-infected cervical epithelial cells that had lost their ability to undergo apoptosis (Kim et al. 1997). Finally, transformed baby mouse kidney (BMK) cells injected into mice were found to die due to hypoxia-induced apoptosis or necrosis, unless apoptosis was blocked by constitutive Bcl-2 expression, in which case the resulting tumors contained giant cells with high polyploidy (Nelson et al. 2004). In vitro, gain of Bcl-2 function or loss of Bak and Bax similarly prevented apoptosis and led to the growth of cells with tetraploid or polyploid DNA content under hypoxic stress (Nelson et al. 2004). Together, these experiments demonstrate that hypoxia can select for cells that have lost their apoptotic potential and are unprotected from hypoxia-induced genomic instability.

Additional studies have suggested that hypoxia may also directly generate apoptosis resistance under certain circumstances. In normal prostate epithelial cells, 24 h of moderate hypoxia (1% oxygen) exposure led to resistance to TRAIL-induced apoptosis by the extrinsic pathway (Walsh et al. 2009). This resistance was attributed to AKT activation and the altered expression of apoptosis-associated proteins, leading to decreased Bax and caspase-8 activity (Walsh et al. 2009). A second study in *Caenorhabditis elegans* described an interesting mechanism in which HIF-1 protects against DNA damage-induced apoptosis in a non-cell-autonomous manner (Sendoel et al. 2010; Powell-Coffman and Coffman 2010). Specifically, HIF-1 induced transcriptional up-regulation and secretion of TYR-2 (mammalian TRP2) from the worm's sensory neurons, which prevented apoptosis in the worm's germ cells by inhibiting CEP-1 (mammalian p53). TRP2 depletion from human melanoma cells leads to increased apoptosis (Sendoel et al. 2010), and it will be interesting to learn if hypoxia affects this pathway in vivo.

Finally, the role of apoptotic resistance in hypoxia-induced genetic instability and cancer progression has been analyzed directly using a range of cells that differed in their apoptotic capacity (Hayashi et al. 2011). Hypoxia or HIF-1 activation depressed DNA repair protein expression and caused transient DNA damage in all cell types. However, prolonged HIF-1 expression led to genetic instability, inactivation of fragile site-associated tumor suppressor genes, and the acquisition of malignant traits only in apoptosis-deficient cells. Therefore, while hypoxic stress leads to cell death from genetic instability in cells with deficient DNA damage checkpoints, it induces genomic instability and malignant progression in cells that have further lost their apoptotic ability.

Conclusion

Hypoxia-induced genetic instability, an integrated part of the hypoxic stress response, involves down-regulation of cellular DNA repair pathways and increased DNA damage rates. In human malignancies, the hypoxic tumor microenvironment likely contributes to the development of a mutator phenotype, which has several parallels with the stress responses in lower organisms. For example, in *Escherichia coli*, activation of the SOS DNA damage stress response and the RpoS starvation stress response results in stress-induced point mutagenesis via error-prone HR DSB repair and limiting levels of the MMR protein MutL (Galhardo et al. 2007). Other bacterial stress responses induce genetic transpositions and a switch to mutagenic NHEJ DSB repair (Galhardo et al. 2007). In addition, hypoxia has recently been found to down-regulate three base excision repair enzymes, Ung, UdgB, and Fpg, in *Mycobacterium smegmatis*, resulting in an altered mutation spectrum (Kurthkoti and Varshney 2010). These phenomena appear analogous to down-regulation of MMR, NER, and HR proteins, Alu-mediated mutagenesis, and the maintenance of NHEJ in the mammalian hypoxic stress response. Cells have also evolved checkpoint signaling and apoptotic mechanisms in order to avoid mutagenesis, and these pathways must be overcome for cells to acquire genetic instability. Hypoxia might induce mutations that inactivate the checkpoint and apoptotic pathways, but can also impact the pathways directly, independent of DNA damage, and can select for cells deficient in these pathways. Hypoxic stress may thus lead to the generation of cells that survive and thrive in the tumor microenvironment, much like the evolution of lower organisms in the response to stress.

References

Beischlag TV, Luis Morales J, Hollingshead BD, Perdew GH (2008) The aryl hydrocarbon receptor complex and the control of gene expression. Crit Rev Eukaryot Gene Expr 18(3):207–250

Bell EL, Klimova TA, Eisenbart J, Moraes CT, Murphy MP, Budinger GR et al (2007) The qo site of the mitochondrial complex III is required for the transduction of hypoxic signaling via reactive oxygen species production. J Cell Biol 177(6):1029–1036, Epub 2007 Jun 11

Bencokova Z, Kaufmann MR, Pires IM, Lecane PS, Giaccia AJ, Hammond EM (2009) ATM activation and signaling under hypoxic conditions. Mol Cell Biol 29(2):526–537, Epub 2008 Nov 3

Beyer S, Kristensen MM, Jensen KS, Johansen JV, Staller P (2008) The histone demethylases JMJD1A and JMJD2B are transcriptional targets of hypoxia-inducible factor HIF. J Biol Chem 283(52):36542–36552, Epub 2008 Nov 4

Bindra RS, Glazer PM (2005) Genetic instability and the tumor microenvironment: towards the concept of microenvironment-induced mutagenesis. Mutat Res 569(1–2):75–85

Bindra RS, Glazer PM (2006) Basal repression of BRCA1 by multiple E2Fs and pocket proteins at adjacent E2F sites. Cancer Biol Ther 5(10):1400–1407, Epub 2006 Oct 30

Bindra RS, Glazer PM (2007a) Co-repression of mismatch repair gene expression by hypoxia in cancer cells: role of the Myc/Max network. Cancer Lett 252(1):93–103, Epub 2007 Feb 1

Bindra RS, Glazer PM (2007b) Repression of RAD51 gene expression by E2F4/p130 complexes in hypoxia. Oncogene 26(14):2048–2057, Epub 2006 Sep 25

Bindra RS, Schaffer PJ, Meng A, Woo J, Maseide K, Roth ME et al (2004) Down-regulation of Rad51 and decreased homologous recombination in hypoxic cancer cells. Mol Cell Biol 24(19):8504–8518

Bindra RS, Schaffer PJ, Meng A, Woo J, Maseide K, Roth ME et al (2005a) Alterations in DNA repair gene expression under hypoxia: elucidating the mechanisms of hypoxia-induced genetic instability. Ann N Y Acad Sci 1059:184–195

Bindra RS, Gibson SL, Meng A, Westermark U, Jasin M, Pierce AJ et al (2005b) Hypoxia-induced down-regulation of BRCA1 expression by E2Fs. Cancer Res 65(24):11597–11604

Bouquet F, Ousset M, Biard D, Fallone F, Dauvillier S, Frit P et al (2011) A DNA-dependent stress response involving DNA-PK occurs in hypoxic cells and contributes to cellular adaptation to hypoxia. J Cell Sci 124:1943–1951

Brown JM (1979) Evidence for acutely hypoxic cells in mouse tumours, and a possible mechanism of reoxygenation. Br J Radiol 52(620):650–656

Brunelle JK, Chandel NS (2002) Oxygen deprivation induced cell death: an update. Apoptosis 7(6):475–482

Cairns RA, Kalliomaki T, Hill RP (2001) Acute (cyclic) hypoxia enhances spontaneous metastasis of KHT murine tumors. Cancer Res 61(24):8903–8908

Camps C, Buffa FM, Colella S, Moore J, Sotiriou C, Sheldon H et al (2008) Hsa-miR-210 is induced by hypoxia and is an independent prognostic factor in breast cancer. Clin Cancer Res 14(5):1340–1348

Chan N, Koritzinsky M, Zhao H, Bindra R, Glazer PM, Powell S et al (2008) Chronic hypoxia decreases synthesis of homologous recombination proteins to offset chemoresistance and radioresistance. Cancer Res 68(2):605–614

Chan SY, Zhang YY, Hemann C, Mahoney CE, Zweier JL, Loscalzo J (2009) MicroRNA-210 controls mitochondrial metabolism during hypoxia by repressing the iron-sulfur cluster assembly proteins ISCU1/2. Cell Metab 10(4):273–284

Chaplin DJ, Olive PL, Durand RE (1987) Intermittent blood flow in a murine tumor: radiobiological effects. Cancer Res 47(2):597–601

Chen H, Yan Y, Davidson TL, Shinkai Y, Costa M (2006) Hypoxic stress induces dimethylated histone H3 lysine 9 through histone methyltransferase G9a in mammalian cells. Cancer Res 66(18):9009–9016

Chen Q, Moghaddas S, Hoppel CL, Lesnefsky EJ (2008) Ischemic defects in the electron transport chain increase the production of reactive oxygen species from isolated rat heart mitochondria. Am J Physiol Cell Physiol 294(2):C460–C466, Epub 2007 Dec 12

Chen H, Kluz T, Zhang R, Costa M (2010) Hypoxia and nickel inhibit histone demethylase JMJD1A and repress Spry2 expression in human bronchial epithelial BEAS-2B cells. Carcinogenesis 31(12):2136–2144, Epub 2010 Sep 29

Cockman ME, Masson N, Mole DR, Jaakkola P, Chang GW, Clifford SC et al (2000) Hypoxia inducible factor-alpha binding and ubiquitylation by the von hippel-lindau tumor suppressor protein. J Biol Chem 275(33):25733–25741

Coquelle A, Toledo F, Stern S, Bieth A, Debatisse M (1998) A new role for hypoxia in tumor progression: induction of fragile site triggering genomic rearrangements and formation of complex DMs and HSRs. Mol Cell 2(2):259–265

Coquelle A, Rozier L, Dutrillaux B, Debatisse M (2002) Induction of multiple double-strand breaks within an hsr by meganucleaseI-SceI expression or fragile site activation leads to formation of double minutes and other chromosomal rearrangements. Oncogene 21(50):7671–7679

Crabtree HG, Cramer W (1933) The action of radium on cancer cells. II. Some factors determining the susceptibility of cancer cells to radium. Proc R Soc Lond B 113(782):238–250

Crosby ME, Kulshreshtha R, Ivan M, Glazer PM (2009) MicroRNA regulation of DNA repair gene expression in hypoxic stress. Cancer Res 69(3):1221–1229, Epub 2009 Jan 13

Culver C, Sundqvist A, Mudie S, Melvin A, Xirodimas D, Rocha S (2010) Mechanism of hypoxia-induced NF-kappaB. Mol Cell Biol 30(20):4901–4921, Epub 2010 Aug 9

Devlin C, Greco S, Martelli F, Ivan M (2011) miR-210: more than a silent player in hypoxia. IUBMB Life 63(2):94–100. doi:10.1002/iub.427, Epub 2011 Feb 24

Edwards RA, Witherspoon M, Wang K, Afrasiabi K, Pham T, Birnbaumer L et al (2009) Epigenetic repression of DNA mismatch repair by inflammation and hypoxia in inflammatory bowel disease-associated colorectal cancer. Cancer Res 69(16):6423–6429, Epub 2009 Jul 28

Farina AR, Tacconelli A, Cappabianca L, Cea G, Panella S, Chioda A et al (2009) The alternative TrkAIII splice variant targets the centrosome and promotes genetic instability. Mol Cell Biol 29(17):4812–4830, Epub 2009 Jun 29

Fasanaro P, D'Alessandra Y, Di Stefano V, Melchionna R, Romani S, Pompilio G et al (2008) MicroRNA-210 modulates endothelial cell response to hypoxia and inhibits the receptor tyrosine kinase ligand ephrin-A3. J Biol Chem 283(23):15878–15883, Epub 2008 Apr 16

Favaro E, Ramachandran A, McCormick R, Gee H, Blancher C, Crosby M et al (2010) MicroRNA-210 regulates mitochondrial free radical response to hypoxia and krebs cycle in cancer cells by targeting iron sulfur cluster protein ISCU. PLoS One 5(4):e10345

Fischer U, Radermacher J, Mayer J, Mehraein Y, Meese E (2008) Tumor hypoxia: impact on gene amplification in glioblastoma. Int J Oncol 33(3):509–515

Freiberg RA, Hammond EM, Dorie MJ, Welford SM, Giaccia AJ (2006a) DNA damage during reoxygenation elicits a Chk2-dependent checkpoint response. Mol Cell Biol 26(5):1598–1609

Freiberg RA, Krieg AJ, Giaccia AJ, Hammond EM (2006b) Checking in on hypoxia/reoxygenation. Cell Cycle 5(12):1304–1307, Epub 2006 Jun 15
Galhardo RS, Hastings PJ, Rosenberg SM (2007) Mutation as a stress response and the regulation of evolvability. Crit Rev Biochem Mol Biol 42(5):399–435
Gee HE, Camps C, Buffa FM, Patiar S, Winter SC, Betts G et al (2010) Hsa-mir-210 is a marker of tumor hypoxia and a prognostic factor in head and neck cancer. Cancer 116(9):2148–2158
Gibson SL, Bindra RS, Glazer PM (2005) Hypoxia-induced phosphorylation of Chk2 in an ataxia telangiectasia mutated-dependent manner. Cancer Res 65(23):10734–10741
Gibson SL, Bindra RS, Glazer PM (2006) CHK2-dependent phosphorylation of BRCA1 in hypoxia. Radiat Res 166(4):646–651
Graeber TG, Osmanian C, Jacks T, Housman DE, Koch CJ, Lowe SW et al (1996) Hypoxia-mediated selection of cells with diminished apoptotic potential in solid tumours. Nature 379(6560):88–91
Guzy RD, Schumacker PT (2006) Oxygen sensing by mitochondria at complex III: the paradox of increased reactive oxygen species during hypoxia. Exp Physiol 91(5):807–819, Epub 2006 Jul 20
Hall EJ, Bedford JS, Oliver R (1966) Extreme hypoxia; its effect on the survival of mammalian cells irradiated at high and low dose-rates. Br J Radiol 39(460):302–307
Hamanaka RB, Chandel NS (2009) Mitochondrial reactive oxygen species regulate hypoxic signaling. Curr Opin Cell Biol 21(6):894–899, Epub 2009 Sep 24
Hammond EM, Giaccia AJ (2004) The role of ATM and ATR in the cellular response to hypoxia and re-oxygenation. DNA Repair (Amst) 3(8–9):1117–1122
Hammond EM, Denko NC, Dorie MJ, Abraham RT, Giaccia AJ (2002) Hypoxia links ATR and p53 through replication arrest. Mol Cell Biol 22(6):1834–1843
Hammond EM, Dorie MJ, Giaccia AJ (2003) ATR/ATM targets are phosphorylated by ATR in response to hypoxia and ATM in response to reoxygenation. J Biol Chem 278(14):12207–12213, Epub 2003 Jan 7
Hammond EM, Dorie MJ, Giaccia AJ (2004) Inhibition of ATR leads to increased sensitivity to hypoxia/reoxygenation. Cancer Res 64(18):6556–6562
Hammond EM, Freiberg RA, Giaccia AJ (2006) The roles of chk 1 and chk 2 in hypoxia and reoxygenation. Cancer Lett 238(2):161–167, Epub 2005 Aug 8
Hanahan D, Weinberg RA (2011) Hallmarks of cancer: the next generation. Cell 144(5):646–674
Hardee ME, Dewhirst MW, Agarwal N, Sorg BS (2009) Novel imaging provides new insights into mechanisms of oxygen transport in tumors. Curr Mol Med 9(4):435–441
Hayashi M, Yoo YG, Christensen J, Huang LE (2011) Requirement of evading apoptosis for HIF-1alpha-induced malignant progression in mouse cells. Cell Cycle 10(14):2364–2372, Epub 2011 Jul 15
Huang LE, Arany Z, Livingston DM, Bunn HF (1996) Activation of hypoxia-inducible transcription factor depends primarily upon redox-sensitive stabilization of its alpha subunit. J Biol Chem 271(50):32253–32259
Huang LE, Gu J, Schau M, Bunn HF (1998) Regulation of hypoxia-inducible factor 1alpha is mediated by an O2-dependent degradation domain via the ubiquitin-proteasome pathway. Proc Natl Acad Sci USA 95(14):7987–7992
Huang LE, Bindra RS, Glazer PM, Harris AL (2007) Hypoxia-induced genetic instability-a calculated mechanism underlying tumor progression. J Mol Med (Berl) 85(2):139–148, Epub 2006 Dec 20
Huang X, Le QT, Giaccia AJ (2010) MiR-210–micromanager of the hypoxia pathway. Trends Mol Med 16(5):230–237, Epub 2010 Apr 29
Jaakkola P, Mole DR, Tian YM, Wilson MI, Gielbert J, Gaskell SJ et al (2001) Targeting of HIF-alpha to the von hippel-lindau ubiquitylation complex by O2-regulated prolyl hydroxylation. Science 292(5516):468–472, Epub 2001 Apr 5
Johnson AB, Denko N, Barton MC (2008) Hypoxia induces a novel signature of chromatin modifications and global repression of transcription. Mutat Res 640(1–2):174–179, Epub 2008 Jan 16

Kallio PJ, Okamoto K, O'Brien S, Carrero P, Makino Y, Tanaka H et al (1998) Signal transduction in hypoxic cells: inducible nuclear translocation and recruitment of the CBP/p300 coactivator by the hypoxia-inducible factor-1alpha. EMBO J 17(22):6573–6586

Kaluz S, Kaluzova M, Stanbridge EJ (2008) Regulation of gene expression by hypoxia: integration of the HIF-transduced hypoxic signal at the hypoxia-responsive element. Clin Chim Acta 395(1–2):6–13, Epub 2008 May 9

Kang MJ, Jung SM, Kim MJ, Bae JH, Kim HB, Kim JY et al (2008) DNA-dependent protein kinase is involved in heat shock protein-mediated accumulation of hypoxia-inducible factor-1alpha in hypoxic preconditioned HepG2 cells. FEBS J 275(23):5969–5981

Kayyali US, Donaldson C, Huang H, Abdelnour R, Hassoun PM (2001) Phosphorylation of xanthine dehydrogenase/oxidase in hypoxia. J Biol Chem 276(17):14359–14365, Epub 2001 Jan 22

Keysar SB, Trncic N, Larue SM, Fox MH (2010) Hypoxia/reoxygenation-induced mutations in mammalian cells detected by the flow cytometry mutation assay and characterized by mutant spectrum. Radiat Res 173(1):21–26

Khanna KK, Jackson SP (2001) DNA double-strand breaks: signaling, repair and the cancer connection. Nat Genet 27(3):247–254

Kim CY, Tsai MH, Osmanian C, Graeber TG, Lee JE, Giffard RG et al (1997) Selection of human cervical epithelial cells that possess reduced apoptotic potential to low-oxygen conditions. Cancer Res 57(19):4200–4204

Kim BM, Choi JY, Kim YJ, Woo HD, Chung HW (2007) Reoxygenation following hypoxia activates DNA-damage checkpoint signaling pathways that suppress cell-cycle progression in cultured human lymphocytes. FEBS Lett 581(16):3005–3012, Epub 2007 May 29

Kirshenbaum LA, Singal PK (1992) Changes in antioxidant enzymes in isolated cardiac myocytes subjected to hypoxia-reoxygenation. Lab Invest 67(6):796–803

Koong AC, Chen EY, Giaccia AJ (1994) Hypoxia causes the activation of nuclear factor kappa B through the phosphorylation of I kappa B alpha on tyrosine residues. Cancer Res 54(6):1425–1430

Koshiji M, Kageyama Y, Pete EA, Horikawa I, Barrett JC, Huang LE (2004) HIF-1alpha induces cell cycle arrest by functionally counteracting myc. EMBO J 23(9):1949–1956, Epub 2004 Apr 8

Koshiji M, To KK, Hammer S, Kumamoto K, Harris AL, Modrich P et al (2005) HIF-1alpha induces genetic instability by transcriptionally downregulating MutSalpha expression. Mol Cell 17(6):793–803

Krieg AJ, Rankin EB, Chan D, Razorenova O, Fernandez S, Giaccia AJ (2010) Regulation of the histone demethylase JMJD1A by hypoxia-inducible factor 1 alpha enhances hypoxic gene expression and tumor growth. Mol Cell Biol 30(1):344–353

Kulshreshtha R, Ferracin M, Wojcik SE, Garzon R, Alder H, Agosto-Perez FJ et al (2007) A microRNA signature of hypoxia. Mol Cell Biol 27(5):1859–1867, Epub 2006 Dec 28

Kurthkoti K, Varshney U (2010) Detrimental effects of hypoxia-specific expression of uracil DNA glycosylase (ung) in mycobacterium smegmatis. J Bacteriol 192(24):6439–6446, Epub 2010 Oct 22

Kuznetsov AV, Kehrer I, Kozlov AV, Haller M, Redl H, Hermann M et al (2011) Mitochondrial ROS production under cellular stress: comparison of different detection methods. Anal Bioanal Chem 400(8):2383–2390, Epub 2011 Feb 20

Kwok TT, Sutherland RM (1989) The radiation response of cells recovering after chronic hypoxia. Radiat Res 119(2):261–267

Lee JH, Choi IJ, Song DK, Kim DK (2010) Genetic instability in the human lymphocyte exposed to hypoxia. Cancer Genet Cytogenet 196(1):83–88

Lehtonen HJ, Makinen MJ, Kiuru M, Laiho P, Herva R, van Minderhout I et al (2007) Increased HIF1 alpha in SDH and FH deficient tumors does not cause microsatellite instability. Int J Cancer 121(6):1386–1389

Li GM (2008) Mechanisms and functions of DNA mismatch repair. Cell Res 18(1):85–98

Li Q, Costa M (2009) c-myc mediates a hypoxia-induced decrease in acetylated histone H4. Biochimie 91(10):1307–1310, Epub 2009 Jul 14

Li C, Jackson RM (2002) Reactive species mechanisms of cellular hypoxia-reoxygenation injury. Am J Physiol Cell Physiol 282(2):C227–C241

Li CY, Little JB, Hu K, Zhang W, Zhang L, Dewhirst MW et al (2001) Persistent genetic instability in cancer cells induced by non-DNA-damaging stress exposures. Cancer Res 61(2):428–432

Lieber MR (2010) The mechanism of double-strand DNA break repair by the nonhomologous DNA end-joining pathway. Annu Rev Biochem 79:181–211

Loboda A, Jozkowicz A, Dulak J (2010) HIF-1 and HIF-2 transcription factors–similar but not identical. Mol Cells 29(5):435–442, Epub 2010 Apr 12

Loeb LA (2001) A mutator phenotype in cancer. Cancer Res 61(8):3230–3239

Lu Y, Chu A, Turker MS, Glazer PM (2011) Hypoxia-induced epigenetic regulation and silencing of the BRCA1 promoter. Mol Cell Biol 31(16):3339–3350, Epub 2011 Jun 13

Majmundar AJ, Wong WJ, Simon MC (2010) Hypoxia-inducible factors and the response to hypoxic stress. Mol Cell 40(2):294–309

Marotta D, Karar J, Jenkins WT, Kumanova M, Jenkins KW, Tobias JW et al (2011) In vivo profiling of hypoxic gene expression in gliomas using the hypoxia marker EF5 and laser-capture microdissection. Cancer Res 71(3):779–789, Epub 2011 Jan 25

Maxwell PH, Wiesener MS, Chang GW, Clifford SC, Vaux EC, Cockman ME et al (1999) The tumour suppressor protein VHL targets hypoxia-inducible factors for oxygen-dependent proteolysis. Nature 399(6733):271–275

Meng AX, Jalali F, Cuddihy A, Chan N, Bindra RS, Glazer PM et al (2005) Hypoxia down-regulates DNA double strand break repair gene expression in prostate cancer cells. Radiother Oncol 76(2):168–176

Mihaylova VT, Bindra RS, Yuan J, Campisi D, Narayanan L, Jensen R et al (2003) Decreased expression of the DNA mismatch repair gene Mlh1 under hypoxic stress in mammalian cells. Mol Cell Biol 23(9):3265–3273

Miyazaki K, Kawamoto T, Tanimoto K, Nishiyama M, Honda H, Kato Y (2002) Identification of functional hypoxia response elements in the promoter region of the DEC1 and DEC2 genes. J Biol Chem 277(49):47014–47021, Epub 2002 Sep 26

Mottram JC (1936) A factor of importance in the radiosensitivity of tumours. Br J Radiol 9:606–614

Moulder JE, Rockwell S (1987) Tumor hypoxia: its impact on cancer therapy. Cancer Metastasis Rev 5(4):313–341

Murphy BJ, Robin ED, Tapper DP, Wong RJ, Clayton DA (1984) Hypoxic coordinate regulation of mitochondrial enzymes in mammalian cells. Science 223(4637):707–709

Nakada C, Tsukamoto Y, Matsuura K, Nguyen TL, Hijiya N, Uchida T et al (2011) Overexpression of miR-210, a downstream target of HIF1alpha, causes centrosome amplification in renal carcinoma cells. J Pathol 224(2):280–288. doi:10.1002/path.2860, Epub 2011 Apr 4

Nakamura H, Tanimoto K, Hiyama K, Yunokawa M, Kawamoto T, Kato Y et al (2008) Human mismatch repair gene, MLH1, is transcriptionally repressed by the hypoxia-inducible transcription factors, DEC1 and DEC2. Oncogene 27(30):4200–4209, Epub 2008 Mar 17

Nakayama K (2009) Cellular signal transduction of the hypoxia response. J Biochem 146(6):757–765, Epub 2009 Oct 28

Nelson DA, Tan TT, Rabson AB, Anderson D, Degenhardt K, White E (2004) Hypoxia and defective apoptosis drive genomic instability and tumorigenesis. Genes Dev 18(17):2095–2107, Epub 2004 Aug 16

Nouspikel T (2009) DNA repair in mammalian cells: nucleotide excision repair: variations on versatility. Cell Mol Life Sci 66(6):994–1009

Pal A, Srivastava T, Sharma MK, Mehndiratta M, Das P, Sinha S et al (2010) Aberrant methylation and associated transcriptional mobilization of alu elements contributes to genomic instability in hypoxia. J Cell Mol Med 14(11):2646–2654. doi:10.1111/j.1582-4934.2009.00792.x

Pires IM, Bencokova Z, Milani M, Folkes LK, Li JL, Stratford MR et al (2010a) Effects of acute versus chronic hypoxia on DNA damage responses and genomic instability. Cancer Res 70(3):925–935, Epub 2010 Jan 26

Pires IM, Bencokova Z, McGurk C, Hammond EM (2010b) Exposure to acute hypoxia induces a transient DNA damage response which includes Chk1 and TLK1. Cell Cycle 9(13):2502–2507

Plateel M, Dehouck MP, Torpier G, Cecchelli R, Teissier E (1995) Hypoxia increases the susceptibility to oxidant stress and the permeability of the blood-brain barrier endothelial cell monolayer. J Neurochem 65(5):2138–2145

Pocock R (2011) Invited review: decoding the microRNA response to hypoxia. Pflugers Arch 461(3):307–315, Epub 2011 Jan 5

Pollard PJ, Loenarz C, Mole DR, McDonough MA, Gleadle JM, Schofield CJ et al (2008) Regulation of jumonji-domain-containing histone demethylases by hypoxia-inducible factor (HIF)-1alpha. Biochem J 416(3):387–394

Powell-Coffman JA, Coffman CR (2010) Apoptosis: lack of oxygen aids cell survival. Nature 465(7298):554–555

Rankin EB, Rha J, Selak MA, Unger TL, Keith B, Liu Q et al (2009) Hypoxia-inducible factor 2 regulates hepatic lipid metabolism. Mol Cell Biol 29(16):4527–4538, Epub 2009 Jun 15

Reynolds TY, Rockwell S, Glazer PM (1996) Genetic instability induced by the tumor microenvironment. Cancer Res 56(24):5754–5757

Rezvani HR, Mahfouf W, Ali N, Chemin C, Ged C, Kim AL et al (2010) Hypoxia-inducible factor-1alpha regulates the expression of nucleotide excision repair proteins in keratinocytes. Nucleic Acids Res 38(3):797–809, Epub 2009 Nov 24

Rice GC, Hoy C, Schimke RT (1986) Transient hypoxia enhances the frequency of dihydrofolate reductase gene amplification in chinese hamster ovary cells. Proc Natl Acad Sci USA 83(16):5978–5982

Rockwell S, Moulder JE (1990) Hypoxic fractions of human tumors xenografted into mice: a review. Int J Radiat Oncol Biol Phys 19(1):197–202

Rodriguez-Jimenez FJ, Moreno-Manzano V, Lucas-Dominguez R, Sanchez-Puelles JM (2008) Hypoxia causes downregulation of mismatch repair system and genomic instability in stem cells. Stem Cells 26(8):2052–2062, Epub 2008 May 29

Rosenberg SM (2001) Evolving responsively: adaptive mutation. Nat Rev Genet 2(7):504–515

Russo CA, Weber TK, Volpe CM, Stoler DL, Petrelli NJ, Rodriguez-Bigas M et al (1995) An anoxia inducible endonuclease and enhanced DNA breakage as contributors to genomic instability in cancer. Cancer Res 55(5):1122–1128

San Filippo J, Sung P, Klein H (2008) Mechanism of eukaryotic homologous recombination. Annu Rev Biochem 77:229–257

Schults MA, Timmermans L, Godschalk RW, Theys J, Wouters BG, van Schooten FJ et al (2010) Diminished carcinogen detoxification is a novel mechanism for hypoxia-inducible factor 1-mediated genetic instability. J Biol Chem 285(19):14558–14564, Epub 2010 Mar 12

Semenza GL, Roth PH, Fang HM, Wang GL (1994) Transcriptional regulation of genes encoding glycolytic enzymes by hypoxia-inducible factor 1. J Biol Chem 269(38):23757–23763

Sendoel A, Kohler I, Fellmann C, Lowe SW, Hengartner MO (2010) HIF-1 antagonizes p53-mediated apoptosis through a secreted neuronal tyrosinase. Nature 465(7298):577–583

Sermeus A, Michiels C (2011) Reciprocal influence of the p53 and the hypoxic pathways. Cell Death Dis 2:e164

Shahrzad S, Quayle L, Stone C, Plumb C, Shirasawa S, Rak JW et al (2005) Ischemia-induced K-ras mutations in human colorectal cancer cells: role of microenvironmental regulation of MSH2 expression. Cancer Res 65(18):8134–8141

Smith J, Tho LM, Xu N, Gillespie DA (2010) The ATM-Chk2 and ATR-Chk1 pathways in DNA damage signaling and cancer. Adv Cancer Res 108:73–112

Stoler DL, Anderson GR, Russo CA, Spina AM, Beerman TA (1992) Anoxia-inducible endonuclease activity as a potential basis of the genomic instability of cancer cells. Cancer Res 52(16):4372–4378

Tacconelli A, Farina AR, Cappabianca L, Desantis G, Tessitore A, Vetuschi A et al (2004) TrkA alternative splicing: a regulated tumor-promoting switch in human neuroblastoma. Cancer Cell 6(4):347–360

Tacconelli A, Farina AR, Cappabianca L, Gulino A, Mackay AR (2005) TrkAIII. A novel hypoxia-regulated alternative TrkA splice variant of potential physiological and pathological importance. Cell Cycle 4(1):8–9, Epub 2005 Jan 5

Tausendschon M, Dehne N, Brune B (2011) Hypoxia causes epigenetic gene regulation in macrophages by attenuating jumonji histone demethylase activity. Cytokine 53(2):256–262, Epub 2010 Dec 4

Taylor CT, Cummins EP (2009) The role of NF-kappaB in hypoxia-induced gene expression. Ann N Y Acad Sci 1177:178–184

Thomlinson RH, Gray LH (1955) The histological structure of some human lung cancers and the possible implications for radiotherapy. Br J Cancer 9(4):539–549

To KK, Sedelnikova OA, Samons M, Bonner WM, Huang LE (2006) The phosphorylation status of PAS-B distinguishes HIF-1alpha from HIF-2alpha in NBS1 repression. EMBO J 25(20):4784–4794, Epub 2006 Oct 5

Um JH, Kang CD, Bae JH, Shin GG, Kim DW, Kim DW et al (2004) Association of DNA-dependent protein kinase with hypoxia inducible factor-1 and its implication in resistance to anticancer drugs in hypoxic tumor cells. Exp Mol Med 36(3):233–242

Vaupel P (2004) Tumor microenvironmental physiology and its implications for radiation oncology. Semin Radiat Oncol 14(3):198–206

Vaupel P, Mayer A (2007) Hypoxia in cancer: significance and impact on clinical outcome. Cancer Metastasis Rev 26(2):225–239

Vaupel P, Kallinowski F, Okunieff P (1989) Blood flow, oxygen and nutrient supply, and metabolic microenvironment of human tumors: a review. Cancer Res 49(23):6449–6465

Vaupel P, Mayer A, Hockel M (2004) Tumor hypoxia and malignant progression. Methods Enzymol 381:335–354

Walsh S, Gill C, O'Neill A, Fitzpatrick JM, Watson RW (2009) Hypoxia increases normal prostate epithelial cell resistance to receptor-mediated apoptosis via AKT activation. Int J Cancer 124(8):1871–1878

Wang GL, Semenza GL (1993) Characterization of hypoxia-inducible factor 1 and regulation of DNA binding activity by hypoxia. J Biol Chem 268(29):21513–21518

Watanabe Y, Maekawa M (2010) Methylation of DNA in cancer. Adv Clin Chem 52:145–167

Watson JA, Watson CJ, McCrohan AM, Woodfine K, Tosetto M, McDaid J et al (2009) Generation of an epigenetic signature by chronic hypoxia in prostate cells. Hum Mol Genet 18(19):3594–3604, Epub 2009 Jul 7

Watson JA, Watson CJ, McCann A, Baugh J (2010) Epigenetics, the epicenter of the hypoxic response. Epigenetics 5(4):293–296, Epub 2010 May 3

Wellmann S, Bettkober M, Zelmer A, Seeger K, Faigle M, Eltzschig HK et al (2008) Hypoxia upregulates the histone demethylase JMJD1A via HIF-1. Biochem Biophys Res Commun 372(4):892–897, Epub 2008 Jun 4

Wenger RH, Stiehl DP, Camenisch G (2005) Integration of oxygen signaling at the consensus HRE. Sci STKE 2005(306):re12

Wilkinson D, Sandhu JK, Breneman JW, Tucker JD, Birnboim HC (1995) Hprt mutants in a transplantable murine tumour arise more frequently in vivo than in vitro. Br J Cancer 72(5):1234–1240

Wouters BG, Koritzinsky M (2008) Hypoxia signalling through mTOR and the unfolded protein response in cancer. Nat Rev Cancer 8(11):851–864, Epub 2008 Oct 10

Xia X, Kung AL (2009) Preferential binding of HIF-1 to transcriptionally active loci determines cell-type specific response to hypoxia. Genome Biol 10(10):R113, Epub 2009 Oct 14

Xia X, Lemieux ME, Li W, Carroll JS, Brown M, Liu XS et al (2009) Integrative analysis of HIF binding and transactivation reveals its role in maintaining histone methylation homeostasis. Proc Natl Acad Sci USA 106(11):4260–4265, Epub 2009 Mar 2

Young SD, Marshall RS, Hill RP (1988) Hypoxia induces DNA overreplication and enhances metastatic potential of murine tumor cells. Proc Natl Acad Sci USA 85(24):9533–9537

Yuan J, Glazer PM (1998) Mutagenesis induced by the tumor microenvironment. Mutat Res 400(1–2):439–446

Yuan J, Narayanan L, Rockwell S, Glazer PM (2000) Diminished DNA repair and elevated mutagenesis in mammalian cells exposed to hypoxia and low pH. Cancer Res 60(16):4372–4376

Zhou X, Sun H, Chen H, Zavadil J, Kluz T, Arita A et al (2010) Hypoxia induces trimethylated H3 lysine 4 by inhibition of JARID1A demethylase. Cancer Res 70(10):4214–4221, Epub 2010 Apr 20

Ziel KA, Grishko V, Campbell CC, Breit JF, Wilson GL, Gillespie MN (2005) Oxidants in signal transduction: impact on DNA integrity and gene expression. FASEB J 19(3):387–394

Chapter 9
Radiation-Induced Delayed Genome Instability and Hypermutation in Mammalian Cells

Christopher P. Allen, Akira Fujimori, Ryuichi Okayasu, and Jac A. Nickoloff

Abstract Genome instability is a hallmark of cancer cells. Inherited cancer predisposition syndromes typically show defects in DNA repair or DNA damage checkpoint systems, collectively called the DNA damage response (DDR). Several mutations in key genes are required to convert a normal cell to a cancer cell, suggesting that an early step in carcinogenesis is the acquisition of a "genome instability" (mutator) phenotype. DDR proteins suppress cancer by preventing spontaneous damage from causing excessive genome instability, and thus, normal cells display very low mutation rates and stable genomes. Genotoxins such as DNA reactive chemicals and radiation cause DNA damage that results in small- and large-scale genetic change (mutations). Recently it has become clear that radiation, including ionizing radiation (IR) such X-rays and charged particles (heavy ion radiation), as well as nonionizing radiation (UV light) induce genome instability many cell generations after the exposure. These delayed effects are seen after high (1–10 Gy) and very low (0.01–0.1 Gy) IR doses, and include hypermutation, hyper-homologous recombination, chromosome instability, and reduced clonogenic survival (delayed death). Similar to immediate effects of radiation, delayed effects show adaptive responses. Here we focus on potential mechanisms that underlie radiation-induced delayed genome instabilities, and discuss the risks of genome destabilizing effects

C.P. Allen
Department of Environmental and Radiological Health Sciences,
Colorado State University, Fort Collins, CO 80523, USA

Research Center for Charged Particle Therapy, and International Open Laboratory,
National Institute of Radiological Sciences, 263-8555 Chiba, Japan

A. Fujimori • R. Okayasu
Research Center for Charged Particle Therapy, and International Open Laboratory,
National Institute of Radiological Sciences, 263-8555 Chiba, Japan

J.A. Nickoloff (✉)
Department of Environmental and Radiological Health Sciences,
Colorado State University, Fort Collins, CO 80523, USA
e-mail: j.nickoloff@colostate.edu

D. Mittelman (ed.), *Stress-Induced Mutagenesis*, DOI 10.1007/978-1-4614-6280-4_9,

of occupational and accidental radiation exposures, and clinical exposures associated with radiation therapy and diagnostic imaging procedures.

Genome Instability and Its Relationship to Cancer

The most important goal of cell division is to pass accurate copies of the genome to daughter cells. Failures at any stage in DNA replication and chromosome segregation result in small- to large-scale genetic changes that are collectively called genome instability. DNA damage enhances genome instability and is often caused by reactive oxygen species (ROS), which are produced during normal cellular metabolism, and by ionizing radiation (IR). DNA damage is also caused by chemical genotoxins, which include environmental toxins (e.g., pesticides) and the majority of cancer chemotherapeutics. Cells have evolved complex systems to respond to genotoxic stress caused by endogenous or exogenous factors. These stress responses include systems that sense DNA damage, arrest cell cycle progression at specific stages of the cell cycle (checkpoints), and repair DNA damage. When damage is severe, programmed death or anti-proliferative pathways are triggered, such as apoptosis, autophagy, necrosis, and senescence (Stephan and Herman 2006; Gozuacik and Kimchi 2004; Vandenabeele et al. 2010; Galluzzi and Kroemer 2008; Eriksson and Stigbrand 2010). These systems are collectively termed the DNA damage response (DDR) which functions in a highly redundant network to minimize genome changes or eliminate cells that sustain too much damage (Branzei and Foiani 2010; Ciccia and Elledge 2010; Machida et al. 2005; Nyberg et al. 2002; Stojic et al. 2004; Budzowska and Kanaar 2009; Putnam et al. 2009; Mordes and Cortez 2008; Reinhardt and Yaffe 2009; Warmerdam and Kanaar 2010). Cells also exhibit "adaptive" responses, defined operationally, that is, adaptation is said to occur when an initial exposure to a low dose of genotoxic agent is found to minimize the effect of a subsequent higher dose (Bonner 2003; Huang et al. 2006). Thus, adaptive responses are protective mechanisms, however when expressed by a tumor cell, such responses have the negative effect of conferring resistance to treatment.

Genome instability occurs over vast scales, with changes ranging from single-base transition and transversion mutations, small insertion or deletion mutations, short-repeat (microsatellite) expansions and contractions, gene-sized rearrangements in repeated genes, such as gene amplification and gene deletion, to whole chromosome changes such as gross chromosomal rearrangements (GCRs), translocations, and gain or loss of entire chromosomes (aneuploidy) (Nickoloff 2002; Abeysinghe et al. 2003) (Fig. 9.1). All of these changes can alter gene expression or gene function and therefore can be broadly defined as "mutations." While some alterations to the genome are phenotypically silent, most gene mutations are deleterious. Although large-scale changes such as translocations may rearrange the genetic material without affecting gene function within that cell, rearranged chromosomes may not segregate properly during mitosis, yielding mutant or inviable daughter cells. Rearranged chromosomes may also fail to segregate properly in meiosis during germ cell production, resulting in sterility or inherited ("transgenerational") defects.

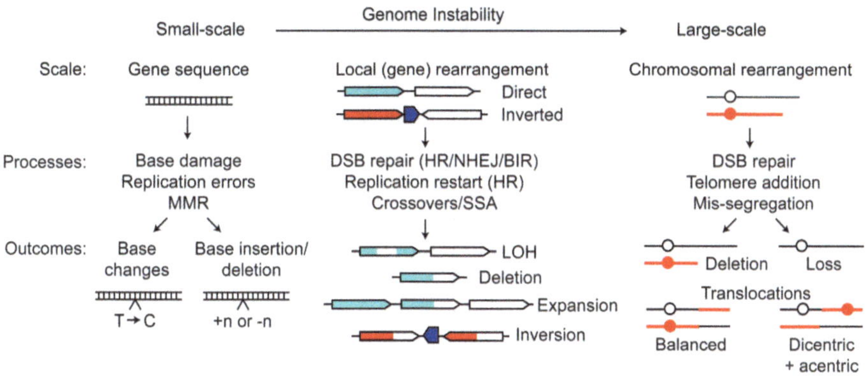

Fig. 9.1 Genome instability occurs over a wide range of scales and is regulated by a variety of DNA dynamic processes. Examples of small-, medium-, and large-scale changes are shown and examples of processes that give rise to specific changes are noted. This list is not exhaustive. For example, NHEJ can result in precise DSB repair, small insertions, or small or large deletions. DNA dynamic processes may directly cause changes (e.g., base-misincorporation during replication) or prevent changes (e.g., mismatch repair, MMR)

It has long been known that cancer cells harbor unstable genomes, but for many years it was unclear whether instability was a cause or an effect of the tumorigenic phenotype. It has been argued that genome instability is an "enabling characteristic" of cancer (Hanahan and Weinberg 2000), and there are now several examples where instability has been shown to precede neoplastic transformation (Hanks et al. 2004; Weaver et al. 2007; Shih et al. 2001; Lengauer et al. 1998). Defects in DNA repair systems increase genome instability and cancer risk, and are associated with many heritable cancer syndromes such as colon cancer (mismatch repair), skin cancer (nucleotide excision repair), and breast cancer (double-strand break (DSB) repair) (Jass 2002; Jiricny 2006; Friedberg 2001; Venkitaraman 2002; Boulton 2006; Gudmundsdottir and Ashworth 2006; Shaheen et al. 2011). Transformation of a normal cell to a cancer cell typically requires mutations in several key genes that regulate cell growth and other properties, but low mutation rates in normal cells preclude acquisition of sufficient mutations within a finite proliferative lifespan. However, cells with unstable genomes have greatly increased mutation rates, and thus, acquisition of a mutator phenotype greatly enhances the probability of transformation to cancer.

Genome Instability Arises from Failed or Inaccurate DNA Repair

Genome instability can arise because of misrepair or failed repair of DNA lesions. For example, DSB repair by homologous recombination (HR) includes subpathways of accurate repair leading to localized loss of heterozygosity (gene conversion), and inaccurate repair by single-strand annealing (SSA), a non-conservative

HR pathway that deletes DNA between linked, direct repeats (Tutt et al. 2001; Schildkraut et al. 2005), and can lead to translocations (Weinstock et al. 2006a, b). Even conservative HR can result in large-scale genome alterations, since gene conversion may occur with an associated crossover event, producing gene-level rearrangements including deletions and inversions, as well as translocations (Nickoloff 2002; Nickoloff et al. 2008). Chromosome-level changes can result from misrepair or failed repair of DNA lesions, as well as errors during chromosome decatenation and segregation (Damelin and Bestor 2007; Yuen et al. 2005).

Defects in HR proteins, such as BRCA2 and BLM, increase genome instability and predispose to cancer (Gudmundsdottir and Ashworth 2006; Cheok et al. 2005). However, if HR is too active, this too can cause genome destabilization and may lead to cancer (Bishop and Schiestl 2001). HR-dependent genome rearrangements have been directly linked to at least 27 human disorders, occurring through interactions between low copy number repeats present on a single chromosome or nonhomologous chromosomes (Stankiewicz and Lupski 2002). The human genome comprises nearly 50% repetitive elements, including large numbers of Alu and other short repeats, that have been implicated in HR-mediated rearrangements in cancer and other human diseases (Abeysinghe et al. 2003; Elliott et al. 2005; Kolomietz et al. 2002). CTG-CAG trinucleotide repeat tract intergenerational expansion contributes to several diseases and has been linked to DNA methylation status as shown by the destabilizing effect of methyltransferase inhibitors (Gorbunova et al. 2004). HSP90 chaperone protein has also been shown to modulate the stability of repeat tracts in response to environmental stress (Mittelman et al. 2010; Mittelman and Wilson 2010). Many labs study HR by using systems comprising repeated genes integrated into the genome. Although these are artificial systems, they are in fact excellent mimics of HR-mediated rearrangements that occur in natural repeats and are therefore highly relevant to the study of human diseases including cancer (Tutt et al. 2001; Nickoloff and Brenneman 2004; D'Anjou et al. 2004; Tremblay et al. 2000).

Human Exposure to Ionizing Radiation

Ionizing radiation (IR) is a "complete carcinogen," that is, it does not require promotion as with certain chemical carincogens. IR causes damage to DNA and other cell components through direct energy absorption, and indirectly by producing ROS. DNA damage can be repaired accurately or misrepaired leading to mutations. DNA repair systems play critical roles in maintaining genome stability in the face of large amounts of spontaneous DNA damage caused by endogenous ROS (byproducts of oxidative metabolism) and external genotoxins such as IR. In addition to low background exposures from cosmic rays and radon, humans are exposed to low doses of IR in various occupations and from diagnostic imaging in the clinic. Very high doses are delivered to cancer patients during radiotherapy, although the highest doses are restricted to tumor tissue, with the surrounding normal tissue receiving low doses. Modern radiotherapy procedures, including intensity modulated radiotherapy, stereotactic radiosurgery, and proton and carbon ion radiotherapy, deliver very high radiation doses to small, well-defined tumor volumes, providing efficient tumor killing while largely sparing normal surrounding tissues. While the risk of secondary tumor induction after radiation therapy is generally low and clearly outweighed by the immediate therapeutic benefits (Newhauser and Durante 2011), some patients, such as children, show significantly increased risk. In fact, for survivors of childhood cancer, the most important risk factor for development of new central nervous system tumors is therapeutic radiation exposure, and in very young children this likely reflects the high sensitivity of the developing brain to radiation (Neglia et al. 2006).

Perhaps even more concerning is the increased use of radiology procedures, such as computed tomography (CT). A single CT scan can result in IR exposures equivalent to >400 normal diagnostic X-rays, spawning significant debate about the safety of CT scans, particularly with regard to wide-spread use in screening and in young children (Rehani and Berry 2000; Wakeford 2008). While MRI scans are considered safer than CT, and MRI can frequently substitute for CT, the much greater cost of MRI has restricted its use. Some types of CT, such as megavoltage CT are used in daily imaging, and doses from a single scan are estimated to range from 1.0 to 3.6 cGy depending on the tissue (Shah et al. 2008). IR doses in the 1–10 cGy range show little or no cytotoxicity, but low doses of IR induce at least two types of delayed genome instability, as discussed below. Thus, an important research goal is to determine the risks associated with low dose exposures in clinical and occupational settings, and then minimize or eliminate exposures that have an appreciable risk of cancer induction. Similarly, an important goal in cancer radiotherapy is to minimize IR exposure of normal tissues, and there is increasing success in this regard with the use of advanced intensity modulated radiotherapy techniques for photon therapy, and by taking advantage of superior dose distributions offered by proton and carbon ion radiotherapy (Allen et al. 2011). Because it is unethical to expose humans to radiation strictly for research purposes, such studies are necessarily performed using epidemiological methods in large patient cohorts, and by using cell and small animal model systems.

Mechanisms of Ionizing Radiation-Induced Delayed Genome Instability

Delayed chromosomal instability (DCI) is assayed using cytogenetic approaches, such as G-banding, chromosome painting, and spectral karyotyping. DCI and delayed death are the best characterized delayed effects of IR (Kim et al. 2006a; Little 1998; Marder and Morgan 1993). DCI and delayed death are induced by high (1–10 Gy) and low (1–10 cGy) doses of low linear energy transfer (LET) IR. IR-induced delayed mutation has been widely studied in cultured cells (Little 1998; Little et al. 1997; Harper et al. 1997; Limoli et al. 1997a; Mothersill et al. 1998; Dahle and Kvam 2003;

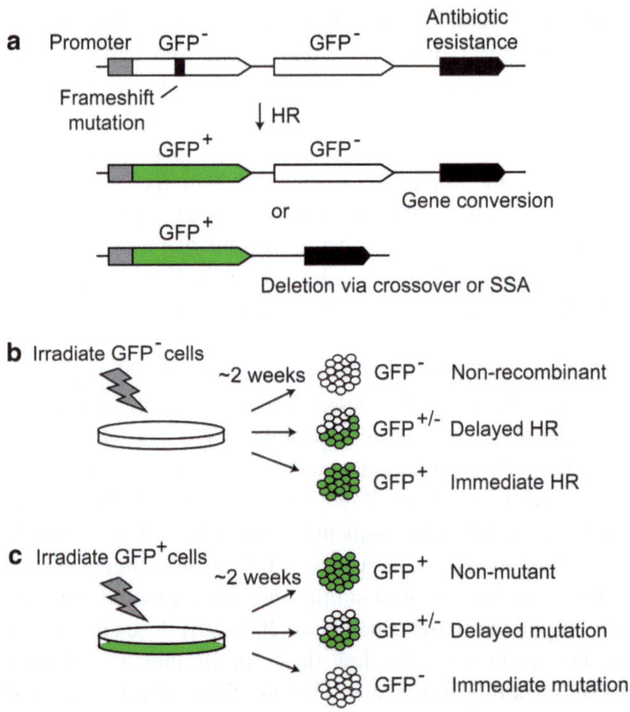

Fig. 9.2 Strategy for detecting delayed instability at an engineered GFP direct repeat. (**a**) The upstream GFP is driven by a promoter but is inactivated by a frameshift mutation; the downstream GFP is inactive because it lacks a promoter. The HR substrate is integrated into the genome using an antibiotic resistance marker. HR products are detected as GFP^+ cells reflecting gene conversion of the frameshift mutation or deletions via crossover or SSA. (**b**) Unirradiated cells (not shown) are stable and typically yield uniform GFP^+ or GFP^- colonies reflecting spontaneous HR products (rare) or non-recombinant, parental cells, respectively. Irradiated cells yield uniform GFP^+ colonies (immediate HR induction or spontaneous HR products), non-recombinant GFP^- colonies, or mixed $GFP^{+/-}$ colonies reflecting DHR. (**c**) Delayed mutation is measured in an analogous manner as DHR, but the assay begins with GFP^+ cells

Evans et al. 2001), and limited data indicates that delayed mutations arise in irradiated mice and are relevant to cancer induction (discussed below). However, none of these studies employed doses less than 1 Gy and most used very high doses (5–12 Gy). A key question is whether low IR doses induce delayed mutation.

DCI and delayed death are highly correlated, suggesting a common underlying mechanism (Huang et al. 2004). In 2004, another type of IR-induced delayed genome instability was discovered. By using human cells harboring a direct repeat GFP HR reporter system, we showed that IR induces delayed HR (DHR) (Huang et al. 2004). An example of a GFP-based HR substrate and its use to assay DHR are shown in Fig. 9.2a, b. In this system, radiation induces HR immediately, producing a homogeneous GFP^+ colony, similar to the well-characterized immediate mutagenic and cytotoxic effects of IR. Some cells will remain GFP^- (non-recombinant) but a third class is mixed $GFP^{+/-}$, reflecting HR events that occur several generations (i.e., several days) after irradiation. Note that once a GFP^{+} recombinant has been isolated, this system can also be used to analyze radiation-induced mutation to GFP^{-}, which can occur either immediately (homogeneous GFP^{-}) or later ($GFP^{+/-}$), reflecting delayed mutation (Fig. 9.2c).

DHR and delayed mutation can also be analyzed with these GFP substrates in real-time using time-lapse fluorescence microscopy or flow cytometry. In preliminary studies we exposed cells with the GFP HR substrate to low doses of high LET carbon ion radiation, which creates complex DNA damage (Hada and Georgakilas 2008). In both time-lapse and flow-cytometric assays, we observed an ~tenfold increase in DHR 6 days after a 1 Gy dose of carbon ion radiation (unpublished results). We hypothesize that the complex damage created by high LET carbon ions underlies this robust DHR response. Time-lapse microscopy also revealed two classes of DHR products that were not detected in colony-based assays including those that appeared to undergo apoptosis, and those that recombined to GFP^{+} but failed to divide and may have senesced. It will be interesting to extend these findings using low doses of both low and high LET IR, and other genotoxic agents.

The Morgan lab showed that DCI is highly correlated with delayed death, and often associated with other delayed instabilities including sister chromatid exchange, micronuclei, and mutation (Limoli et al. 1997a). In a parallel study of DHR and DCI, it was shown that DHR did not correlate with DCI or delayed death, indicating that DHR and DCI/death arise by distinct mechanisms (Huang et al. 2004). Interestingly, although DCI and DHR reflect distinct mechanisms, IR induces both types of delayed instabilities with high efficiency (~2–10% of surviving cells), at very low doses (1–10 cGy), and strikingly these instabilities do not increase in a linear dose-dependent manner (Huang et al. 2006; Huang et al. 2004). For example, IR-induced DHR is induced ~2.5-fold by a 1 cGy dose, but only ~tenfold by a 5 Gy dose which is 500-fold greater (Huang et al. 2006). IR-induced DHR also displays an adaptive response: DHR induction is induced three- to fourfold when cells are irradiated with 1–5 cGy, or 1–5 cGy followed by 5 Gy, but these levels are several fold lower than that seen with a single 5 Gy exposure (Huang et al. 2006). The nonlinear dose dependence, and adaptive response indicate that these delayed effects are not due to direct inactivation of a gene or even members of a large gene family (such as DNA repair genes). Instead, the target appears to be larger than the genome and is probably larger than the nucleus (Huang et al. 2006; Little 1998; Huang et al. 2004). Thus, these IR-induced genome instabilities are often described as non-targeted effects, and in this regard, they are reminiscent of bystander effects (Morgan 2003).

Hu and coworkers addressed the question of whether genomic instability was the result of direct nuclear or cytoplasmic irradiations using microbeam irradiation and found that heritable rearrangements of chromosome 11 were seen at significantly increased frequencies in progeny of cells that were irradiated in nuclear or cytoplasmic regions compared to non-irradiated controls (Hu et al. 2011). Ionizing radiation produces DSBs with "ragged" ends and many other types of DNA damage including single-strand breaks and base damage. "Clean" DSBs, such as those created by nucleases, do not induce DCI (Limoli et al. 1997b), but IR is sufficient and its

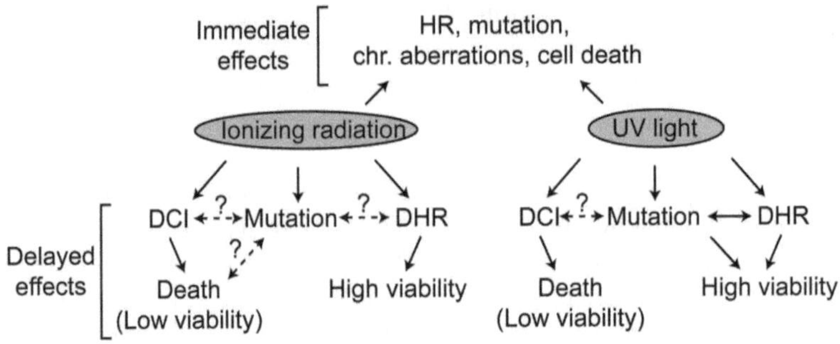

Fig. 9.3 Delayed genomic instabilities result in different cell fates. IR and UV induce a shared set of immediate genetic changes (above), but different types of delayed effects. IR induces DHR, delayed mutation and DCI, cells that display DHR show high viability but DCI is associated with delayed death. UV-induced DHR is associated with delayed mutation that yields a complex mutation spectrum, and high viability. UV induces DCI and delayed death and perhaps distinct types of mutations associated with DCI and delayed death

effects are enhanced by DSB repair deficiencies (Okayasu et al. 2000; Ponnaiya et al. 1997; Barber et al. 2002), suggesting that DCI and delayed death may be triggered by complex and/or poorly repaired DSBs (Little 1998; Evans et al. 2001; Limoli et al. 1997b). Parallel studies to determine if nuclease-induced DSBs fail to induce DHR have not yet been performed.

Like IR, chemical genotoxins and UV induce mutations immediately after exposure, and these agents also induce delayed mutations and DCI with high efficiency (Dahle and Kvam 2003; Dahle et al. 2005; Stamato et al. 1995; Stamato and Perez 1998; Bielfeld et al. 1989). UV also enhances DHR, and delayed mutation and DHR were highly correlated, suggesting that these instabilities are triggered by a common underlying factor. Interestingly, unlike immediate UV mutagenesis, which typically produces a spectrum of simple point mutations that lie in poly-pyrimidine tracts (Tornaletti et al. 1993), the UV-induced delayed mutation spectrum was very unusual, with a high fraction of mutated *HPRT* genes showing multiple, and often complex mutations that may reflect processing of multiple DNA lesions caused by bursts of ROS (Durant et al. 2006). This model is consistent with the association of DHR with delayed mutation, and the observation that HR is greatly stimulated by the ROS generating agent phenytoin (Winn et al. 2003). UV also induces delayed death (O'Reilly and Mothersill 1997) and delayed chromosomal instability (Dahle and Kvam 2003), but as seen with IR-induced DHR, UV-induced DHR was not associated with delayed death (Durant et al. 2006). Thus, as with IR, there appears to be at least two distinct types of delayed genomic instabilities induced by UV (Fig. 9.3). UV-induced DHR is associated with delayed mutation and high viability (Durant et al. 2006), and UV-induced DCI is associated with delayed mutation (Dahle and Kvam 2003), and probably delayed death. It's possible that the delayed mutations associated with DCI arise by a distinct mechanism from those associated with DHR, but this is unknown because only mutation frequencies were measured in the DCI study (Dahle and Kvam 2003).

There are two key questions that remain in the IR-induced delayed instability field: how is genome instability propagated through many generations, and for how long? Because delayed effects are seen 20 or more generations after irradiation, the original damage is diluted through generations and is therefore unlikely to underlie persistent delayed effects. Studies in X-irradiated CHO cells showed that spontaneous mutations at the *HPRT* locus persisted for 95–100 population doublings, producing a heritable mutator phenotype that was two orders of magnitude higher than in non-irradiated controls (Chang and Little 1992). There has been speculation that delayed instability depends on changes in methylation patterns, gene expression, or intracellular signaling/cellular homeostasis (Pogribny et al. 2004; Tawa et al. 1998; Barcellos-Hoff and Brooks 2001; Baverstock 2000; Snyder and Morgan 2004), although the Morgan lab ruled out coordinated gene expression changes (Snyder and Morgan 2005). DCI is known to persist for more than a year, but no studies have addressed the persistence of DHR.

There is growing evidence that persistent oxidative stress underlies at least some forms of delayed genome instability such as DCI, potentially reflecting mitochondrial dysfunction (Kim et al. 2006b), and this idea is supported by the observation that DCI is induced by chronic exposure to H_2O_2 and glucose oxidase (Limoli and Giedzinski 2003). One way to explore the mechanism of IR-induced delayed mutation is to generate mutation spectra. There have been many studies of delayed mutations after IR, typically with high IR doses, but most were limited to mutation frequency analysis (Harper et al. 1997; Limoli et al. 1997a; Dahle and Kvam 2003; Evans et al. 2001; Barber et al. 2002; Carls and Schiestl 1999). Mutation spectra were generated in only two studies. Little and colleagues (1997) used a Southern blot approach to show that delayed *HPRT* mutations were more similar to spontaneous mutations (few large deletions) than to mutations directly induced by IR (frequent large deletions). Ullrich and Ponnayia (1998) analyzed IR-induced delayed mutations in *p53 at* the nucleotide level, but only three mutants were characterized making it difficult to infer mechanism. Next-generation sequencing technology will likely prove more tractable to generate expanded or genome-wide mutation spectra in the future.

As mentioned above, DSB repair capacity appears to modulate IR-induced DCI, and at least at high doses, delayed mutation: cells and mice with DNA-PKcs defects are more prone to these IR-induced instabilities (Okayasu et al. 2000; Ponnaiya et al. 1997; Barber et al. 2002). The idea that repair modulates delayed instability is also supported by experiments showing that delayed instability is suppressed when cell cycle progression is blocked by contact inhibition, presumably reflecting increased time for repair prior to DNA replication (Roy et al. 1999). It is not known whether DSB repair deficiency enhances IR-induced DHR. It is possible that DSB repair deficiency enhances IR-induced delayed genome instabilities because complex, long-lived damage enhances cellular stress responses, or at the least, causes these responses to be more persistent. Further studies are required to resolve these important questions.

Ionizing Radiation-Induced Delayed Instability in Animal Tissues

Several studies have focused on delayed instability in mice after whole-body irradiation. Ullrich and colleagues (Ullrich and Ponnaiya 1998) isolated cells from mouse mammary tissue after a 1 Gy whole body X-ray exposure. Genomically unstable clones were pleiomorphic, preneoplastic, and capable of progressing to neoplasia when cultured in vitro or in vivo, and subclones harbored a variety of p53 mutations. Whole-body irradiation of male mice results in delayed instability in progeny mice, transmitted via damaged sperm but occurring in tissues many cell generations after the original exposure, a clear example of transgenerational instability (Barber et al. 2002; Carls and Schiestl 1999). An acute 7.5 Gy whole body exposure induced HR between direct YFP repeats in cutaneous tissues of transgenic "recombomice," a measure of direct HR induction by IR (Kovalchuk et al. 2004). The same dose delivered chronically at 28 cGy/day had the opposite effect: HR was suppressed below control (spontaneous) levels (Kovalchuk et al. 2004), although follow-up histological studies suggest that HR suppression may actually reflect preferential death of the acinar cells in which HR was monitored (B. Engelward, personal communication). The Engelward lab also demonstrated bystander effects, including epigenetic changes such as gene methylation and altered gene expression in tissue in an unirradiated half of a mouse monitored after the other half was irradiated with 1 Gy, although this study did not evaluate delayed genome instability (Koturbash et al. 2006).

Stepwise Cancer Progression and Radiation Carcinogenesis

Understanding the role of genetic instability in tumorigenesis is a longstanding goal. For years it was argued that destabilization of the genome was an effect not a cause of cancer. In 1976, Peter Nowell formalized the hypothesis whereby a single cell evolved from normal to tumorigenic through an iterative process of mutation, clonal expansion, and selection of increasingly aggressive phenotypes (Nowell 1976). The driver was hypothesized to be genetic instability. The finite lifespan of a cell is believed to suppress tumorigenesis by limiting the replicative lifespan before cells acquire the mutations necessary for tumor formation. Therefore immortalization was recognized as a trait that was acquired extremely rarely, but once acquired became complicit in tumor evolution by providing additional time for a cell to accumulate mutations. Many studies have used viral proteins to generate genetic instability and subsequent tumorigenesis, but these must be interpreted in light of potential interactions between cell and viral proteins. Genome instability generated by low dose IR is the most relevant model for assessing risks associated with medical or other radiation exposures.

Studies of IR-induced tumors in mice and other model organisms have provided insight into the mechanistic basis for radiation induction of cancer. The influence of genetic background in susceptibility to radiation-induced cancer is clear, with the tumor types induced and their incidence being determined by the strain of mouse irradiated (Okayasu et al. 2000; Storer et al. 1988; Szymanska et al. 1999; Yu et al. 2001). Radiation-induced genomic instability, frequently revealed as cytogenetic abnormalities, may account for the high frequency of neoplastic transformation observed in certain model systems. Cancer may be induced at very low, but non-zero frequencies, by low doses of IR, as described by the linear no-threshold model, a model that has been the subject of lively debate (BEIR-VII 2006). Whether or not cancer is induced by extremely low doses of IR, it is clear that cells show robust delayed effects to very low IR doses, and it remains unclear whether mechanisms of tumorigenesis induced by high doses, such as direct inactivation of tumor suppressor genes via large-scale deletions or chromosomal aberrations, are biologically significant at very low doses. Low doses of IR might induce cancer by distinct mechanisms, involving delayed effects such as DHR. This is of concern because DHR has the potential for causing large-scale genome rearrangements, yet it is not associated with low viability like DCI (Huang et al. 2004). Cancer arises within complex tissue microenvironments, and radiation may have many nongenetic effects such as induction of inflammation in stromal cells or other effects that contribute to neoplastic transformation and tumor progression (Barcellos-Hoff et al. 2005; Fujimori et al. 2005). These mechanisms are not mutually exclusive and each may contribute to tumorigenesis in various tissues to different degrees (BEIR-VII 2006).

Acknowledgments We thank Joel Bedford, Howard Liber, Susan Bailey, Michael Weil, and William Morgan for many helpful discussions. Research in the Nickoloff laboratory is supported by NIH grant R01 GM084020, and by the Japan National Institute of Radiological Sciences (NIRS) International Open Laboratory program. Research in the Okayasu and Fujimori laboratories is supported by the NIRS, and Okayasu also receives support from a Grant-in-Aid award from the Japan Society for the Promotion of Science.

References

Abeysinghe SS, Chuzhanova N, Krawczak M, Ball EV, Cooper DN (2003) Translocation and gross deletion breakpoints in human inherited disease and cancer I: nucleotide composition and recombination-associated motifs. Hum Mutat 22:229–244

Allen CP, Borak TB, Tsujii H, Nickoloff JA (2011) Heavy charged particle radiobiology: using enhanced biological effectiveness and improved beam focusing to advance cancer therapy. Mutat Res 711:150–157

Barber R, Plumb MA, Boulton E, Roux I, Dubrova YE (2002) Elevated mutation rates in the germ line of first- and second-generation offspring of irradiated male mice. Proc Natl Acad Sci USA 99:6877–6882

Barcellos-Hoff MH, Brooks AL (2001) Extracellular signaling through the microenvironment: a hypothesis relating carcinogenesis, bystander effects, and genomic instability. Radiat Res 156: 618–627

Barcellos-Hoff MH, Park C, Wright EG (2005) Radiation and the microenvironment—tumorigenesis and therapy. Nat Rev Cancer 5:867–875

Baverstock K (2000) Radiation-induced genomic instability: a paradigm-breaking phenomenon and its relevance to environmentally induced cancer. Mutat Res 454:89–109

BEIR-VII (2006) Exposure to low levels of ionizing radiation. In Committee to Assess Health Risks from Exposure to Low Levels of Ionizing Radiation, Washington, DC

Bielfeld V, Weichenthal M, Roser M, Breitbart E, Berger J, Seemanova E, Rudiger HW (1989) Ultraviolet-induced chromosomal instability in cultured fibroblasts of heterozygote carriers for xeroderma pigmentosum. Cancer Genet Cytogenet 43:219–226

Bishop AJ, Schiestl RH (2001) Homologous recombination as a mechanism of carcinogenesis. Biochim Biophys Acta 1471:M109–121

Bonner WM (2003) Low-dose radiation: thresholds, bystander effects, and adaptive responses. Proc Natl Acad Sci USA 100:4973–4975

Boulton SJ (2006) Cellular functions of the BRCA tumour-suppressor proteins. Biochem Soc Trans 34:633–645

Branzei D, Foiani M (2010) Maintaining genome stability at the replication fork. Nat Rev Mol Cell Biol 11:208–219

Budzowska M, Kanaar R (2009) Mechanisms of dealing with DNA damage-induced replication problems. Cell Biochem Biophys 53:17–31

Carls N, Schiestl RH (1999) Effect of ionizing radiation on transgenerational appearance of p(un) reversions in mice. Carcinogenesis 20:2351–2354

Chang WP, Little JB (1992) Persistently elevated frequency of spontaneous mutations in progeny of CHO clones surviving X-irradiation: association with delayed reproductive death. Mutat Res 270:191–199

Cheok CF, Bachrati CZ, Chan KL, Ralf C, Wu L, Hickson ID (2005) Roles of the Bloom's syndrome helicase in the maintenance of genome stability. Biochem Soc Trans 33:1456–1459

Ciccia A, Elledge SJ (2010) The DNA damage response: making it safe to play with knives. Mol Cell 40:179–204

D'Anjou H, Chabot C, Chartrand P (2004) Preferential accessibility to specific genomic loci for the repair of double-strand breaks in human cells. Nucleic Acids Res 32:6136–6143

Dahle J, Kvam E (2003) Induction of delayed mutations and chromosomal instability in fibroblasts after UVA-, UVB-, and X-radiation. Cancer Res 63:1464–1469

Dahle J, Noordhuis P, Stokke T, Svendsrud DH, Kvam E (2005) Multiplex polymerase chain reaction analysis of UV-A- and UV-B-induced delayed and early mutations in V79 Chinese hamster cells. Photochem Photobiol 81:114–119

Damelin M, Bestor TH (2007) The decatenation checkpoint. Br J Cancer 96:201–205

Durant ST, Paffett KS, Shrivastav M, Timmins GS, Morgan WF, Nickoloff JA (2006) UV radiation induces delayed hyperrecombination associated with hypermutation in human cells. Mol Cell Biol 26:6047–6055

Elliott B, Richardson C, Jasin M (2005) Chromosomal translocation mechanisms at intronic alu elements in mammalian cells. Mol Cell 17:885–894

Eriksson D, Stigbrand T (2010) Radiation-induced cell death mechanisms. Tumour Biol 31:363–372

Evans HH, Horng MF, Ricanati M, Diaz-Insua M, Jordan R, Schwartz JL (2001) Diverse delayed effects in human lymphoblastoid cells surviving exposure to high-LET ^{56}Fe particles or low-LET ^{137}Cs gamma radiation. Radiat Res 156:259–271

Friedberg EC (2001) How nucleotide excision repair protects against cancer. Nat Rev Cancer 1:22–33

Fujimori A, Okayasu R, Ishihara H, Yoshida S, Eguchi-Kasai K, Nojima K, Ebisawa S, Takahashi S (2005) Extremely low dose ionizing radiation up-regulates CXC chemokines in normal human fibroblasts. Cancer Res 65:10159–10163

Galluzzi L, Kroemer G (2008) Necroptosis: a specialized pathway of programmed necrosis. Cell 135:1161–1163
Gorbunova V, Seluanov A, Mittelman D, Wilson JH (2004) Genome-wide demethylation destabilizes CTG.CAG trinucleotide repeats in mammalian cells. Hum Mol Genet 13: 2979–2989
Gozuacik D, Kimchi A (2004) Autophagy as a cell death and tumor suppressor mechanism. Oncogene 23:2891–2906
Gudmundsdottir K, Ashworth A (2006) The roles of BRCA1 and BRCA2 and associated proteins in the maintenance of genomic stability. Oncogene 25:5864–5874
Hada M, Georgakilas AG (2008) Formation of clustered DNA damage after high-LET irradiation: a review. J Radiat Res (Tokyo) 49:203–210
Hanahan D, Weinberg RA (2000) The hallmarks of cancer. Cell 100:57–70
Hanks S, Coleman K, Reid S, Plaja A, Firth H, FitzPatrick D, Kidd A, Méhes K, Nash R, Robin N et al (2004) Constitutional aneuploidy and cancer predisposition caused by biallelic mutations in BUB1B. Nat Genet 36:1159–1161
Harper K, Lorimore SA, Wright EG (1997) Delayed appearance of radiation-induced mutations at the *HPRT* locus in murine hematopoietic cells. Exp Hematol 25:263–269
Hu B, Grabham P, Nie J, Balajee AS, Zhou H, Hei TK, Geard CR (2012) Intrachromosomal changes and genomic instability in site-specific microbeam-irradiated and bystander human-hamster hybrid cells. Radiat Res 177:25–34
Huang L, Grimm S, Smith LE, Kim PM, Nickoloff JA, Goloubeva OG, Morgan WF (2004) Ionizing radiation induces delayed hyperrecombination in mammalian cells. Mol Cell Biol 24: 5060–5068
Huang L, Kim PM, Nickoloff JA, Morgan WF (2006) Targeted and non-targeted effects of low-dose ionizing radiation on delayed genomic instability in human cells. Cancer Res 67: 1099–1104
Jass JR (2002) Pathogenesis of colorectal cancer. Surg Clin North Am 82:891–904
Jiricny J (2006) The multifaceted mismatch-repair system. Nat Rev Mol Cell Biol 7:335–346
Kim GJ, Chandrasekaran K, Morgan WF (2006a) Mitochondrial dysfunction, persistently elevated levels of reactive oxygen species and radiation-induced genomic instability: a review. Mutagenesis 21:361–367
Kim GJ, Fiskum GM, Morgan WF (2006b) A role for mitochondrial dysfunction in perpetuating radiation-induced genomic instability. Cancer Res 66:10377–10383
Kolomietz E, Meyn MS, Pandita A, Squire JA (2002) The role of Alu repeat clusters as mediators of recurrent chromosomal aberrations in tumors. Genes Chromosomes Cancer 35:97–112
Koturbash I, Rugo RE, Hendricks CA, Loree J, Thibault B, Kutanzi K, Pogribny I, Yanch JC, Engelward BP, Kovalchuk O (2006) Irradiation induces DNA damage and modulates epigenetic effectors in distant bystander tissue in vivo. Oncogene 25:4267–4275
Kovalchuk O, Hendricks CA, Cassie S, Engelward AJ, Engelward BP (2004) In vivo recombination after chronic damage exposure falls to below spontaneous levels in "recombomice". Mol Cancer Res 2:567–573
Lengauer C, Kinzler KW, Vogelstein B (1998) Genetic instabilities in human cancers. Nature 396:643–649
Limoli CL, Giedzinski E (2003) Induction of chromosomal instability by chronic oxidative stress. Neoplasia (New York, NY) 5:339–346
Limoli CL, Kaplan MI, Corcoran J, Meyers M, Boothman DA, Morgan WF (1997a) Chromosomal instability and its relationship to other end-points of genomic instability. Cancer Res 57:5557–5563
Limoli CL, Kaplan MI, Phillips JW, Adair GM, Morgan WF (1997b) Differential induction of chromosomal instability by DNA strand-breaking agents. Cancer Res 57:4048–4056
Little JB (1998) Radiation-induced genomic instability. Int J Radiat Biol 74:663–671
Little JB, Nagasawa H, Pfenning T, Vetrovs H (1997) Radiation-induced genomic instability: delayed mutagenic and cytogenetic effects of X-rays and a-particles. Radiat Res 148:299–307

Machida YJ, Hamlin JL, Dutta A (2005) Right place, right time, and only once: replication initiation in metazoans. Cell 123:13–24

Marder BA, Morgan WF (1993) Delayed chromosomal instability induced by DNA damage. Mol Cell Biol 13:6667–6677

Mittelman D, Wilson JH (2010) Stress, genomes, and evolution. Cell Stress Chaperones 15: 463–466

Mittelman D, Sykoudis K, Hersh M, Lin Y, Wilson JH (2010) Hsp90 modulates CAG repeat instability in human cells. Cell Stress Chaperones 15:753–759

Mordes DA, Cortez D (2008) Activation of ATR and related PIKKs. Cell Cycle 7:2809–2812

Morgan WF (2003) Is there a common mechanism underlying genomic instability, bystander effects and other nontargeted effects of exposure to ionizing radiation? Oncogene 22: 7094–7099

Mothersill C, Crean M, Lyons M, McSweeney J, Mooney R, O'Reilly J, Seymour CB (1998) Expression of delayed toxicity and lethal mutations in the progeny of human cells surviving exposure to radiation and other environmental mutagens. Int J Radiat Biol 74:673–680

Neglia JP, Robison LL, Stovall M, Liu Y, Packer RJ, Hammond S, Yasui Y, Kasper CE, Mertens AC, Donaldson SS et al (2006) New primary neoplasms of the central nervous system in survivors of childhood cancer: a report from the Childhood Cancer Survivor Study. J Natl Cancer Inst 98:1528–1537

Newhauser WD, Durante M (2011) Assessing the risk of second malignancies after modern radiotherapy. Nat Rev Cancer 11:438–448

Nickoloff JA (2002) Recombination: mechanisms and roles in tumorigenesis. In: Bertino JR (ed) Encyclopedia of cancer, vol 4, 2nd edn, Elsevier Science. San Diego, USA, pp 49–59

Nickoloff JA, Brenneman MA (2004) Analysis of recombinational repair of DNA double-strand breaks in mammalian cells with I-*Sce*I nuclease. In: Waldman AS (ed) Genetic recombination—reviews and protocols. Humana Press, Totowa, NJ, pp 35–52

Nickoloff JA, De Haro LP, Wray J, Hromas R (2008) Mechanisms of leukemia translocations. Curr Opin Hematol 15:338–345

Nowell PC (1976) The clonal evolution of tumor cell populations. Science 194:23–28

Nyberg KA, Michelson RJ, Putnam CW, Weinert TA (2002) Toward maintaining the genome: DNA damage and replication checkpoints. Annu Rev Genet 36:617–656

Okayasu R, Suetomi K, Yu Y, Silver A, Bedford JS, Cox R, Ullrich RL (2000) A deficiency in DNA repair and DNA-PKcs expression in the radiosensitive BALB/c mouse. Cancer Res 60:4342–4345

O'Reilly JP, Mothersill C (1997) Comparative effects of UV A and UV B on clonogenic survival and delayed cell death in skin cell lines from humans and fish. Int J Radiat Biol 72: 111–119

Pogribny I, Raiche J, Slovack M, Kovalchuk O (2004) Dose-dependence, sex- and tissue-specificity, and persistence of radiation-induced genomic DNA methylation changes. Biochem Biophys Res Commun 320:1253–1261

Ponnaiya B, Cornforth MN, Ullrich RL (1997) Radiation-induced chromosomal instability in balb/c and C57bl/6 mice: the difference is as clear as black-and-white. Radiat Res 147:121–125

Putnam CD, Jaehnig EJ, Kolodner RD (2009) Perspectives on the DNA damage and replication checkpoint responses in Saccharomyces cerevisiae. DNA Repair 8:974–982

Rehani MM, Berry M (2000) Radiation doses in computed tomography. The increasing doses of radiation need to be controlled. BMJ 320:593–594

Reinhardt HC, Yaffe MB (2009) Kinases that control the cell cycle in response to DNA damage: Chk1, Chk2, and MK2. Curr Opin Cell Biol 21:245–255

Roy K, Kodama S, Suzuki K, Watanabe M (1999) Delayed cell death, giant cell formation and chromosome instability induced by X-irradiation in human embryo cells. J Radiat Res 40: 311–322

Schildkraut E, Miller CA, Nickoloff JA (2005) Gene conversion and deletion frequencies during double-strand break repair in human cells are controlled by the distance between direct repeats. Nucleic Acids Res 33:1574–1580

Shah AP, Langen KM, Ruchala KJ, Cox A, Kupelian PA, Meeks SL (2008) Patient dose from megavoltage computed tomography imaging. Int J Radiat Oncol Biol Phys 70:1579–1587

Shaheen M, Allen C, Nickoloff JA, Hromas R (2011) Synthetic lethality: exploiting the addiction of cancer to DNA repair. Blood 117:6074–6082

Shih IM, Zhou W, Goodman SN, Lengauer C, Kinzler KW, Vogelstein B (2001) Evidence that genetic instability occurs at an early stage of colorectal tumorigenesis. Cancer Res 61:818–822

Snyder AR, Morgan WF (2004) Radiation-induced chromosomal instability and gene expression profiling: searching for clues to initiation and perpetuation. Mutat Res 568:89–96

Snyder AR, Morgan WF (2005) Lack of consensus gene expression changes associated with radiation-induced chromosomal instability. DNA Repair 4:958–970

Stamato TD, Perez ML (1998) EMS and UV-light-induced colony sectoring and delayed mutation in Chinese hamster cells. Int J Radiat Biol 74:739–745

Stamato TD, Richardson E, Perez ML (1995) UV-light induces delayed mutations in Chinese hamster cells. Mutat Res 328:175–181

Stankiewicz P, Lupski JR (2002) Genome architecture, rearrangements and genomic disorders. Trends Genet 18:74–82

Stephan JS, Herman PK (2006) The regulation of autophagy in eukaryotic cells: do all roads pass through Atg1? Autophagy 2:146–148

Stojic L, Brun R, Jiricny J (2004) Mismatch repair and DNA damage signalling. DNA Repair 3:1091–1101

Storer JB, Mitchell TJ, Fry RJ (1988) Extrapolation of the relative risk of radiogenic neoplasms across mouse strains and to man. Radiat Res 114:331–353

Szymanska H, Sitarz M, Krysiak E, Piskorowska J, Czarnomska A, Skurzak H, Hart AA, de Jong D, Demant P (1999) Genetics of susceptibility to radiation-induced lymphomas, leukemias and lung tumors studied in recombinant congenic strains. Int J Cancer 83:674–678

Tawa R, Kimura Y, Komura J, Miyamura Y, Kurishita A, Sasaki MS, Sakurai H, Ono T (1998) Effects of X-ray irradiation on genomic DNA methylation levels in mouse tissues. J Radiat Res 39:271–278

Tornaletti S, Rozek D, Pfeifer GP (1993) The distribution of UV photoproducts along the human p53 gene and its relation to mutations in skin cancer. Oncogene 8:2051–2057

Tremblay A, Jasin M, Chartrand P (2000) A double-strand break in a chromosomal LINE element can be repaired by gene conversion with various endogenous LINE elements in mouse cells. Mol Cell Biol 20:54–60

Tutt A, Bertwistle D, Valentine J, Gabriel A, Swift S, Ross G, Griffin C, Thacker J, Ashworth A (2001) Mutation in Brca2 stimulates error-prone homology-directed repair of DNA double-strand breaks occurring between repeated sequences. EMBO J 20:4704–4716

Ullrich RL, Ponnaiya B (1998) Radiation-induced instability and its relation to radiation carcinogenesis. Int J Radiat Biol 74:747–754

Vandenabeele P, Galluzzi L, Vanden Berghe T, Kroemer G (2010) Molecular mechanisms of necroptosis: an ordered cellular explosion. Nat Rev Mol Cell Biol 11:700–714

Venkitaraman AR (2002) Cancer susceptibility and the functions of BRCA1 and BRCA2. Cell 108:171–182

Wakeford R (2008) Childhood leukaemia following medical diagnostic exposure to ionizing radiation in utero or after birth. Radiat Prot Dosimetry 132:166–174

Warmerdam DO, Kanaar R (2010) Dealing with DNA damage: relationships between checkpoint and repair pathways. Mutat Res 704:2–11

Weaver BA, Silk AD, Montagna C, Verdier-Pinard P, Cleveland DW (2007) Aneuploidy acts both oncogenically and as a tumor suppressor. Cancer Cell 11:25–36

Weinstock DM, Elliott B, Jasin M (2006a) A model of oncogenic rearrangements: differences between chromosomal translocation mechanisms and simple double-strand break repair. Blood 107:777–780

Weinstock DM, Richardson CA, Elliott B, Jasin M (2006b) Modeling oncogenic translocations: distinct roles for double-strand break repair pathways in translocation formation in mammalian cells. DNA Repair 5:1065–1074

Winn LM, Kim PM, Nickoloff JA (2003) Oxidative stress-induced homologous recombination as a novel mechanism for phenytoin-initiated toxicity. J Pharmacol Exp Ther 306:523–527

Yu Y, Okayasu R, Weil MM, Silver A, McCarthy M, Zabriskie R, Long S, Cox R, Ullrich RL (2001) Elevated breast cancer risk in irradiated BALB/c mice associates with unique functional polymorphism of the Prkdc (DNA-dependent protein kinase catalytic subunit) gene. Cancer Res 61:1820–1824

Yuen KW, Montpetit B, Hieter P (2005) The kinetochore and cancer: what's the connection? Curr Opin Cell Biol 17:576–582

Chapter 10
Radiation-Induced Bystander Effects and Stress-Induced Mutagenesis

Carmel Mothersill and Colin Seymour

Abstract Two very exciting scientific revolutions have occurred in biology in the last 20 or so years. For many years, the world believed that mutations occur at random. The discovery of stress-induced mutagenesis has changed ideas about mutation and evolution and revealed mutagenic programs that differ from standard spontaneous mutagenesis in rapidly proliferating cells. In an equally exciting paradigm shift in the seemingly unrelated field of radiation biology, we have uncovered a potential driver for stress-induced mutagenesis: the bystander effect. Stress-induced mutations occur during growth-limiting stress: nutrient stress for bacteria, and radiation damage for bystander effects. Stress-induced mutations can include adaptive mutations that allow survival in the otherwise hostile environment. The stress responses increase mutagenesis specifically when cells are maladapted to their environments, potentially accelerating evolution. Stress-induced mutation mechanisms may provide important models for the genomic instability underlying some cancers and genetic diseases, for rapid adaptation of biota in changed environments, for resistance to chemotherapeutic and antibiotic drugs, for pathogenicity of microbes, and for many other important evolutionary processes. This chapter reviews our knowledge of radiation-induced bystander effects, and how they may contribute to our mechanistic understanding of mutagenesis and adaptive evolution. Implications for disease therapy, protection of species—including humans—from stressors such as radiation, and mechanisms of evolutionary change are discussed.

C. Mothersill (✉)
Medical Physics and Applied Radiation Sciences,
McMaster University, Hamilton, ON L8S 4K1, Canada
e-mail: mothers@mcmaster.ca

C. Seymour
McMaster University, Hamilton, ON L8S 4K1, Canada

D. Mittelman (ed.), *Stress-Induced Mutagenesis*, DOI 10.1007/978-1-4614-6280-4_10,

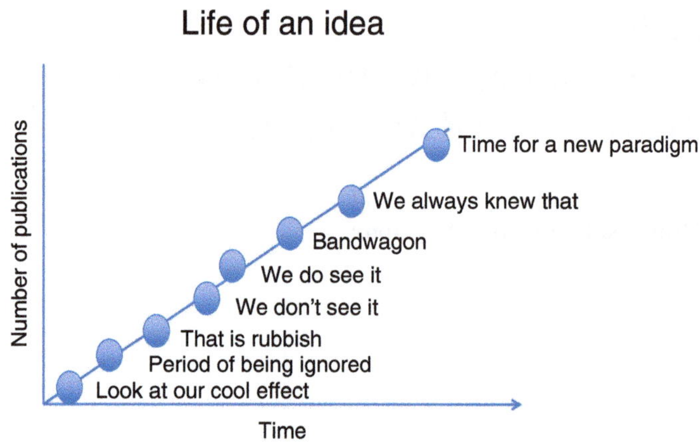

Fig. 10.1 A cynical but often true view of the phases through which a paradigm breaking idea goes as evidence mounts up

Introduction

Scientific ideas have a life similar to that shown in Fig. 10.1—they start as revolutions, gradually gain acceptance—usually in the face of overwhelming data—and are finally accepted as having been obvious all along. Two exciting and seemingly unrelated revolutions occurred within the last 15–20 years, which are just now coming together in a way which will have major implications for our understanding of biology, from basic molecular biology to ecosystems biology. The first revolution concerns the subject of this book: stress-induced mutagenesis (SIM). The second concerns the subject of this chapter: radiation-induced bystander effects (RIBE). Although the fields developed independently, they both have overturned a central paradigm of biology: mutagenesis is a random process, and any change resulting from the process is due to post-mutation selection of traits that confer a survival advantage. Darwin and neo-Darwinian science were considered to have proven wrong the idea that the environment could influence the rate or fate of mutations. However, in the case of SIM, the environment was shown to induce changes in mutation rate, while with RIBE, cells and organisms receiving low doses of radiation were shown to signal to their neighbors, leading to induction of genomic instability—i.e., a greater than normal tolerance of mutation by the system. The aim of this chapter is to introduce RIBE and discuss the mechanisms and implications, while drawing parallels between the SIM and RIBE processes.

RIBE Phenomenology

Background

The "non-targeted effects" of ionizing radiation—including bystander effects and genomic instability—are unique in that no classic mutagenic event occurs in the cell showing the effect (Wright 2010; Ilnytskyy and Kovalchuk 2011; Mothersill and

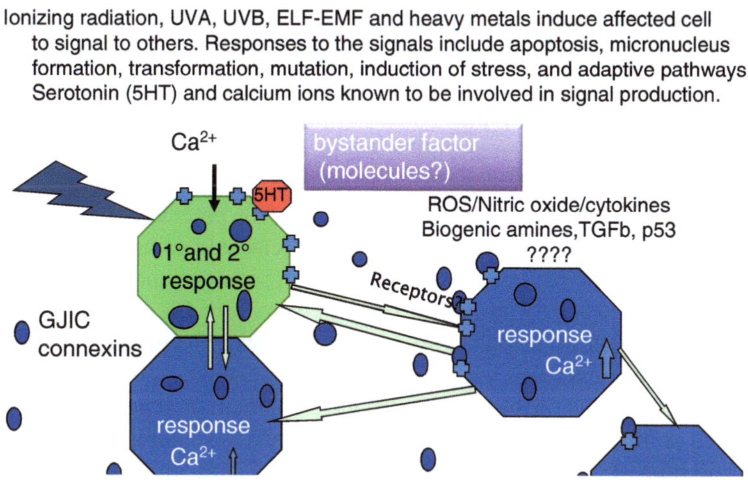

Fig. 10.2 Schematic showing the bystander effect and currently known mechanisms

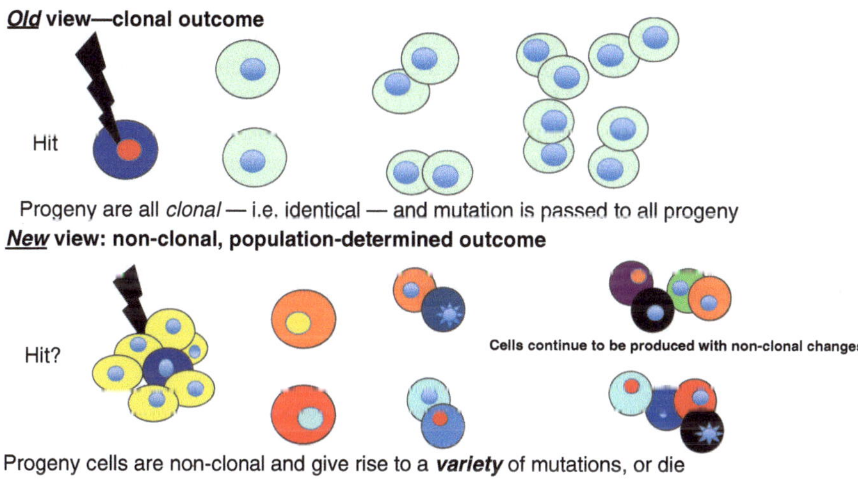

Fig. 10.3 Schematic showing the process of genomic instability generation in progeny after irradiation of progenitors

Seymour 2012). In the case of bystander effects, cells which were not in the field affected by the radiation show high levels of mutations, chromosomal aberrations, and membrane signaling changes (horizontal transmission of mutations and information, which may be damaging or adaptive); see Fig. 10.2 (Nagasawa and Little 1992; Mothersill and Seymour 1997, 2001; Hei et al. 2008). In the case of genomic instability, generations of cells derived from an irradiated progenitor appear normal, but then lethal and non-lethal mutations appear in distant progeny (a phenomenon called "vertical transmission"); see Fig. 10.3 (Kadhim et al. 1992; Seymour and Mothersill 1997; Aypar et al. 2011). Both phenomena are characterized by high yields of non-clonal mutations and distant de novo occurrence of events both in space and time. The occurrence of several normal cell divisions followed by the abrupt appearance of high yields of aberrant cells precludes a mutator phenotype or other conventional explanations (Christians et al. 1995; Dayal et al. 2008), and appear to indicate a threshold for induction of a generalized form of stress-induced mutagenesis—the subject of this book—which is well documented in bacteria (Hersh et al. 2004; Galhardo et al. 2007). It is implicit in considerations of RIBE and genomic instability that there must be a "driver," i.e., a substance, signal, or process that leads to sustained generation of mutations once the process is induced, without these mutations diminishing with time or being selected against, even when they result in the death of the cells containing them (Seymour and Mothersill 1997; Mothersill et al. 2000a, b; Rugo et al. 2011). In fact, a key discovery leading to this scientific revolution was published by our group in 1986 (Seymour et al. 1986): the high yield of lethal mutations in cells that survived radiation exposure. Later studies showed that, depending on the cell line, the yield of mutations was dose independent and ranged from zero to 90% cell death per population doubling (Seymour and Mothersill 1988). For a given cell line, the yield was constant, and our conclusion was that we were seeing a reset mutation tolerance in post-irradiation survivors. The data challenged conventional views about selection since the survivors should have repopulated the cell cultures with progeny that did not carry the division failure trait. Later, as research on RIBE began to emerge, our conclusion was that the bystander process generated the lethal mutations in the population, which explained why it did not go away (Seymour and Mothersill 1997; Lyng et al. 2002a, b). This, in effect, was a demonstration of stress-induced mutagenesis. In the radiobiology language and mindset of the time, however, the best guess as to mechanism was that the cell stress led to toxic free radical production (Kashino et al. 2007; Ermakov et al. 2009; Azzam et al. 2012; Widel et al. 2012), which broke DNA, leading to random mutations. Modern science recognizes cell stress as an active process, first recognized in 1962 as a response to heat shock (Milkman 1962; Haak and Kregel 2008).

Cell Stress

Stress is a word with multiple meanings. In the context discussed here, it refers to the ways cells can respond to stress, ranging from the activation of survival pathways to the initiation of cell death that eventually eliminates damaged cells. Whether cells mount a protective response or succumb to death depends to a large extent on the nature and duration of the stress, as well as the cell type. For example, milder stresses can lead to protection through activation of the heat shock response, or the unfolded protein response (UPR). The literature describes several types of cell death (e.g., apoptosis, necrosis, pyroptosis, or autophagic cell death) and the mechanism by which a cell dies often depends on various exogenous factors, as well as the cell's ability to handle the stress to which it is exposed. There are many molecules and cellular processes that play critical roles in normal cell signaling and survival responses, while also having a dual role in inducing cell death, including endoplasmic reticulum (ER) stress (Moretti et al. 2007; Johnson et al. 2011) and Ca^{2+} signaling

(Walleczek 1992; Kim et al. 2006). The RIBE literature has identified calcium as critical in the radiation response pathway (Lyng et al. 2000, 2006; Shao et al. 2006), suggesting that RIBE falls into the category of an active cellular stress response whose outcome depends on environmental conditions in and around the irradiated cell. However, the alternative theory is that radiation generates the toxic radicals—among which oxygen radicals are most important in radiation damage—and that by some type of futile cycle, these are perpetually regenerated in cells long after the actual period of irradiation. This phenomenon is known as persistent oxidative stress (Ermakov et al. 2009; Martinez-Outschoorn et al. 2010; Buonanno et al. 2011).

Oxidative Stress

Oxidative stress describes cell damage caused by an overabundance of oxidants, including reactive oxygen species (ROS)—e.g., free radicals, oxygen ions, and peroxide. ROS are harmful in excess, but are also necessary for important cellular functions such as cell signaling (Lyng et al. 2011). Also, oxidative stress plays a role in cellular processes such as aging and apoptosis. In a normal cell, ROS are produced as a by-product of metabolic processes. Enzymes, including superoxide dismutase and catalase, control the level in cells, as do certain dietary vitamins and cofactors (Azzam et al. 2002; Spitz et al. 2004; Dahle et al. 2005; Decourscy and Ligeti 2005). However, if the level of oxidants outweighs the level of antioxidants, damage is caused to nuclear and mitochondrial DNA, proteins, and lipids. If this damage is irreparable, then injury, mutagenesis, carcinogenesis, accelerated senescence, and cell death can occur (Haak and Kregel 2008; Sedelnikova et al. 2010; Widel et al. 2012). Oxidative stress has been linked to diseases, including some allergic and inflammatory skin diseases (Martins et al. 1986; Fariello 1988; Simpson et al. 2003; Lankin et al. 2005; Moreira et al. 2005; Okayama 2005; kovacic and Somanathan 2010; Chiavaroli et al. 2011; Cotella et al. 2012).

A state of oxidative stress can be induced by a number of factors, including chemical agents and radiation. Radiation-induced damage and oxidative stress are closely tied. Irradiated cells produce damaging ROS. Treatment with antioxidants can dampen the detrimental effects of high dose radiation exposure (Mothersill et al. 2000a, b; Preta et al. 2009). Cytoplasmic irradiation can result in damage to nuclear DNA, and experiments with free radical scavengers support the idea that this DNA damage is dependent on ROS generation (Huang and Sheridan 1981; Gillies 1987; Slade and Radman 2011). Evidence suggests that a cell's oxidative state not only plays a role at the time of radiation exposure, but also has effects long after exposure. As the result of irradiation, cells can produce ROS for many days after irradiation, and results by Lyng et al. (1996) and Clutton et al. (1996) in completely different systems show that the elevated ROS effect can be transmitted to progeny. In addition to ROS production, irradiated cells are stimulated to increase their expression of antioxidants (Lee et al. 2001; Kojima et al. 2011), which may be a mechanism leading to the perpetuation of the effect. The elevation of antioxidant defenses may play a role in the radioadaptive response (Fan et al. 2007; Carbone et al. 2009; McDonald et al. 2010); that is, the triggering of cellular effects upon radiation exposure that protect a cell from a subsequent radiation challenge.

Following ionizing radiation exposure, several signal transduction pathways (e.g., ERK1/2, JNK, p38, and ATM) and transcription factors (e.g., AP1, NFκB, GADD153, and p53) are activated (Hei et al. 2011; Lagerwerf et al. 2011). Many of these signaling and gene expression pathways are also sensitive to changes in intracellular oxidation/reduction reactions (Hitchler and Domann 2007; Chen et al. 2008; Ozden et al. 2011). This again highlights the dual nature of RIBE responses where the outcome is not a clear consequence of dose or exposure, but can be modulated by the environment in which the initial exposure and the consequent response take place.

What Constitutes a Stressor?

This raises the question: what actually is a stressor? A dictionary definition is "An agent, condition, or other stimulus that causes stress to an organism" (http://medical-dictionary.thefreedictionary.com/stress); however, this raises the question of what is meant by "stress." To quote Wikipedia, "Stress is a term that is commonly used today but has become increasingly difficult to define. It shares, to some extent, common meanings in both the biological and psychological sciences. Stress typically describes a negative concept that can have an impact on one's mental and physical well-being, but it is unclear what exactly defines stress and whether or not stress is a cause, an effect, or the process connecting the two. With organisms as complex as humans, stress can take on entirely concrete or abstract meanings with highly subjective qualities, satisfying definitions of both cause and effect in ways that can be both tangible and intangible." While this definition is written to define human stress, it is difficult to know where the boundary exists in non-human species, as well as cells and tissues, in relation to the link between the physical experience of a stressor and the associated neural, hormonal and homeostatic response to a stressful condition. This question is fundamental to the understanding of both radiation harm and risk, and mutagenic risk from stress. As stated, above stress has negative connotations at the level of the individual, but at the level of the population or ecosystem it may be an essential driver of adaptation. An attempt to conceptualize this is shown in Fig. 10.4. Here, stress-induced mutations are the result of induced genomic instability and provide abundant material for adaptation and selection.

Generalized "Stress"

Perhaps the most useful definition of biological stress is disruption to normal environmental conditions. This idea was first developed by Cannon in 1926

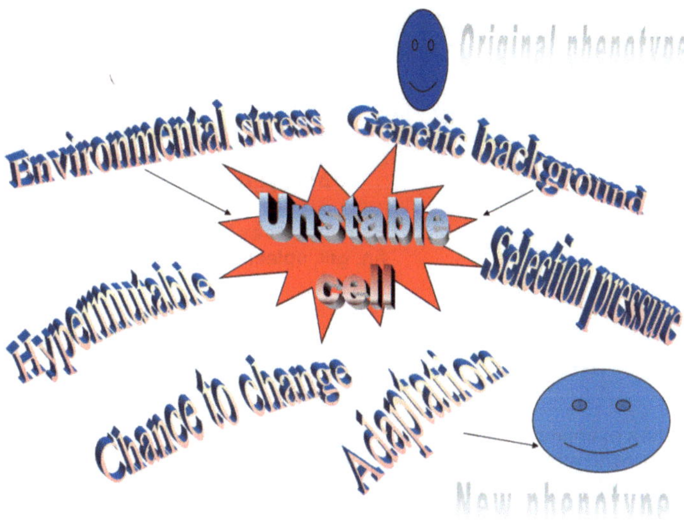

Fig. 10.4 A conceptual depiction of stress-induced mutagenesis in evolution

(Cannon 1926) who coined the term "homeostasis" to describe the normal state in the cell, tissue, or organism. Stress was then supposed to be any change that disrupted homeostasis (Selye 1956). More recently, Koolhaas et al. (2011) defined stress as disruption that exceeded the ability of the system to cope. In other words, it is only stress if homeostatic mechanisms are saturated. The problem with this idea is that it defines stress as only occurring after the dose response curve has entered the damaging region, while proponents of theories such as hormesis, adaptive response, and directed mutagenesis claim that the key role of stressors is before the region of the dose response curve where actual damage is seen. In fact, the inter-animal and inter-plant signaling now known to occur following low dose radiation exposure to selected individuals in a population (Surinov et al. 2001, 2004; Mothersill et al. 2006a, b; Mothersill et al. 2007a, b; Mothersill et al. 2009; Audette-Stuart and Yankovich 2011; Isaeva and Surinov 2011; Smith et al. 2011) would suggest that the "stress response"—including stress-induced mutagenesis—can be activated in advance of exposure to the stressor. Figure 10.5 shows a typical experimental setup used to demonstrate this point. Fascinating questions needing attention include the nature of mechanisms and signals involved, the specificity of the response within or between species, and the purpose of such preemptive mutagenesis, which in some systems leads to the production of lethal genetic damage in organisms that never were exposed to radiation (Seymour and Mothersill 1997; Chai and Hei 2008; Shao et al. 2008; Hanot et al. 2009). Are these "casualties of war," or hypersensitive individuals at the extreme end of a sensitivity range? The concept is illustrated in Fig. 10.6.

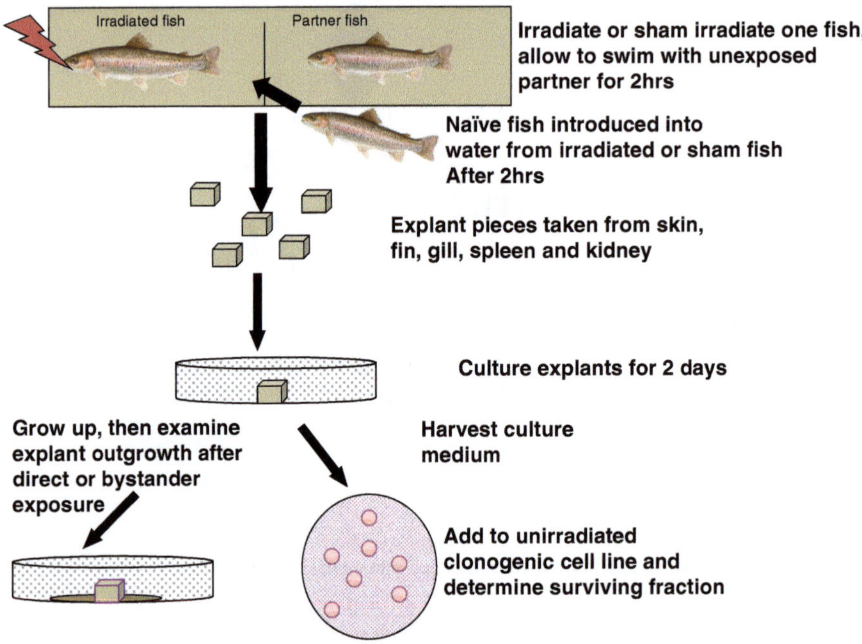

Fig. 10.5 Measuring bystander response to radiation in vivo

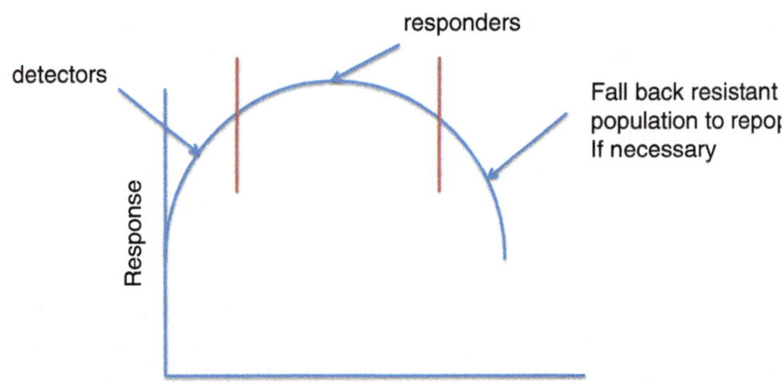

Fig. 10.6 Concept of first responders/reactors which are very sensitive to the stressor and alert others

Stressors and Innate Immune Response

The casualties of war concept is reminiscent of the innate immune response, which is highly important in lower organisms (invertebrates and fish) that do not have well-developed inducible immune systems (Plouffe et al. 2005; Magnadóttir 2006). Organisms relying on innate immune systems for defense use signaling similar to RIBE signals to induce protective responses, such as toxic secretions that aim to destroy "non-self"—i.e., cells or organisms perceived as different (Ottaviani et al. 1998). Many colonial organisms such as tunicates and lichens use the production of antibiotics or other cytotoxins to prevent other colonies from encroaching on their territory (Rinkevich and Weissman 1992; Ausubel 2005). This similarity between RIBE and innate immunity extends to the perception of stress, as well as the spatial and temporal relationship between exposure and response (Matzinger 1998; McBride et al. 2004). It is likely that studies of the mechanisms involved in innate immunity could help us understand the relationship between the stressor and the transduction of the stress response in general, including the specific situation of stress-induced mutagenesis.

Multiple Stressors

The reality in biology is that stressors seldom act alone. In the environment or in the individual, multiple and complex factors act to produce stress. Often an initiating stressor leads to other secondary stresses in the system. When trying to determine either the mechanisms or the implications of stress exposure, it is critical to bear this in mind. The text box below lists some of the issues in this field under the headings of "theoretical issues" and "practical issues." There is no agreement about how to deal with multiple stressors. Current opinion is that iso-effect should be used to try to add up the impacts of mixed exposures rather than iso-dose (Mothersill and Seymour 2009a, b; Mothersill et al. 2007a, b), but both approaches neglect to consider possible synergistic interactions between stressors, which could increase or decrease the potential additive effect. Also disregarded in this approach is the likely existence of response thresholds, which can be tipping points for the initiation of programs such as stress-induced mutagenesis, death, or other catastrophic change in response. Hormetic or adaptive effects are also not accommodated, even though they are well established in toxicology (Calabrese et al. 2011; Gori and Münzel 2012) and involve the concept of low doses of toxins, including radiation, being good for organisms, while high doses are toxic. How these non-linear dose response relationships might be altered or affected by the presence of multiple stressors is not known, but the data that does exist suggests that complex interactions do occur (Mothersill et al. 1998; Salbu et al. 2008; Souidi et al. 2009; Vanhoudt et al. 2010). This is clearly an area needing attention both from the theoretical and practical perspective.

Issues of concern in the multiple stressor field
Theoretical issues
• Is the time sequence of exposure important?
• Is the relative dose of each stressor important?
• Are there stressors that dominate?
• Can pre-exposure to one stressor confer resistance to a subsequent stressor?
Practical issues
• Compliance can only be measured if there are dose limits, or other quantitative parameters to meet
• Multiple stressor "doses" in the chemical world can be added together, but how can we "add" radiation dose and chemical dose?
• What about adaptation, hormetic or synergistic responses, and antagonistic effects?
• Might we need a different number for populations from pristine environments with no stress adaptation?

Randomness and Emergent Properties

A final point to discuss before dealing with mechanisms is the role of randomness and emergent properties of irradiated systems. Clearly, unstable systems such as cell populations showing non-clonal genomic instability are demonstrating randomness. These mutations are unpredictable, both in where and when they occur. An emergent property of the system is, however, that the random mutation burden triggers a higher order control at the tissue level, such as signaling to coordinate apoptosis, growth, or differentiation, as appropriate, which ensures tissue function is preserved (Cox 2006; Damiani et al. 2011; Ichinose and Kobayashi 2011). At the level of the ecosystem or population, the randomness of individual death/mutation leads to the emergent property of adaptation benefiting the higher order survival of the species or biotype.

Mechanisms

Dose Responses

The most striking thing about bystander dose response curves is that they rarely show a continuous increase in effect with dose. Rather they show very low dose thresholds and then saturate (Mothersill and Seymour 1998a, b; Prise et al. 2002; Liu et al. 2006). This means that the greatest relative contribution of the bystander effect to the total effect is in the low dose region (Fig. 10.7). In some systems there is a binary effect (Schettino et al. 2003; Schettino et al. 2005), meaning that for any given cell, the expression of the effect is either on or off. In terms of mechanisms, this type of dose response suggests that a trigger level of radiation is necessary to activate the cellular pathway(s) involved in signal production. This is very exciting, since calcium ion-gated channels (5HT type 3A) are thought to be implicated in signal production (Miyachi et al. 1994; Dobrev 2007; Poon et al. 2007; Saroya et al. 2009; Sreevidya et al. 2010), and, as radiation is ionizing and affects cell membranes, it is probably possible to calculate the energy

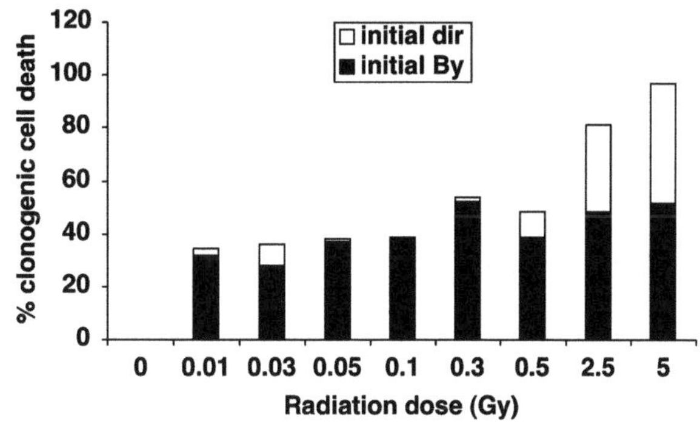

Fig. 10.7 Relative contribution of direct (dir) and bystander (by) cell death to the total amount of cell death caused by a radiation dose

needed to open a channel by using biophysical modeling. Work using specific inhibitors of the type 3A receptor involves ondansetron and kitryl, (Sagrada et al. 1991; Andrews et al. 1992; Aapro and Blower 2005), both of which prevent bystander effects in irradiated cells. This is an exciting area in radiotherapy.

Much of the mechanistic research in the bystander field concerns the use of microbeams. Proper review of this field is outside the scope of this chapter, but it is well reviewed (Matsumoto et al. 2009; Prise and Schettino 2011). Here, the energy is deposited in a cell or in part of a cell, and effects are measured in adjacent or distant cells. This allows very precise quantification of the dose to a cell or part of a cell, and precise evaluation of the distance the signal can travel (Schettino et al. 2003; Prise et al. 2006). Thus, both spatial and temporal information can be obtained. Problems limiting the usefulness of microbeams are that it is almost impossible to get data for whole organisms—some experiments have been done with very small fish (http://www.raraf.org/in%20vivo.htm)—and there is concern about the confounding effects of the fluorescent dyes and UV imaging needed to visualize the results (Gault et al. 2007). No system is perfect, though, and there are also many criticisms of in vivo models where part of the animal is shielded by neural and hormonal influences, as well as by circulating blood and lymph, neither of which can be controlled.

Another aspect of dose response is the dose of bystander medium required to induce the response in a recipient cell. Mothersill and Seymour (1997) and Ryan et al. (2008a, b) looked at this and found that a cut off cell number for producing enough bystander signal in 1 h into 5 ml of culture medium was 50,000 cells. Cell numbers below this did not produce enough signal for the clonogenic reporter assay to be able to detect. However, in Ryan's work, it was found that a 1:1 dilution of bystander medium with unused medium completely removed the effect. Lower dilutions gave a dose response. This strongly suggests that rather than containing a mystery bystander factor, bystander medium may be deficient in some factor, which may have been bound or used by the irradiated cells. Our candidate factor, which we have found to be bound by irradiated cells, is serotonin (5HT) (Mothersill et al. 2010a, b), and we now suggest that it is bound to the 5HT 3A receptor to allow the calcium channel to open, thus permitting calcium influx into the cells. Calcium influx is a known very early event in stress response using the MAPK pathway, and serotonin is a neurotransmitter known to have other important functions and receptors in non-neural cells (Poon et al. 2007; Müller and Meineke 2011). This hypothesis assumes that, at least in the system used in our laboratory, serotonin is necessary for the normal function of the un-irradiated cells that receive bystander medium, but we have yet to confirm this.

Cell Line Differences

A key early question in the bystander field was whether all cells showed bystander effects. Much of the confusion and problems with early acceptance of data revolved around the fact that many cell lines did not show bystander-like effects (Groesser et al. 2008; Sowa et al. 2010), or they did sometimes, or only in some labs (Fournier et al. 2009). Most of these early problems have not been resolved as being due to serum batch variability, light sensitivity of bystander medium, or actual cell line variation (Mothersill et al. 2004; Ryan et al. 2008a, b; Vines et al. 2008; Ryan et al. 2009a, b; Mothersill et al. 2010a, b). There is now a growing consensus that radioresistant cell lines with large survival curve shoulders do not show toxic bystander effects, while radiosensitive cell lines with small or no shoulders have large bystander effects. A screen of 13 cell lines also suggested that cell lines demonstrating low dose hypersensitivity (HRS) did not have bystander effects, but the question remains as to whether these lines show bystander effects in response to low doses but not high doses (Mothersill et al. 2002). Further research aimed at teasing out the relationship between toxic bystander effects and adaptive responses surprisingly suggest that the phenomena have unrelated mechanisms and may actually be mutually exclusive (Ryan et al. 2009b). This is strange because a key theory about why bystander effects exist is that they may lead to adaptive responses and aid the evolution of fitter populations. Perhaps the specific radiobiological technique for demonstrating adaptive responses (a very small dose of radiation administered 4–6 h before a challenge dose) is actually revealing a mechanism that is not true adaptation in the evolutionary sense. Support for this comes from the data published by Audette-Stuart et al. (2011) showing that frogs living in ponds contaminated with tritium were better able to withstand a challenge dose of radiation than frogs from pristine ponds. They also conferred the resistance to the tadpoles with which they were swimming (an inter-animal bystander effect) (Audette-Stuart and Yankovich 2011). In cell culture, however, it is likely that the effects are separate because fractionation of the direct dose of radiation leads to greater survival than if the total dose is administered at once, but the medium harvested from cultures treated with fractionated irradiation is actually more toxic than that from once-irradiated cells (Mothersill and Seymour 2002; Little et al. 2005). This is consistent with the "on/off" nature of the bystander mechanism, but further suggests that fractionating the dose gives a double production of bystander effect. Perhaps the reason cell lines showing bystander effects, but do not show obvious adaptive responses after a small dose followed by a challenge dose, is because in the bystander flasks there was a double dose of the toxic effect, and no inter-dose recovery.

Some experiments have been done using specific knockout pairs of cell lines to elucidate pathways that are important in bystander effects (Nagasawa et al. 2008; Zhang et al. 2008; Mothersill et al. 2009; Tsai et al. 2009). Some interesting data from our own laboratory includes the demonstration that the medium from repair-deficient cell lines had more toxic effects than medium from repair deficient cell lines (Mothersill et al. 2004). Additionally, if medium from irradiated repair-proficient cell lines was placed on repair-deficient cells, they acquired resistance. The reverse was also true (Mothersill et al. 2006a, b). Experiments with p53 knockout cell lines revealed a role for p53 in response to the signal, but not in production of the signal (Mothersill et al. 2011).

In Vivo Versus In Vitro Non-targeted Effects

A key argument used in the early days of the non-targeted effects field was that they were not relevant in vivo, and they were often considered artifacts of cell culture. Efforts to establish in vivo relevance had two aspects: first, to confirm occurrence in in vivo models, and second, to argue that the mechanisms were relevant to radiation risk or radiation therapy outcome. The first aspect has been well-proven, and is reviewed in (Mothersill and Seymour 2009a, b; Choi et al. 2010; Mothersill et al. 2010a, b; Mancuso et al. 2012; Rastogi et al. 2012). The second is highly controversial and remains to be decided. In vivo models for studying bystander effects and instability include attempts to partially irradiate animals and look for effects in un-irradiated organs (Mancuso et al. 2008; Koturbash et al. 2011), while ex vivo approaches where irradiation takes place in vivo in bystander assays are done using blood, urine, or tissue explants in vitro (Mothersill et al. 1999; Belyakov et al. 2003; O'Dowd et al. 2006), and models where organisms are irradiated and then put close with non-irradiated organisms that show effects (Mothersill et al. 2006a, b; Mothersill et al. 2007a, b; Surinov 2007; Biedrzycki et al. 2010; Audette-Stuart and Yankovich 2011). In regard to the relevance of the effects, both adaptive and adverse effects can be demonstrated (Yu 2010; Mobbs et al. 2011; Saxena et al. 2011; Klokov et al. 2012; Rödel et al. 2012), suggesting that downstream mechanisms probably process the stress signal produced by the irradiation, and that differences in response are due to genetic and epigenetic factors determining the expression of relevant pathways.

Dealing with Randomness

Random mutations resulting from genomic instability or bystander stress-induced mutagenesis complicate radiation risk modeling because they introduce uncertainties into the models. At the environment, population, or ecosystem level in biology, such randomness is not an issue, as population survival is the goal, but in the area of human risk, where the goal is protection of the individual, these uncertainties lead to major headaches and the use of the precautionary principle. The elucidation of mechanisms involved in the induction and expression of non-targeted effects is thus of great importance, as is the understanding of the contribution of genetic background and environmental factors discussed earlier.

Epigenetics, System Genetics and NTEs

The randomness issue has led to the development of new branches of radiation biology, among them system radiation biology, cloud radiobiology and system radiation genetics (Mothersill and Seymour 2009a, b; Gierer 2012; Mothersill and Seymour 2012). These sciences attempt to model chaotic and random behavior of systems to understand the tipping points, thresholds, and emergent properties that lead to functional outcomes and behaviors that are not predictable from the study of the function of system components in isolation (Noble 2003; Bateman and Wood 2009; Lilly et al. 2010; Zhao and Ricci 2010; Kupfer 2012; Prins et al. 2012; Meyer et al. 2012). These sciences aim to develop tools for the accurate prediction of outcomes. Key to the success of such approaches is the link between experimentalists and modelers (Mariotti et al. 2010; Kundrát and Friedland 2012), because models are only as good as the data used to develop them.

Implications

There are three obvious and distinct areas where the ability of bystander effects to induce mutagenesis is of importance. These are radiation and environmental protection and risk evaluation, radiotherapy for treatment and understanding carcinogenesis for prevention of cancer, and, finally, understanding mechanisms of evolution and the importance of environmental drivers in inducing mutations. Many of these issues have already been addressed, but some summary thoughts are discussed here.

Protection

The key issues for protection are the shape of the dose response curve at low doses, the mechanisms underlying targeted and non-targeted effects, the relative impacts of genetics and environment on the outcome of multiple stressors, and for environmental protection—how to develop endpoints indicating harm or risk at organizational levels higher than the cell or individual that are sufficient for human protection (Bréchignac et al. 2012). The reason this is important is that we may need to protect different things in ecosystems versus individuals—e.g., the structure and function of the system versus the level of damage in the individual. The whole field of radiation protection is controversial at present, mainly due to uncertainty about the relevance of non-targeted effects and adaptive responses. These disturb the assumed linear relationship between DNA damage (leading to the possibility of a carcinogenic mutation) and energy deposition in the nucleus. Even though the experimental data used to support this hypothesis was from high dose exposures to the Japanese, it was considered acceptable to extrapolate from the high-dose-associated incidence of cancer to predict low-dose-associated incidence (Kellerer 2000; Puskin 2009; Little 2010). On one side of the argument, proponents of hormesis and adaptive response argue that low doses are less dangerous, while those working on genomic instability and some bystander researchers consider low doses to be more dangerous and to underlie other pathologies in addition to those involving carcinogenesis; for example, cardiac damage, immune insufficiency, and neurodegenerative disease (Manton et al. 2004; Hug et al. 2006; Lowe et al. 2009; Barjaktarovic et al. 2011; Sriharshan et al. 2012). Clearly, answers are badly needed in this area.

Therapy

The issues for radiotherapy also concern genetics and stress-induced mutation. Here the issues concern predicting outcome after therapy, e.g., tumor recurrence, or risk of second cancers; optimizing the dose to the tumor given it may be more radioresistant than surrounding normal tissue which may show toxic bystander responses; investigating whether genomic instability and stress-induced mutagenesis in the patient may make the malignancy or aggression of the tumor worse; and investigating whether there are novel drug targets to be found that might regulate bystander mechanisms to the benefit of the patient (Mothersill and Seymour 2004; Prise and O'Sullivan 2009; Sjostedt and Bezak 2010). The resolution of these questions also requires an understanding of non-targeted effects mechanisms, but the doses of direct radiation involved are much higher, and so it is easier to see effects. However, the variation in results obtained due to human genetic variation, and the problems of confounding factors such as smoking and other lifestyle choices, makes it difficult to obtain large enough cohorts of patients to obtain meaningful results. Harnessing bystander mechanisms for therapeutic gain using targeted radiotherapy (Dadachova 2010; Sorensen et al. 2012) or inhibitors of serotonin (Wang 2008; Salvo et al. 2012) seem to be the most promising avenues to improve therapy at present.

Evolutionary Mechanisms

The most exciting theoretical implication of non-targeted effects is clearly that they may be important in evolution. This is discussed at length earlier in the chapter. It is possible that stress-induced mutagenesis is perpetuated in systems by bystander mechanisms, providing accelerated change in direct response to the stressor. This hypothesis neatly links Lamarckian ideas with Darwinian ideas to make a logical framework whereby Darwin's random mutations are accelerated and more directed in response to environmental change (stress), leading to the more rapid and responsive adaptation of the organism to the new environment. Investigation of these mechanisms is sure to lead to exciting science over the next few years.

References

Aapro M, Blower P (2005) 5-hydroxytryptamine type-3 receptor antagonists for chemotherapy-induced and radiotherapy-induced nausea and emesis: can we safely reduce the dose of administered agents? Cancer 104:1–18

Andrews PL, Bhandari P, Davey PT, Bingham S, Marr HE, Blower PR (1992) Are all 5-HT3 receptor antagonists the same? Eur J Cancer 28A(Suppl 1):S2–S6

Audette-Stuart M, Kim SB, McMullin D, Festarini A, Yankovich TL, Carr J, Mulpuru S (2011) Adaptive response in frogs chronically exposed to low doses of ionizing radiation in the environment. J Environ Radioact 102:566–573

Audette-Stuart M, Yankovich T (2011) Bystander effects in bullfrog tadpoles. Radioprotection 46:S497–S502

Ausubel FM (2005) Are innate immune signaling pathways in plants and animals conserved? Nat Immunol 6:973–979

Aypar U, Morgan WF, Baulch JE (2011) Radiation-induced genomic instability: are epigenetic mechanisms the missing link? Int J Radiat Biol 87:179–191

Azzam EI, De Toledo SM, Spitz DR, Little JB (2002) Oxidative metabolism modulates signal transduction and micronucleus formation in bystander cells from alpha-particle-irradiated normal human fibroblast cultures. Cancer Res 62:5436–5442

Azzam EI, Jay-Gerin JP, Pain D (2012) Ionizing radiation-induced metabolic oxidative stress and prolonged cell injury. Cancer Lett 327:48–60

Barjaktarovic Z, Schmaltz D, Shyla A, Azimzadeh O, Schulz S, Haagen J, Dörr W, Sarioglu H, Schäfer A, Atkinson MJ, Zischka H, Tapio S (2011) Radiation-induced signaling results in mitochondrial impairment in mouse heart at 4 weeks after exposure to X-rays. PLoS One 6:e27811

Bateman A, Wood M (2009) Cloud computing. Bioinformatics 25:1475

Belyakov OV, Folkard M, Mothersill C, Prise KM, Michael BD (2003) A proliferation-dependent bystander effect in primary porcine and human urothelial explants in response to targeted irradiation. Br J Cancer 88:767–774

Biedrzycki ML, Jilany TA, Dudley SA, Bais HP (2010) Root exudates mediate kin recognition in plants. Commun Integr Biol 3:28–35

Bréchignac F, Bradshaw C, Carroll S, Fuma S, Håkanson L, Jaworska A, Kapustka L, Kawaguchi I, Monte L, Oughton D, Sazykina T, Strand P (2012) Towards an ecosystem approach for environment protection with emphasis on radiological hazards. IUR Report no 7, IUR Report series, 103 pages, http://www.iur-uir.org

Buonanno M, de Toledo SM, Pain D, Azzam EI (2011) Long-term consequences of radiation-induced bystander effects depend on radiation quality and dose and correlate with oxidative stress. Radiat Res 175:405–415

Calabrese V, Cornelius C, Cuzzocrea S, Iavicoli I, Rizzarelli E, Calabrese EJ (2011) Hormesis, cellular stress response and vitagenes as critical determinants in aging and longevity. Mol Aspects Med 32:279–304

Cannon WB (1926) Physiological regulation of normal states: some tentative postulates concerning biological homeostatics. In: Pettit A (ed) A Charles Richet: ses amis, ses collègues, ses élèves. Éditions Médicales, Paris, p 91

Carbone MC, Pinto M, Antonelli F, Amicarelli F, Balata M, Belli M, Conti Devirgiliis L, Ioannucci L, Nisi S, Sapora O, Satta L, Simone G, Sorrentino E, Tabocchini MA (2009) The cosmic silence experiment: on the putative adaptive role of environmental ionizing radiation. Radiat Environ Biophys 48:189–196

Chai Y, Hei TK (2008) Radiation induced bystander effect in vivo. Acta Med Nagasaki 53:S65–S69

Chen H, Xu G, Zhao Y, Tian B, Lu H, Yu X, Xu Z, Ying N, Hu S, Hua Y (2008) A novel OxyR sensor and regulator of hydrogen peroxide stress with one cysteine residue in Deinococcus radiodurans. PLoS One 3:e1602

Chiavaroli V, Giannini C, De Marco S, Chiarelli F, Mohn A (2011) Unbalanced oxidant-antioxidant status and its effects in pediatric diseases. Redox Rep 16:101–107

Choi VW, Cheng SH, Yu KN (2010) Radioadaptive response induced by alpha-particle-induced stress communicated in vivo between zebrafish embryos. Environ Sci Technol 44:8829–8834

Christians FC, Newcomb TG, Loeb LA (1995) Potential sources of multiple mutations in human cancers. Prev Med 24(4):329–332

Clutton SM, Townsend KM, Walker C, Ansell JD, Wright EG (1996) Radiation-induced genomic instability and persisting oxidative stress in primary bone marrow cultures. Carcinogenesis 17:1633–1639

Cotella D, Hernandez-Enriquez B, Wu X, Li R, Pan Z, Leveille J, Link CD, Oddo S, Sesti F (2012) Toxic role of K+channel oxidation in mammalian brain. J Neurosci 32:4133–4144

Cox LA Jr (2006) Universality of J-shaped and U-shaped dose-response relations as emergent properties of stochastic transition systems. Dose Response 3:353–368

Dahle J, Kvam E, Stokke T (2005) Bystander effects in UV-induced genomic instability: antioxidants inhibit delayed mutagenesis induced by ultraviolet A and B radiation. J Carcinog 4:11

Damiani C, Serra R, Villani M, Kauffman SA, Colacci A (2011) Cell-cell interaction and diversity of emergent behaviours. IET Syst Biol 5:137–144

Dadachova E (2010) Cancer therapy with alpha-emitters labeled peptides. Semin Nucl Med 40:204–208

Dayal D, Martin SM, Limoli CL, Spitz DR (2008) Hydrogen peroxide mediates the radiation-induced mutator phenotype in mammalian cells. Biochem J 413:185–191

Decoursey TE, Ligeti E (2005) Regulation and termination of NADPH oxidase activity. Cell Mol Life Sci 62:2173–2193

Dobrev D (2007) 5-hydroxytryptamine and atrial arrhythmogenesis: a "culprit mechanism" or bystander in patients with chronic atrial fibrillation? J Mol Cell Cardiol 42:51–53

Ermakov AV, Konkova MS, Kostyuk SV, Egolina NA, Efremova LV, Veiko NN (2009) Oxidative stress as a significant factor for development of an adaptive response in irradiated and nonirradiated human lymphocytes after inducing the bystander effect by low-dose X-radiation. Mutat Res 669:155–161

Fan M, Ahmed KM, Coleman MC, Spitz DR, Li JJ (2007) Nuclear factor-kappaB and manganese superoxide dismutase mediate adaptive radioresistance in low-dose irradiated mouse skin epithelial cells. Cancer Res 67:3220–3228

Fariello RG (1988) Experimental support for the implication of oxidative stress in the genesis of parkinsonian syndromes. Funct Neurol 3:407–412

Fournier C, Barberet P, Pouthier T, Ritter S, Fischer B, Voss KO, Funayama T, Hamada N, Kobayashi Y, Taucher-Scholz G (2009) No evidence for DNA and early cytogenetic damage in bystander cells after heavy-ion microirradiation at two facilities. Radiat Res 171:530–540

Galhardo RS, Hastings PJ, Rosenberg SM (2007) Mutation as a stress response and the regulation of evolvability. Crit Rev Biochem Mol Biol 42:399–435

Gault N, Rigaud O, Poncy JL, Lefaix JL (2007) Biochemical alterations in human cells irradiated with alpha particles delivered by macro- or microbeams. Radiat Res 167:551–562

Gierer A (2012) The hydra model - a model for what? Int J Dev Biol 55:437–445

Gillies NE (1987) Effects of radiations on cells. Br Med J (Clin Res Ed) 295:1390–1391

Gori T, Münzel T (2012) Biological effects of low-dose radiation: of harm and hormesis. Eur Heart J 33:292–295

Groesser T, Cooper B, Rhydberg B (2008) Lack of bystander effects from high-LET radiation for early cytogenetic end points. Radiat Res 170:794–802

Haak J, Kregel KC (2008) 1962-2007: a cell stress odyssey. Novartis Found Symp 291:3–15

Hanot M, Hoarau J, Carrière M, Angulo JF, Khodja H (2009) Membrane-dependent bystander effect contributes to amplification of the response to alpha-particle irradiation in targeted and nontargeted cells. Int J Radiat Oncol Biol Phys 75:1247–1253

Hei TK, Zhou H, Ivanov VN, Hong M, Lieberman HB, Brenner DJ, Amundson SA, Geard CR (2008) Mechanism of radiation-induced bystander effects: a unifying model. J Pharm Pharmacol 60:943–950

Hei TK, Zhou H, Chai Y, Ponnaiya B, Ivanov VN (2011) Radiation induced non-targeted response: mechanism and potential clinical implications. Curr Mol Pharmacol 4:96–105

Hersh MN, Ponder RG, Hastings PJ, Rosenberg SM (2004) Adaptive mutation and amplification in Escherichia coli: two pathways of genome adaptation under stress. Res Microbiol 155:352–359

Hitchler MJ, Domann FE (2007) An epigenetic perspective on the free radical theory of development. Free Radic Biol Med 43:1023–1036

http://medical-dictionary.thefreedictionary.com/stress

http://www.raraf.org/in%20vivo.htm

Folker Meyer, Argonne National Laboratory, Susan Gregurick, DOE-BER, Peg Folta, LLNL Workshop: using clouds for parallel computations in systems biology

Huang PC, Sheridan RB III (1981) Genetic and biochemical studies with ataxia telangiectasia. A review. Hum Genet 59:1–9

Hug K, Röösli M, Rapp R (2006) Magnetic field exposure and neurodegenerative diseases–recent epidemiological studies. Soz Praventivmed 51:210–220

Ichinose G, Kobayashi M (2011) Emergence of cooperative linkages by random intensity of selection on a network. Biosystems 105:1–9

Ilnytskyy Y, Kovalchuk O (2011) Non-targeted radiation effects-an epigenetic connection. Mutat Res 714:113–125

Isaeva VG, Surinov BP (2011) Effect of natural and postradiation volatile secretions of mice on the immune reactivity and blood cellularity of irradiated animals. Radiats Biol Radioecol 51:444–450

Johnson GG, White MC, Grimaldi M (2011) Stressed to death: targeting endoplasmic reticulum stress response induced apoptosis in gliomas. Curr Pharm Des 17:284–292

Kadhim MA, Macdonald DA, Goodhead DT, Lorimore SA, Marsden SJ, Wright EG (1992) Transmission of chromosomal instability after plutonium alpha-particle irradiation. Nature 355:738–740

Kashino G, Prise KM, Suzuki K, Matsuda N, Kodama S, Suzuki M, Nagata K, Kinashi Y, Masunaga S, Ono K, Watanabe M (2007) Effective suppression of bystander effects by DMSO treatment of irradiated CHO cells. J Radiat Res (Tokyo) 48:327–333

Kellerer AM (2000) Risk estimates for radiation-induced cancer–the epidemiological evidence. Radiat Environ Biophys 39:17–24

Kim R, Emi M, Tanabe K (2006) Role of mitochondria as the gardens of cell death. Cancer Chemother Pharmacol 57:545–553

Klokov D, Leskov K, Araki S, Zou Y, Goetz EM, Luo X, Willson D, Boothman DA (2012) Low dose IR-induced IGF-1-sCLU expression: a p53-repressed expression cascade that interferes with TGFβ1 signaling to confer a pro-survival bystander effect. Oncogene. 2012 Mar 5. doi: 10.1038/onc.2012.64. [Epub ahead of print]

Kojima S, Takai E, Tsukimoto M (2011) ATP released from low-dose gamma ray-irradiated cells activates intracellular antioxidant systems via purine receptors. Anti-Aging Med 8:108–113

Koolhaas JM, Bartolomucci A, Buwalda B, de Boer SF, Flügge G, Korte SM, Meerlo P, Murison R, Olivier B, Palanza P, Richter-Levin G, Sgoifo A, Steimer T, Stiedl O, van Dijk G, Wöhr M, Fuchs E (2011) Stress revisited: a critical evaluation of the stress concept. Neurosci Biobehav Rev 35:1291–1301

Koturbash I, Zemp F, Kolb B, Kovalchuk O (2011) Sex-specific radiation-induced microRNAome responses in the hippocampus, cerebellum and frontal cortex in a mouse model. Mutat Res 722:114–118

Kovacic P, Somanathan R (2010) Dermal toxicity and environmental contamination: electron transfer, reactive oxygen species, oxidative stress, cell signaling, and protection by antioxidants. Rev Environ Contam Toxicol 203:119–138

Kundrát P, Friedland W (2012) Track structure calculations on intracellular targets responsible for signal release in bystander experiments with transfer of irradiated cell-conditioned medium. Int J Radiat Biol 88:98–102

Kupfer DM (2012) Cloud computing in biomedical research. Aviat Space Environ Med 83:152–153
Lagerwerf S, Vrouwe MG, Overmeer RM, Fousteri MI, Mullenders LH (2011) DNA damage response and transcription. DNA Repair (Amst) 10:743–750
Lankin VZ, Lisina MO, Arzamastseva NE, Konovalova GG, Nedosugova LV, Kaminnyi AI, Tikhaze AK, Ageev FT, Kukharchuk VV, Belenkov YN (2005) Oxidative stress in atherosclerosis and diabetes. Bull Exp Biol Med 140:41–43
Lee SA, Dritschilo A, Jung M (2001) Role of ATM in oxidative stress-mediated c-Jun phosphorylation in response to ionizing radiation and CdCl2. J Biol Chem 276:11783–11790
Lilly GD, Lee J, Kotov NA (2010) "Cloud" assemblies: quantum dots form electrostatically bound dynamic nebulae around large gold nanoparticles. Phys Chem Chem Phys 12:11878–11884
Little MP, Filipe JA, Prise KM, Folkard M, Belyakov OV (2005) A model for radiation-induced bystander effects, with allowance for spatial position and the effects of cell turnover. J Theor Biol 232:329–338
Little MP (2010) Do non-targeted effects increase or decrease low dose risk in relation to the linear-non-threshold (LNT) model? Mutat Res 687:17–27
Liu Z, Mothersill CE, McNeill FE, Lyng FM, Byun SH, Seymour CB, Prestwich WV (2006) A dose threshold for a medium transfer bystander effect for a human skin cell line. Radiat Res 166:19–23
Lowe XR, Bhattacharya S, Marchetti F, Wyrobek AJ (2009) Early brain response to low-dose radiation exposure involves molecular networks and pathways associated with cognitive functions, advanced aging and Alzheimer's disease. Radiat Res 171:53–65
Lyng FM, O'Reilly S, Cottell DC, Seymour CB, Mothersill C (1996) Persistent expression of morphological abnormalities in the distant progeny of irradiated cells. Radiat Environ Biophys 35:273–283
Lyng FM, Seymour CB, Mothersill C (2000) Production of a signal by irradiated cells which leads to a response in unirradiated cells characteristic of initiation of apoptosis. Br J Cancer 83:1223–1230
Lyng FM, Seymour CB, Mothersill C (2002a) Early events in the apoptotic cascade initiated in cells treated with medium from the progeny of irradiated cells. Radiat Prot Dosimetry 99:169–172
Lyng FM, Seymour CB, Mothersill C (2002b) Initiation of apoptosis in cells exposed to medium from the progeny of irradiated cells: a possible mechanism for bystander-induced genomic instability? Radiat Res 157:365–370
Lyng FM, Maguire P, McClean B, Seymour C, Mothersill C (2006) The involvement of calcium and MAP kinase signaling pathways in the production of radiation-induced bystander effects. Radiat Res 165:400–409
Lyng FM, Howe OL, McClean B (2011) Reactive oxygen species-induced release of signalling factors in irradiated cells triggers membrane signalling and calcium influx in bystander cells. Int J Radiat Biol 87:683–695
Magnadóttir B (2006) Innate immunity of fish (overview). Fish Shellfish Immunol 20:137–151
Mancuso M, Pasquali E, Leonardi S, Tanori M, Rebessi S, Di Majo V, Pazzaglia S, Toni MP, Pimpinella M, Covelli V, Saran A (2008) Oncogenic bystander radiation effects in Patched heterozygous mouse cerebellum. Proc Natl Acad Sci USA 105:12445–12450
Mancuso M, Pasquali E, Giardullo P, Leonardi S, Tanori M, Di Majo V, Pazzaglia S, Saran A (2012) The radiation bystander effect and its potential implications for human health. Curr Mol Med 12:613–624
Manton KG, Volovik S, Kulminski A (2004) ROS effects on neurodegeneration in Alzheimer's disease and related disorders: on environmental stresses of ionizing radiation. Curr Alzheimer Res 1:277–293
Mariotti L, Facoetti A, Alloni D, Bertolotti A, Ranza E, Ottolenghi A (2010) Effects of ionizing radiation on cell-to-cell communication. Radiat Res 174:280–289
Martinez-Outschoorn UE, Balliet RM, Rivadeneira DB, Chiavarina B, Pavlides S, Wang C, Whitaker-Menezes D, Daumer KM, Lin Z, Witkiewicz AK, Flomenberg N, Howell A, Pestell RG, Knudsen ES, Sotgia F, Lisanti MP (2010) Oxidative stress in cancer associated fibroblasts

drives tumor-stroma co-evolution: a new paradigm for understanding tumor metabolism, the field effect and genomic instability in cancer cells. Cell Cycle 9:3256–3276
Martins RN, Harper CG, Stokes GB, Masters CL (1986) Increased cerebral glucose-6-phosphate dehydrogenase activity in Alzheimer's disease may reflect oxidative stress. J Neurochem 46:1042–1045
Matsumoto H, Tomita M, Otsuka K, Hatashita M (2009) A new paradigm in radioadaptive response developing from microbeam research. J Radiat Res (Tokyo) 50(Suppl A):A67–A79
Matzinger P (1998) An innate sense of danger. Semin Immunol 10:399–415
McBride WH, Chiang CS, Olson JL, Wang CC, Hong JH, Pajonk F, Dougherty GJ, Iwamoto KS, Pervan M, Liao YP (2004) A sense of danger from radiation. Radiat Res 162:1–19
McDonald JT, Kim K, Norris AJ, Vlashi E, Phillips TM, Lagadec C, Della Donna L, Ratikan J, Szelag H, Hlatky L, McBride WH (2010) Ionizing radiation activates the Nrf2 antioxidant response. Cancer Res 70:8886–8895
Milkman R (1962) Temperature effects on day old Drosophila pupae. J Gen Physiol 45:777–799
Miyachi Y, Kasai H, Ohyama H, Yamada T (1994) Changes of aggressive behavior and brain serotonin turnover after very low-dose X-irradiation of mice. Neurosci Lett 175:92–94
Mobbs SF, Muirhead CR, Harrison JD (2011) Risks from ionising radiation: an HPA viewpoint paper for Safegrounds. J Radiol Prot 31:289–307
Moreira PI, Honda K, Liu Q, Santos MS, Oliveira CR, Aliev G, Nunomura A, Zhu X, Smith MA, Perry G (2005) Oxidative stress: the old enemy in Alzheimer's disease pathophysiology. Curr Alzheimer Res 2:403–408
Moretti L, Cha YI, Niermann KJ, Lu B (2007) Switch between apoptosis and autophagy: radiation-induced endoplasmic reticulum stress? Cell Cycle 6:793–798
Mothersill C, Seymour C (1997) Medium from irradiated human epithelial cells but not human fibroblasts reduces the clonogenic survival of unirradiated cells. Int J Radiat Biol 71:421–427
Mothersill C, Crean M, Lyons M, McSweeney J, Mooney R, O'Reilly J, Seymour CB (1998) Expression of delayed toxicity and lethal mutations in the progeny of human cells surviving exposure to radiation and other environmental mutagens. Int J Radiat Biol 74:673–680
Mothersill C, Seymour CB (1998a) Mechanisms and implications of genomic instability and other delayed effects of ionizing radiation exposure. Mutagenesis 13:421–426
Mothersill C, Seymour CB (1998b) Cell-cell contact during gamma irradiation is not required to induce a bystander effect in normal human keratinocytes: evidence for release during irradiation of a signal controlling survival into the medium. Radiat Res 149:256–262
Mothersill CE, O'Malley KJ, Murphy DM, Seymour CB, Lorimore SA, Wright EG (1999) Identification and characterization of three subtypes of radiation response in normal human urothelial cultures exposed to ionizing radiation. Carcinogenesis 20:2273–2278
Mothersill C, Stamato TD, Perez ML, Cummins R, Mooney R, Seymour CB (2000a) Involvement of energy metabolism in the production of 'bystander effects' by radiation. Br J Cancer 82:1740–1746
Mothersill C, Kadhim MA, O'Reilly S, Papworth D, Marsden SJ, Seymour CB, Wright EG (2000b) Dose- and time-response relationships for lethal mutations and chromosomal instability induced by ionizing radiation in an immortalized human keratinocyte cell line. Int J Radiat Biol 76:799–806
Mothersill C, Seymour C (2001) Radiation-induced bystander effects: past history and future directions. Radiat Res 155:759–767
Mothersill C, Seymour CB (2002) Bystander and delayed effects after fractionated radiation exposure. Radiat Res 158:626–633
Mothersill C, Seymour CB, Joiner MC (2002) Relationship between radiation-induced low-dose hypersensitivity and the bystander effect. Radiat Res 157:526–532
Mothersill C, Seymour RJ, Seymour CB (2004) Bystander effects in repair-deficient cell lines. Radiat Res 161:256–263
Mothersill C, Seymour RJ, Seymour CB (2006a) Increased radiosensitivity in cells of two human cell lines treated with bystander medium from irradiated repair-deficient cells. Radiat Res 165:26–34

Mothersill C, Bucking C, Smith RW, Agnihotri N, O'Neill A, Kilemade M, Seymour CB (2006b) Communication of radiation-induced stress or bystander signals between fish in vivo. Environ Sci Technol 40:6859–6864

Mothersill C, Smith RW, Agnihotri N, Seymour CB (2007a) Characterization of a radiation-induced stress response communicated in vivo between zebrafish. Environ Sci Technol 41:3382–3387

Mothersill C, Mosse, I, Seymour, C (eds) (2007) Multiple stressors: a challenge for the future. Proceedings of the NATO advanced research workshop on multipollution exposure and risk assessment - a challenge for the future, Minsk, Belarus, Oct 1–5, Series: NATO science for peace and security series C: environmental security

Mothersill C, Seymour C (2009a) Implications for environmental health of multiple stressors. J Radiol Prot 29:A21–A28

Mothersill C, Smith RW, Hinton TG, Aizawa K, Seymour CB (2009) Communication of radiation-induced signals in vivo between DNA repair deficient and proficient medaka (Oryzias latipes). Environ Sci Technol 43:3335–3342

Mothersill C, Seymour C (2009b) Communication of ionising radiation signals–a tale of two fish. Int J Radiat Biol 85:909–919

Mothersill C, Seymour CB (2004) Radiation-induced bystander effects–implications for cancer. Nat Rev Cancer 4:158–164

Mothersill C, Saroya R, Smith RW, Singh H, Seymour CB (2010a) Serum serotonin levels determine the magnitude and type of bystander effects in medium transfer experiments. Radiat Res 174:119–123

Mothersill C, Smith RW, Saroya R, Denbeigh J, Rowe B, Banevicius L, Timmins R, Moccia R, Seymour CB (2010b) Irradiation of rainbow trout at early life stages results in legacy effects in adults. Int J Radiat Biol 86:817–828

Mothersill C, Bristow RG, Harding SM, Smith RW, Mersov A, Seymour CB (2011) A role for p53 in the response of bystander cells to receipt of medium borne signals from irradiated cells. Int J Radiat Biol 87:1120–1125

Mothersill C, Seymour C (2012) Changing paradigms in radiobiology. Mutat Res 750:85–95

Müller K, Meineke V (2011) Radiation-induced mast cell mediators differentially modulate chemokine release from dermal fibroblasts. J Dermatol Sci 61:199–205

Nagasawa H, Little JB (1992) Induction of sister chromatid exchanges by extremely low doses of alpha-particles. Cancer Res 52:6394–6396

Nagasawa H, Wilson PF, Chen DJ, Thompson LH, Bedford JS, Little JB (2008) Low doses of alpha particles do not induce sister chromatid exchanges in bystander Chinese hamster cells defective in homologous recombination. DNA Repair (Amst) 7:515–522

Noble D (2003) The future: putting humpty-dumpty together again. Biochem Soc Trans 31:156–158

O'Dowd C, Mothersill CE, Cairns MT, Austin B, McClean B, Lyng FM, Murphy JE (2006) The release of bystander factor(s) from tissue explant cultures of rainbow trout (Onchorhynchus mykiss) after exposure to gamma radiation. Radiat Res 166:611–617

Okayama Y (2005) Oxidative Stress in Allergic and Inflammatory Skin Diseases. Curr Drug Targets Inflamm Allergy 4:517–519

Ottaviani E, Valensin S, Franceschi C (1998) The neuro-immunological interface in an evolutionary perspective: the dynamic relationship between effector and recognition systems. Front Biosci 3:d431–d435

Ozden O, Park SH, Kim HS, Jiang H, Coleman MC, Spitz DR, Gius D (2011) Acetylation of MnSOD directs enzymatic activity responding to cellular nutrient status or oxidative stress. Aging (Albany NY) 3:102–107

Plouffe DA, Hanington PC, Walsh JG, Wilson EC, Belosevic M (2005) Comparison of select innate immune mechanisms of fish and mammals. Xenotransplantation 12:266–277

Poon RC, Agnihotri N, Seymour C, Mothersill C (2007) Bystander effects of ionizing radiation can be modulated by signaling amines. Environ Res 105:200–211

Preta G, de Klark R, Glas R (2009) A role for nuclear translocation of tripeptidyl-peptidase II in reactive oxygen species-dependent DNA damage responses. Biochem Biophys Res Commun 389:575–579

Prins P, Belhachemi D, Möller S, Smant G (2012) Scalable computing for evolutionary genomics. Methods Mol Biol 856:529–545

Prise KM, Belyakov OV, Newman HC, Patel S, Schettino G, Folkard M, Michael BD (2002) Non-targeted effects of radiation: bystander responses in cell and tissue models. Radiat Prot Dosimetry 99:223–226

Prise KM, Folkard M, Kuosaite V, Tartier L, Zyuzikov N, Shao C (2006) What role for DNA damage and repair in the bystander response? Mutat Res 597:1–4

Prise KM, O'Sullivan JM (2009) Radiation-induced bystander signalling in cancer therapy. Nat Rev Cancer 9:351–360

Prise KM, Schettino G (2011) Microbeams in radiation biology: review and critical comparison. Radiat Prot Dosimetry 143:335–339

Puskin JS (2009) Perspective on the use of LNT for radiation protection and risk assessment by the U.S. Environmental Protection Agency. Dose Response 7:284–291

Rastogi S, Coates PJ, Lorimore SA, Wright EG (2012) Bystander-type effects mediated by long-lived inflammatory signaling in irradiated bone marrow. Radiat Res 177:244–250

Rinkevich B, Weissman IL (1992) Allogeneic resorption in colonial protochordates: consequences of nonself recognition. Dev Comp Immunol 16:275–286

Rödel F, Frey B, Gaipl US, Keilholz L, Fournier C, Manda K, Schöllnberger H, Hildebrandt G, Rödel C (2012) Modulation of inflammatory immune reactions by low-dose ionizing radiation: molecular mechanisms and clinical application. Curr Med Chem 19:1741–1750

Rugo RE, Mutamba JT, Mohan KN, Yee T, Chaillet JR, Greenberger JS, Engelward BP (2011) Methyltransferases mediate cell memory of a genotoxic insult. Oncogene 30:751–756

Ryan LA, Seymour CB, O'Neill-Mehlenbacher A, Mothersill CE (2008a) Radiation-induced adaptive response in fish cell lines. J Environ Radioact 99:739–747

Ryan LA, Smith RW, Seymour CB, Mothersill CE (2008b) Dilution of irradiated cell conditioned medium and the bystander effect. Radiat Res 169:188–196

Ryan LA, Seymour CB, Mothersill CE (2009a) Investigation of non-linear adaptive responses and split dose recovery induced by ionizing radiation in three human epithelial derived cell lines. Dose Response 7:292–306

Ryan LA, Seymour CB, Joiner MC, Mothersill CE (2009b) Radiation-induced adaptive response is not seen in cell lines showing a bystander effect but is seen in lines showing HRS/IRR response. Int J Radiat Biol 85:87–95

Salbu B, Denbeigh J, Smith RW, Heier LS, Teien HC, Rosseland BO, Oughton D, Seymour CB, Mothersill C (2008) Environmentally relevant mixed exposures to radiation and heavy metals induce measurable stress responses in Atlantic salmon. Environ Sci Technol 42:3441–3446

Sagrada A, Turconi M, Bonali P, Schiantarelli P, Micheletti R, Montagna E, Nicola M, Algate DR, Rimoldi EM, Donetti A (1991) Antiemetic activity of the new 5-HT3 antagonist DAU 6215 in animal models of cancer chemotherapy and radiation. Cancer Chemother Pharmacol 28:470–474

Salvo N, Doble B, Khan L, Amirthevasar G, Dennis K, Pasetka M, Deangelis C, Tsao M, Chow E (2012) Prophylaxis of radiation-induced nausea and vomiting using 5-hydroxytryptamine-3 serotonin receptor antagonists: a systematic review of randomized trials. Int J Radiat Oncol Biol Phys 82:408–417

Saroya R, Smith R, Seymour C, Mothersill C (2009) Injection of resperpine into zebrafish, prevents fish to fish communication of radiation-induced bystander signals: confirmation in vivo of a role for serotonin in the mechanism. Dose Response 8:317–330

Saxena P, Bhatnagar A, Nishad DK, Tyagi R, Rana P, Ali S, Mittal G (2011) Transient cell function disruption by low dose acute exposure of ionizing radiation. Indian J Exp Biol 49:895–903

Schettino G, Folkard M, Michael BD, Prise KM (2005) Low-dose binary behavior of bystander cell killing after microbeam irradiation of a single cell with focused c(k) x rays. Radiat Res 163:332–336

Schettino G, Folkard M, Prise KM, Vojnovic B, Held KD, Michael BD (2003) Low-dose studies of bystander cell killing with targeted soft X rays. Radiat Res 160:505–511

Sedelnikova OA, Redon CE, Dickey JS, Nakamura AJ, Georgakilas AG, Bonner WM (2010) Role of oxidatively induced DNA lesions in human pathogenesis. Mutat Res 704:152–159

Selye H (1956) The stress of life. McGraw-Hill, New York

Seymour CB, Mothersill C, Alper T (1986) High yields of lethal mutations in somatic mammalian cells that survive ionizing radiation. Int J Radiat Biol Relat Stud Phys Chem Med 50:167–179

Seymour CB, Mothersill C (1988) Re. Clonogenicity of the progeny of surviving cells after irradiation. Int J Radiat Biol 54:497–504

Seymour CB, Mothersill C (1997) Delayed expression of lethal mutations and genomic instability in the progeny of human epithelial cells that survived in a bystander-killing environment. Radiat Oncol Investig 5:106–110

Shao C, Lyng FM, Folkard M, Prise KM (2006) Calcium fluxes modulate the radiation-induced bystander responses in targeted glioma and fibroblast cells. Radiat Res 166:479–487

Shao C, Prise KM, Folkard M (2008) Signaling factors for irradiated glioma cells induced bystander responses in fibroblasts. Mutat Res 638:139–145

Sjostedt S, Bezak E (2010) Non-targeted effects of ionising radiation and radiotherapy. Australas Phys Eng Sci Med 33:219–231

Simpson EP, Yen AA, Appel SH (2003) Oxidative stress: a common denominator in the pathogenesis of amyotrophic lateral sclerosis. Curr Opin Rheumatol 15:730–736

Slade D, Radman M (2011) Oxidative stress resistance in Deinococcus radiodurans. Microbiol Mol Biol Rev 75:133–191

Smith RW, Mothersill C, Hinton T, Seymour CB (2011) Exposure to low level chronic radiation leads to adaptation to a subsequent acute X-ray dose and communication of modified acute X-ray induced bystander signals in medaka (Japanese rice fish, Oryzias latipes). Int J Radiat Biol 87:1011–1022

Sorensen A, Mairs RJ, Braidwood L, Joyce C, Conner J, Pimlott S, Brown M, Boyd M (2012) In vivo evaluation of a cancer therapy strategy combining HSV1716-mediated oncolysis with gene transfer and targeted radiotherapy. J Nucl Med 53:647–654

Souidi M, Tissandie E, Racine R, Ben Soussan H, Rouas C, Grignard E, Dublineau I, Gourmelon P, Lestaevel P, Gueguen Y (2009) Uranium: properties and biological effects after internal contamination. Ann Biol Clin (Paris) 67:23–38

Sowa MB, Goetz W, Baulch JE, Pyles DN, Dziegielewski J, Yovino S, Snyder AR, de Toledo SM, Azzam EI, Morgan WF (2010) Lack of evidence for low-LET radiation induced bystander response in normal human fibroblasts and colon carcinoma cells. Int J Radiat Biol 86:102–113

Spitz DR, Azzam EI, Li JJ, Gius D (2004) Metabolic oxidation/reduction reactions and cellular responses to ionizing radiation: a unifying concept in stress response biology. Cancer Metastasis Rev 23:311–322

Sreevidya CS, Fukunaga A, Khaskhely NM, Masaki T, Ono R, Nishigori C, Ullrich SE (2010) Agents that reverse UV-Induced immune suppression and photocarcinogenesis affect DNA repair. J Invest Dermatol 130:1428–1437

Sriharshan A, Boldt K, Sarioglu H, Barjaktarovic Z, Azimzadeh O, Hieber L, Zitzelsberger H, Ueffing M, Atkinson MJ, Tapio S (2012) Proteomic analysis by SILAC and 2D-DIGE reveals radiation-induced endothelial response: four key pathways. J Proteomics 75:2319–2330

Surinov BP, Isaeva VG, Tokarev OI (2001) Allelopathic activity of volatile secretions in irradiated animals. Radiats Biol Radioecol 41:645–649

Surinov BP, Isaeva VG, Dukhova NN (2004) Postirradiation volatile secretions of mice: syngeneic and allogeneic immune and behavioral effects. Bull Exp Biol Med 138:384–386

Surinov BP (2007) Mice with radiation or toxic damage or malignant tumors produce aversive chemosignals repelling intact animals. Dokl Biol Sci 414:199–201

Tsai KK, Stuart J, Chuang YY, Little JB, Yuan ZM (2009) Low-dose radiation-induced senescent stromal fibroblasts render nearby breast cancer cells radioresistant. Radiat Res 172:306–313

Vanhoudt N, Vandenhove H, Horemans N, Wannijn J, Van Hees M, Vangronsveld J, Cuypers A (2010) The combined effect of uranium and gamma radiation on biological responses and oxidative stress induced in Arabidopsis thaliana. J Environ Radioact 101:923–930

Vines AM, Lyng FM, McClean B, Seymour C, Mothersill CE (2008) Bystander signal production and response are independent processes which are cell line dependent. Int J Radiat Biol 84:83–90

Walleczek J (1992) Electromagnetic field effects on cells of the immune system: the role of calcium signaling. FASEB J 6:3177–3185

Wang XS (2008) Pathophysiology of cancer-related fatigue. Clin J Oncol Nurs 12(5 Suppl):11–20

Widel M, Przybyszewski WM, Cieslar-Pobuda A, Saenko YV, Rzeszowska-Wolny J (2012) Bystander normal human fibroblasts reduce damage response in radiation targeted cancer cells through intercellular ROS level modulation. Mutat Res 731:117–124

Wright EG (2010) Manifestations and mechanisms of non-targeted effects of ionizing radiation. Mutat Res 687:28–33

Yu CC (2010) Radiation safety in the neonatal intensive care unit: too little or too much concern? Pediatr Neonatol 51:311–319

Zhang Y, Zhou J, Held KD, Redmond RW, Prise KM, Liber HL (2008) Deficiencies of double-strand break repair factors and effects on mutagenesis in directly gamma-irradiated and medium-mediated bystander human lymphoblastoid cells. Radiat Res 169:197–206

Zhao Y, Ricci PF (2010) Modeling dose-response at low dose: a systems biology approach for ionization radiation. Dose Response 8:456–477

Chapter 11
Stress Induced Mutagenesis, Genetic Diversification, and Cell Survival via Anastasis, the Reversal of Late Stage Apoptosis

Ho Lam Tang, Ho Man Tang, and Denise J. Montell

Abstract Changes in genomic DNA are critical for evolution as they generate genetic diversity, which is the substrate for natural selection. However, most mutations are deleterious, so protective mechanisms have evolved such as apoptotic cell death, to eliminate damaged cells. Apoptosis is generally assumed to be irreversible once massive destruction of structural and functional cellular components occurs. Recent surprising studies reveal that dying cells can reverse the apoptotic process, survive, and proliferate, even after sustaining DNA damage. This process has been named anastasis. While most cells repair their damaged DNA, residual genetic alterations persist in some cells and can result in oncogenic transformation. Although proliferation of transformed cells is a negative consequence, anastasis may serve useful purposes as well. For example, such a cell survival mechanism could serve to salvage postmitotic cells, which are difficult to replace, and thereby limit permanent tissue damage due to transient stresses. The DNA mutations that persist following anastasis represent a form of stress-induced mutagenesis, increasing genetic and phenotypic diversity in response to environmental or physiological stresses that initiate apoptosis. Negative side effects of this otherwise beneficial process may include carcinogenesis and evolution of drug resistance following chemotherapy.

Ho Lam Tang and Ho Man Tang have contributed equally to this chapter.

H.L. Tang • H.M. Tang • D.J. Montell (✉)
Department of Biological Chemistry, Center for Cell Dynamics, Johns Hopkins University School of Medicine, Baltimore, MD 21205, USA
e-mail: dmontell@jhmi.edu

D. Mittelman (ed.), *Stress-Induced Mutagenesis*, DOI 10.1007/978-1-4614-6280-4_11,

Introduction

Anastasis is a Greek word that means “rising to life” and refers to a recently recognized mechanism that rescues dying cells by arresting and reversing the process of apoptosis (Tang et al. 2012). While the physiological significance and molecular mechanisms of this new phenomenon remain to be fully elucidated, the observation that cells can survive and proliferate following DNA damage and repair immediately suggests a wide range of physiological and pathological implications. These include mechanisms for tissue repair following stress or injury, tumorigenesis, evolution of drug resistance, and preventing, arresting or reversing the progression of degenerative diseases. Programmed cell death is generally assumed to be the last decision of a living cell, so survival by anastasis is an unexpected phenomenon. Here, we first provide a brief review of the physiological functions and biochemical mechanisms of apoptosis. Then we describe the unexpected discovery of anastasis in both cancer cell lines and a variety of normal cell types, both primary cells and cell lines. We define the features of anastasis and some of its documented consequences. We discuss the implications of these observations for tissue homeostasis and recovery from injury, the etiology and treatment of diseases, stress-induced genetic diversity, and evolution. We also discuss strategies to detect, track, and study anastasis and indicate exciting avenues for future research in this emerging field.

Importance of Controlled Cell Death

The discovery of the cell suicide process known as apoptosis (Kerr et al. 1972), though extremely important, did not initially generate great excitement. The notion that cell death is a key feature of organismal life seemed counter intuitive and was slow to attract attention (Glücksmann 1951). However, once it became clear that apoptosis is essential for normal animal development (Horvitz 1999; Lettre and Hengartner 2006) and that insufficient apoptosis causes cancer (Brown and Attardi 2005), research in this field exploded (Jacobson et al. 1997; Lam 2004; Fuchs and Steller 2011). Apoptosis describes a particular type of programmed cell death that is characterized by a series of morphological changes cells exhibit as they die (Kerr et al. 1972). These include cell shrinkage and plasma membrane blebbing, nuclear condensation and DNA fragmentation. The term “programmed cell death” describes stereotypical deaths that occur at predictable places and times during development (Jacobson et al. 1997; Lockshin and Williams 1964). However, apoptosis can also be induced in response to drugs, irradiation, and other environmental insults as well as in response to specific cytokines such as tumor necrosis factor α (TNFα). Other types of cell death include autophagy and necrosis (Galluzzi et al. 2012) as well as an unusual non-apoptotic cell death process in *Caenorhabditis elegans* (Abraham et al. 2007; Blum et al. 2012). During autophagy, cells degrade their own organelles via lysosomes (Chen and Klionsky 2011; Codogno and Meijer 2005). This process allows cells to survive nutrient deprivation but can ultimately also result in death.

In contrast to apoptosis and autophagy, necrosis is characterized by swelling of organelles and cells, and ultimately loss of plasma membrane integrity (Golstein and Kroemer 2007). Whereas apoptotic cells emit signals that stimulate phagocytic cells to engulf them, necrotic cells can accumulate, for example forming a necrotic core in tumors (Byrne 2010). For many years, necrosis was believed to be an accidental and uncontrolled process, but recent evidence suggests that necrosis is actually a regulated form of cell death (Vandenabeele et al. 2010). While other forms of cell death are attracting increasing attention, apoptosis remains the best characterized type of cell death, proper regulation of which is clearly critical for normal development and tissue homeostasis and to prevent and combat intractable diseases such as autoimmunity, cancer, and neurodegeneration (Fuchs and Steller 2011).

Apoptosis plays essential roles in embryo development by removing unneeded cells (Jacobson et al. 1997; Fuchs and Steller 2011; Baehrecke 2002). For example, apoptosis sculpts fingers and toes from paddle-shaped limb buds when cells between the developing digits die. Apoptosis eliminates unnecessary structures such as the tadpole's tail during metamorphosis as well as excess cells. During nervous system development for example, excess cells form; however, trophic factors such as nerve growth factor are produced in limiting quantities so that only the appropriate number of cells survive. Injured or mutated cells also undergo apoptosis (Jacobson et al. 1997; Fuchs and Steller 2011; Baehrecke 2002). In vivo, apoptotic cells are engulfed and consumed by neighboring cells or lymphocytes for clearance.

Dysfunction of apoptosis can have fatal consequences. For example, cancer cells that acquire resistance to cell death usually kill their host (Mrozek et al. 2003; Vazquez et al. 2008; Stratton et al. 2009). Failure to eliminate excess neurons can also be lethal to a developing embryo (Kuida et al. 1996). Excess cell death is not good either and can contribute to degenerative diseases like Parkinson's and Alzheimer's (Mattson 2000; Friedlander 2003). Elucidating the regulatory mechanisms governing apoptotic cell death has enhanced and will continue to enhance our understanding of how cell death shapes life, and provide new strategies for the treatment of these intractable diseases.

While apoptosis is essential in multicellular organisms, it is becoming clear that apoptosis also benefits unicellular organisms particularly in adverse conditions. For example, during nutrient stress, "old" yeast cells undergo apoptosis so that the "young" yeast cells have a better chance to survive (Gourlay et al. 2006; Buttner et al. 2006; Carmona-Gutierrez et al. 2010). This enhances the overall fitness of the population, and leads to stress induced mutagenesis, which increases both genetic and phenotypic diversity. In this way microorganisms can better adapt to their changing environment (Rosenberg 2001).

Life and Death Decisions

Eukaryotic cells face a continual struggle between life and death, fought at the biochemical level between pro-apoptotic and anti-apoptotic stimuli. Whether a cell lives or dies depends on the balance of competing signals. Both extrinsic and intrinsic stimuli can initiate cell execution events by triggering mitochondrial translocation of pro-apoptotic Bcl-2 family proteins such as BAX, BAD, and BID. External pro-survival signals include growth factors and cytokines, which inhibit death by binding to their corresponding receptors and activating pro-survival pathways. For example, growth factors can inhibit BAD by Akt-dependent phosphorylation (Datta et al. 1997; Zhu et al. 2002). In contrast, extracellular death ligands such as TNF-α and Fas ligand promote apoptosis by binding to death receptors to trigger activation of caspase-8 (Wang et al. 2008; Juo et al. 1998), an apoptotic protease that cleaves BID, which then translocates to mitochondria to trigger the release of execution proteins (Luo et al. 1998; Li et al. 1998; Schug et al. 2011). This initiator caspase also triggers the caspase cascade by cleaving and activating caspase-3 (Stennicke et al. 1998; Riedl and Shi 2004), which is an effector caspase that plays important roles in cellular demolition during apoptosis. In addition to responding to external signals, cells continually assess their internal health, checking for integrity of their organelles and DNA. For example, mitochondrial damage can increase the sensitivity of cells to apoptosis (Tait and Green 2010). DNA damage can trigger p53, which activates pathways ranging from DNA repair for mild damage to apoptosis for irreparable damage (Murray-Zmijewski et al. 2008). During early apoptosis, pro-apoptotic BCL-2 family proteins such as BAX and BAD translocate to mitochondria to activate mitochondria-dependent apoptotic pathways (Wang 2001), triggering mitochondrial fragmentation and permeabilization, which can nevertheless still be inhibited by the pro-survival BCL-2 family members such as BCL-2 and BCL-XL (Chipuk et al. 2010). Thus, the decision as to whether to live or die during apoptosis appears to be made at the mitochondrial surface. Once mitochondria-dependent apoptotic pathways are fully activated, including release of cytochrome c and activation of executioner caspase-3, apoptosis is generally believed to be irreversible (Galluzzi et al. 2012; Riedl and Shi 2004; Tait and Green 2010; Chipuk et al. 2010).

Comprehensive and Rapid Destruction During the Apoptotic Process

Once mitochondrial permeabilization and caspase activation occur, massive and rapid destruction of functional and structural components typically ensues. As described above, during apoptosis intrinsic and extrinsic pro-apoptotic signals converge on mitochondria to trigger permeabilization (Wang 2001; Taylor et al. 2008). Since mitochondria are the bioenergetic and metabolic centers of the cell, widespread mitochondrial damage could in principle be sufficient to cause cell death. The permeabilization of mitochondria is a critical apoptotic event that releases cell executioners to the cytosol, initiating both caspase-dependent and -independent events, all of which can lead to cell death (Wang 2001; Green and Kroemer 2004; Kroemer and Martin 2005). These executioners include apoptosis-inducing factor (AIF), endonuclease G (EndoG), and cytochrome c (Wang 2001; Riedl and Shi 2004). AIF and EndoG translocate to the nucleus to destroy DNA (Susin et al. 1999; Li et al. 2001). Cytochrome c release activates the caspase cascade which then destroys structural and functional proteins (Luthi and Martin 2007), resulting in the morphological manifestations of apoptosis such as nuclear condensation, cell shrinkage, and membrane blebbing (Taylor et al. 2008). Activated caspases also trigger DNA destruction by activating DNA fragmentation factor/caspase-activated DNAse (DFF40/CAD) by cleaving its inhibitor DFF45/ICAD (Liu et al. 1997; Enari et al. 1998). Caspases also damage DNA repair systems, for example by cleaving and inactivating Poly(ADP)-ribose polymerase-1 (PARP), which is required for genomic stability (Lazebnik et al. 1994; Wang et al. 1997). The mitochondrial release of the Second Mitochondria-derived Activator of Caspase (Smac), which is also known as Direct IAP Binding Protein with Low pI (DIABLO) strengthens the caspase cascade by directly binding and inactivating Inhibitor of Apoptosis Proteins (IAPs) (Du et al. 2000; Verhagen et al. 2000; Chai et al. 2000). In addition to cleaving structural and functional components of the cell (Luthi and Martin 2007), caspases also cleave BID to form tBID, which translocates to mitochondria to reinforce mitochondrial permeabilization and release cell executioners in a positive feedback loop (Luo et al. 1998; Li et al. 1998; Kuwana et al. 2002).

Apoptosis can be a rapid process. While it could take minutes to hours or even days to commit to apoptosis, depending on the cell type and stimulus, time-lapse studies show that once activated, the mitochondrial release of cytochrome c to the cytosol can be completed in about 5 min. This timing was observed in a variety of cell lines and in response to a variety of stimuli (Goldstein et al. 2000, 2005). Cytosolic cytochrome c then triggers the caspase cascade, which once initiated can be completed within 5 min (Tyas et al. 2000; Takemoto et al. 2003). Within 10 min of caspase activation, cells display morphological hallmarks of apoptosis, including nuclear condensation and cell shrinkage (Tyas et al. 2000; Takemoto et al. 2003). Therefore, after mitochondrial fragmentation and activation of executioner caspases, apoptosis typically follows rapidly (Chipuk et al. 2010; Taylor et al. 2008; Riedl and Shi 2004; Takemoto et al. 2003). Although the precise "point of no return" is still controversial (Galluzzi et al. 2012), the dying process is generally considered to be irreversible at the execution stage.

Reversal of Apoptosis in Cell Cultures

Accumulating evidence demonstrates that apoptotic cells can reverse the dying process well beyond the point generally assumed to be irrevocable. A variety of human cancer cell lines including cervical, skin, liver, breast, and prostate that are chemically induced to undergo apoptosis can reverse the process and survive (Fig. 11.1) (Tang et al. 2009). Cells do so even after displaying multiple hallmarks of execution stage apoptosis including mitochondrial fragmentation

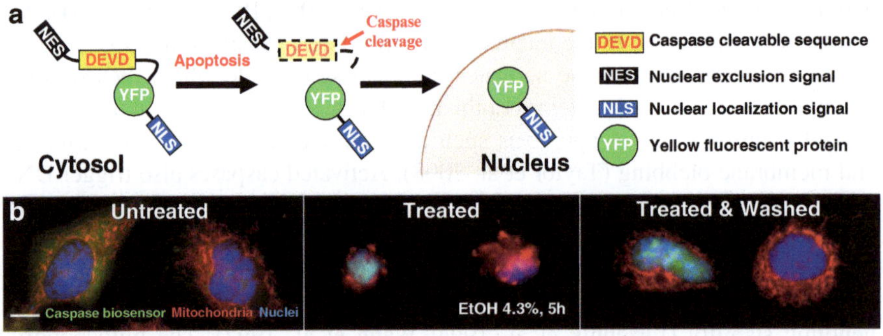

Fig. 11.1 Reversal of apoptosis after caspase activation. (**a**) Schematic diagram of a caspase biosensor, a fusion protein composed of a nuclear exclusion sequence (NES), a caspase cleavage site (DEVD), yellow fluorescent protein (YFP), and a nuclear localization signal (NLS). Upon caspase activation, YFP accumulates in the nucleus until the protein is degraded. (**b**) Fluorescence microscopy of HeLa cells expressing the caspase biosensor before (Untreated), during (Treated, 5 h), and after (Treated and Washed, 3 h) exposure to 4.3% ethanol. *Merged images*: caspase biosensor (YFP, *green*), mitochondria (*red*), and nuclei (*blue*). Cell at *left* expresses the caspase biosensor, which translocates to the nucleus where it remains even as the cell recovers normal morphology. The cell at *right* does not express the biosensor. Scale bar, 10 μm

and dysfunction, nuclear condensation, cytoplasmic shrinkage, membrane blebbing, and caspase activation (Tang et al. 2009, 2012). Although these signs were previously believed to represent commitment to death, washing away of chemical inducers is sufficient to allow dying cells to regain normal morphology and even to divide. However, dying cells have failed as yet to reverse apoptosis once they display nuclear fragmentation, which appears to be a truly terminal event (Tang et al. 2009).

A recent study also demonstrates that reversibility of late stage apoptosis can occur in a variety of normal cell types (Tang et al. 2012). Primary liver and heart cells, macrophages, NIH 3T3 fibroblasts, and brain cells, and cancer cells are all able to recover following removal of chemicals that induce apoptosis, even after they shrink and their mitochondria fragment, nuclei condense, and plasma membrane blebs (Tang et al. 2012). Interestingly, cells can reverse apoptosis even after activation of caspase 3 and apoptotic nucleases such as AIF, EndoG, and CAD, after inactivation of DNA repair systems, and after DNA is damaged. Yet when cells reverse apoptosis and survive, most DNA damage is repaired, suggesting the reactivation of repair systems. However, permanent genetic alterations occur in some of the cells (Fig. 11.2) (Tang et al. 2012), possibly because of incomplete DNA repair, consistent with an earlier hypothesis that arresting apoptosis in DNA-damaged cells should result in mutations (Vaughan et al. 2002; Sheridan and Martin 2008; Tubio and Estivill 2011). The ability of cells to reverse late stage apoptosis and survive, even if a bit damaged, could be a mechanism to allow cells and tissues to survive and recover from transient insults that would otherwise be fatal (Fig. 11.3).

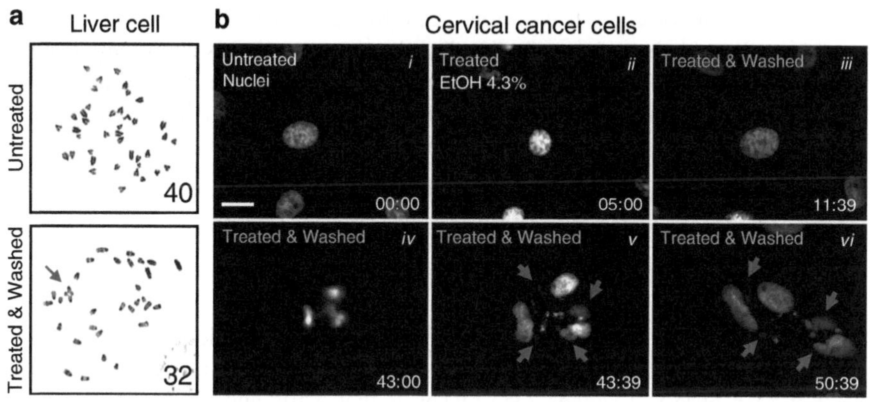

Fig. 11.2 Genetic alterations in cells after anastasis. (**a**) Inverted DAPI-banding image of metaphase spreads of untreated mouse primary liver cells (Untreated) compared to cells that were treated with 4.5% ethanol for 5 h, washed and then cultured with fresh medium for 3 days (Treated and Washed). Abnormal chromosomal configuration (indicated by an *arrow*) and number (indicated on the corresponding images) were observed in a small fraction of washed cells. (**b**) Monochromatic nuclear images of a single HeLa cell before ethanol induction (Untreated, *i*), after induction with 4.3% ethanol in culture medium for 5 h (Treated, *ii*), and then after washing and further culture in fresh medium (Treated and Washed, *iii–vi*). *Arrows* indicate some micronuclei that form. Time scale h:min. Scale bar, 30 µm. Images are modified from Tang et al. (Tang et al. 2012)

Fig. 11.3 Proposed model for the cause of mutation after anastasis

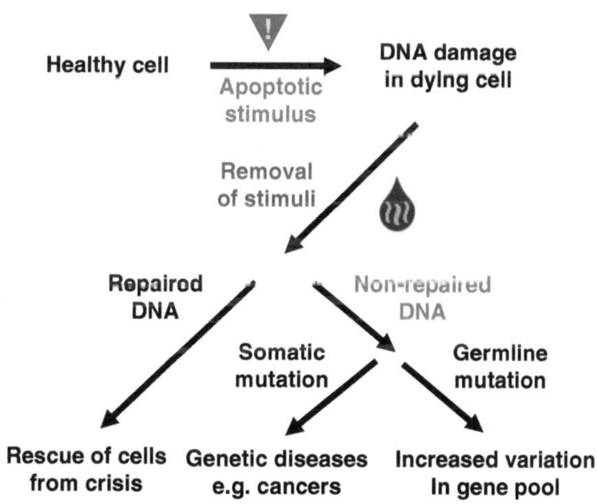

Anastasis Nomenclature

The word "Anastasis" (Ανασtάσης) means "rising to life" in Greek and has been proposed to describe this process of reversing apoptosis (Tang et al. 2012). The word apoptosis is a compound word made up of a Greek prefix ("apo") attached to a Greek stem ("ptosis"), and is used to describe the "dropping off" or "falling off," as of petals from flowers or leaves from trees (Kerr et al. 1972). "Anastasis", a compound word consisting of a prefix "ana" and a stem "stasis," is commonly used in Greek to denote the meaning of standing again, or rising to life. Since the key concept in "ptosis" is "falling to death" and that in "anastasis" is "rising to life," "anastasis" is an apt antonym to the term "apoptosis."

Detection of Anastasis

The observation that simply washing away an inducer is sufficient to allow cells to reverse apoptosis and survive raises the question as to why the process of anastasis has not been observed sooner. One reason is that while apoptosis was originally described based on the morphological appearance of cells in fixed tissue (Kerr et al. 1972), cells that have undergone anastasis resemble non-apoptotic cells and therefore are not easily recognized morphologically. Hallmarks of apoptosis include cytoplasmic and chromatin condensation, nuclear fragmentation, plasma membrane blebbing, and engulfment by resident phagocytes (Kerr et al. 1972; Jacobson et al. 1997). During anastasis, dying cells regain normal morphology and largely repair their damage (Tang et al. 2009, 2012). Therefore, in fixed tissue, cells in the process of anastasis resemble apoptotic cells and those that have completed anastasis are difficult to distinguish from normal cells.

To detect anastasis, one of the most reliable strategies is to continuously monitor cells throughout the processes of apoptosis and anastasis by time-lapse fluorescence microscopy (Tang et al. 2009, 2012). Overall cellular morphology can be monitored with DIC optics, and mitochondrial fragmentation, which is one of the features of early apoptosis (Taylor et al. 2008), can be observed by fluorescence microscopy. Biochemical markers of apoptosis including exposure of phosphatidylserine (PS) on the outer leaflet of the plasma membrane, release of mitochondrial cytochrome c after mitochondrial permeabilization, activation of proteases such as caspases, and DNA fragmentation can also be observed microscopically (Galluzzi et al. 2012; Taylor et al. 2008). Direct observation of cells following treatment with an inducer of apoptosis and then washing, reveals that they can recover normal morphology and go on to survive and divide even after undergoing nuclear condensation, cytoplasmic shrinkage, mitochondrial fragmentation and membrane blebbing. Moreover, time-lapse imaging of cells expressing a caspase activity biosensor shows that the same cells that display executioner caspase activation can recover normal morphology, survive, and proliferate after removal of the inducer (Tang et al. 2012).

It is important to note that although caspase-3 activation is generally considered to be a point of no return for dying cells, there are nevertheless examples of cells that activate caspases and then arrest the process and survive. One well-characterized example occurs during spermatid maturation in *Drosophila melanogaster* (Arama et al. 2003). In this instance, maturing sperm destroy most of their cytoplasmic contents in a limited demolition that requires apoptosome activity and effector caspase activation but does not result in cell death. This is an interesting example of arrested apoptosis; however, it is clearly distinct from anastasis in that the process does not reverse. In addition, although DNA damage is another hallmark of apoptosis, DNA is presumably safe-guarded from damage during sperm development, and therefore that DNAses are not activated and DNA repair is not inactivated. In contrast, time-lapse and cytogenetic studies show formation of micronuclei and other genetic alterations following anastasis (Tang et al. 2012). These observations are consistent with an earlier hypothesis that arrest of true apoptosis should result in DNA damage (Vaughan et al. 2002; Sheridan and Martin 2008; Tubio and Estivill 2011), although it is important to distinguish between the notion of arrest of apoptosis, which has been known for some time, and actual reversal of the process that takes place in cells undergoing anastasis (Tang et al. 2012), which is a newly described phenomenon.

Fluorescently labeled annexin V can also be used to follow anastasis. Annexin V binds efficiently to phosphatidylserine (PS), which moves from the inner to the outer leaflet of the plasma membrane during early apoptosis (Logue et al. 2009). Both ethanol-induced primary rat heart cells and jasplakinolide-induced brain cells bind fluorescein isothiocyanate-conjugated annexin V (annexin V-FITC) (Tang et al. 2012). These cells also exhibit other hallmarks of apoptosis including mitochondrial fragmentation, nuclear condensation, and membrane blebbing. After removal of the inducers, cells retain annexin V-FITC labeling as they recover, providing one way to label and track cells that reverse apoptosis. Another strategy is to sort out the annexin labeled cells for further study. Using a cell sorting approach, studies show that a fraction of apoptosis-induced, Annexin V-FITC-labeled B cell lymphoma cells and mouse mammary carcinoma cells with defective p53 protein continue to proliferate after removal of apoptotic stimuli (Geske et al. 2001; Hammill et al. 1999). However, Annexin V labels cells at a relatively early stage of apoptosis from which it is not as surprising that cells can recover, compared to later stages. An additional limitation of this approach is that annexin V-FITC labeling does not last long once the apoptotic stimulus is removed and the cells begin to recover. In the future it will be important to develop more robust methods to permanently label cells that have undergone anastasis.

Potential Examples and Consequences of Anastasis

Does anastasis naturally occur in animals? A variety of examples of tissue recovery following stresses that are known to induce apoptosis suggest that it is certainly possible. For example, embryos can survive radiation that causes massive apoptosis (Abrams et al. 1993). Neuronal apoptotic cell death can be induced in photoreceptor cells of flies, rabbits and rats by exposure to excess light (Milligan et al. 1997; McKechnie and Foulds 1980; Gordon et al. 2002) and yet the signs of apoptosis vanish quickly when the stressed animals are reexposed to a normal light and dark cycle (Milligan et al. 1997; McKechnie and Foulds 1980; Gordon et al. 2002). During heart failure, hallmarks of apoptosis including mitochondrial release of

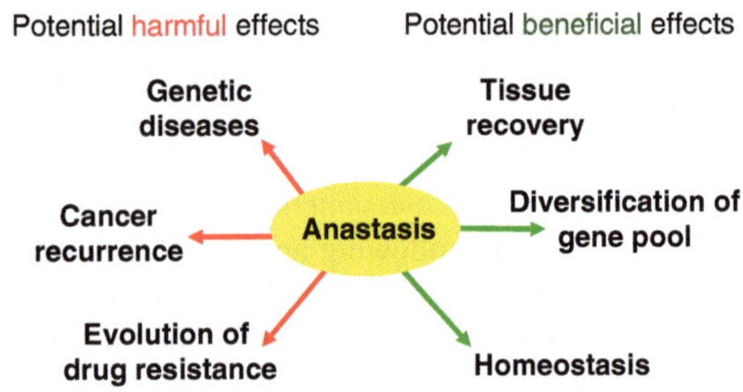

Fig. 11.4 Potential consequences of anastasis

cytochrome c and activation of caspase-3 are observed in significant numbers of cardiomyocytes, but these cells maintain normal nuclear morphology (Narula et al. 1999; Reed and Paternostro 1999), suggesting possible arrest of the apoptotic process (Narula et al. 2006), and reversal of apoptosis in isolated primary rat heart cells has also been observed (Tang et al. 2012). Importantly, advanced heart failure can be reversed after unloading the heart with a ventricular assist device (Dandel et al. 2011; Demirozu and Frazier 2012), and the recovering heart tissue displays increased expression of pro-survival genes (Milting et al. 1999; Haider et al. 2002), which are also induced in primary liver cells undergoing anastasis (Tang et al. 2012). Moreover, inhibition of these proteins prevents successful reversal of apoptosis (Tang et al. 2012). Although it is possible that each of these dramatic examples of in vivo tissue repair occurs by removal of apoptotic cells by phagocytosis and their replacement by compensatory cell division, which is stimulated by apoptotic cells (Bergmann and Steller 2010), an alternative or additional explanation may be that anastasis is a heretofore unrecognized mechanism that limits permanent damage and promotes healing and survival following transient stresses. The clearest potential example of reversible apoptosis in vivo to date appears to occur in the heart, following brief ischemic events (Kenis et al. 2010). Cardiomyocytes in both rabbits and mice can be labeled with Annexin V following transient ischemia and inhibition of caspases prevents this labeling, demonstrating that the cells enter the apoptotic program. Yet the cells go on to survive, internalize Annexin V, and reestablish phosphatidylserine asymmetry across the plasma membrane. These studies are suggestive that anastasis can occur in vivo in a very important physiological setting.

Just as apoptosis has both positive and negative consequences, anastasis may have two sides (Fig. 11.4). Anastasis could be a mechanism by which cancer cells initially survive chemotherapy and radiation treatments and later evolve resistance. In addition to the direct mutagenic effects of radiation and chemotherapy drugs (Ross 1999; Johnstone et al. 2002; Fu et al. 2012), the process of apoptosis causes new mutations some of which persist in cells that reverse the dying process (Tang et al. 2012). Interestingly many chemo- and radiation therapies induce apoptosis and are delivered episodically with recovery periods in between successive treatments. This approach is similar to applying an apoptotic inducer and washing it away (Tang et al. 2009, 2012). Therefore, cancer cells may undergo anastasis between anticancer treatments and could acquire new mutations and thus transform to more aggressive and metastatic phenotypes as a result. Acquisition of mutations in normal cells as a consequence of anastasis might also cause new tumors after cycles of anticancer therapy.

Recent next generation sequencing reveals preferred DNA cleavage sites in apoptotic cells (Fullwood et al. 2011). Therefore, it is possible that cells acquire specific mutations at these sites when the DNA damage is not fully repaired after anastasis, leading to specific diseases. In fact, studies of leukemia show that sublethal apoptotic induction results in cleavage within the *MLL* gene on chromosome 11q23 within a breakpoint cluster region (bcr) (Stanulla et al. 1997). Sublethal apoptotic induction promotes *MLL* gene translocations (Betti et al. 2001), *TEL* breaks, and also formation of *TEL-AML1* fusion (Eguchi-Ishimae et al. 2001), which is related to acute lymphoblastic leukemia (Zelent et al. 2004). These studies suggest a possible link between anastasis and carcinogenesis, and that suppression of anastasis could function synergistically with conventional chemo- and radiation therapies, preventing relapse, emergence of drug resistance, and initiation of new cancers.

Anastasis could also theoretically be a mechanism to introduce DNA mutations that promote genetic diversity and evolution. Environmental stresses can trigger apoptosis in germ cells in a variety of organisms (Aitken et al. 2011). For example in mice, germ cells undergo apoptosis in response to temperature stress (Perez-Crespo et al. 2008), and in fruit flies and *C. elegans* germ cells undergo apoptosis in response to starvation (Drummond-Barbosa and Spradling 2001; Salinas et al. 2006). Upon refeeding, fertility is restored. An interesting possibility is that germ cells might undergo anastasis in response to such transient stresses and the progeny produced from germ cells that have undergone anastasis might acquire new mutations as a consequence. Such stress-induced mutagenesis could in principle promote genetic diversity, improving the likelihood that some individuals out of a large population would be better adapted to the adverse environment. While relatively well established in single cell organisms such as bacteria and yeast (Rosenberg 2001), this phenomenon has not yet been explored in multicellular organisms. It is likely that only organisms with very large populations would benefit from such a strategy. Insects are therefore likely candidates for multicellular organisms that might benefit from stress-induced mutagenesis. Populations of mosquitoes, for example, appear to evolve resistance to insecticides rapidly (Maxmen 2012), so it will be of interest to determine whether insect germ cells undergo anastasis in response to various stresses and whether this ultimately enhances their ability to adapt to a changing environment. In other multicellular organisms, stress-induced mutagenesis might be an evolutionary vestige even if the mechanism is not currently beneficial; alternatively this mechanism may have been co-opted for other beneficial purposes such as salvaging cells that are difficult to replace, as discussed earlier. Further study of the full range of cell types and stresses that induce anastasis will help clarify these open questions.

Molecular Mechanisms of Anastasis

During apoptosis, multiple cell execution pathways are activated, and massive destruction occurs in a dying cell, raising the question as to how a dying cell reverses apoptosis at the execution stage. The molecular mechanisms of anastasis have only just begun to be investigated. Not surprisingly new gene transcription is required and a time course microarray analysis reveals that pro-survival genes such as members of the Bcl-2 and heat shock protein (HSP) families, Xiap and Mdm2, are up-regulated during the process (Fig. 11.5) (Tang et al. 2012). Clearly cells undergoing anastasis need to (1) suppress the expression or activities of proteins that induce apoptosis, (2) remove proteins that have been cleaved during induction of apoptosis, most likely via protein degradation pathways, and (3) repair damage caused in the dying process, including DNA and organelle damage (Fig. 11.6).

To suppress activated execution pathways, removal or inactivation of pro-apoptotic factors would seem to be the first critical step. During apoptosis, upstream initiator caspases (e.g., Caspase-8 and -9) cleave and activate downstream effector caspases (e.g., Caspase-3 and -7). So to reverse apoptosis, activated caspases should be degraded. In fact, during anastasis cleaved, active caspase-3 disappears and full-length caspase-3 returns to levels comparable with healthy cells (Tang et al. 2012). X-linked inhibitor of apoptosis protein (XIAP) may contribute to reverse apoptosis because it is a potent inhibitor of caspases (Deveraux et al. 1997, 1998; Scott et al. 2005; Ditzel et al. 2008) that promotes degradation of caspase-3 (Suzuki et al. 2001). Transcription of XIAP increases during anastasis (Tang et al. 2012). Suppression of XIAP in knockout mice or by a specific inhibitor in cell

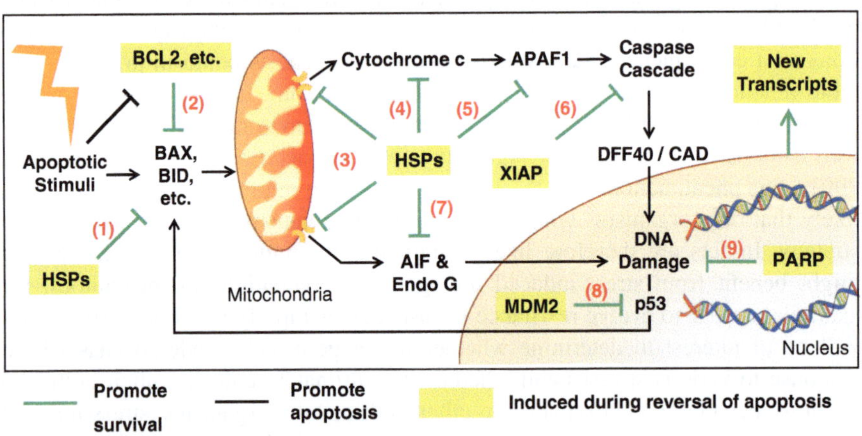

Fig. 11.5 Proposed model for the mechanisms of anastasis based on time course microarray analysis. Schematic representation of pro-apoptotic and pro-survival pathways. *Black arrows* indicate pro-apoptotic pathways. *Green lines* indicate pro-survival signals that could suppress pro-apoptotic pathways during anastasis. Primary references are cited in *parentheses* in the figure as supporting information [(*1*) Gabai et al. 2002; (*2*) Yin et al. 1994; (*3*) Paul et al. 2002; (*4*) Bruey et al. 2000; (*5*) Pandey et al. 2000; (*6*) Deveraux et al. 1997; (*7*) Ravagnan et al. 2001; (*8*) Momand et al. 1992; (*9*) de Murcia et al. 1997]. Modified from Tang et al. 2012

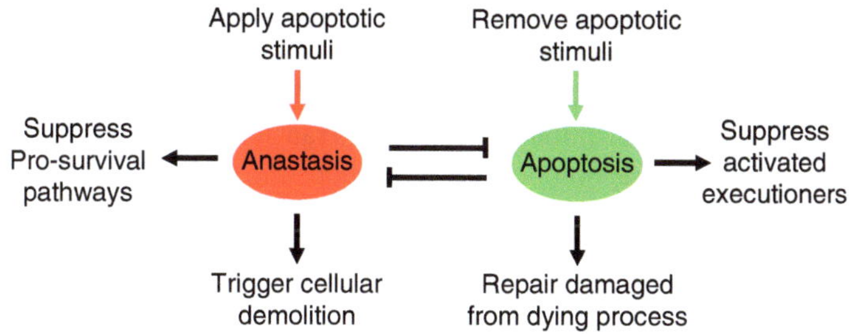

Fig. 11.6 Proposed model for the relationships between apoptosis and anastasis

culture does not by itself cause cell death (Tang et al. 2012; Harlin et al. 2001). However, inhibition of XIAP prevents anastasis (Tang et al. 2012), suggesting that XIAP plays an important role in the process.

In order to stably reverse apoptosis, the pro-apoptotic signals upstream of caspases should also be inhibited. As described above, during apoptosis, pro-apoptotic Bcl-2 family members such as BAX and BID translocate to mitochondria to cause fragmentation and outer membrane permeabilization (MOMP) by forming supramolecular openings on mitochondrial outer membranes (Luo et al. 1998; Kuwana et al. 2002; Wei et al. 2000). This causes release of mitochondrial executioners such as cytochrome c, which trigger caspase cascades (Wang 2001). In addition, mitochondrial release of Smac/Diablo unleashes caspases by binding to IAPs (Chai et al. 2000; Liu et al. 2000; Wu et al. 2000). So pro-survival Bcl-2 family proteins could contribute to anastasis as they suppress cytochrome c release by neutralizing factors such as Bax and Bid, which induce MOMP (Yang et al. 1997; Dlugosz et al. 2006; Peng et al. 2006). Heat shock protein 90 (HSP90) and HSP27 promote survival by preventing apoptosome formation (Pandey et al. 2000; Bruey et al. 2000), and HSP27 suppresses the release of cytochrome c (Paul et al. 2002). XIAP suppresses the mitochondrial release of Smac and promotes its degradation (MacFarlane et al. 2002; Flanagan et al. 2010). Importantly, the transcription of the pro-survival Bcl-2 family members, Hsp90 and Hsp27, and Xiap increases during early anastasis and by suppressing the functions of Bcl-2, HSP90, and XIAP prevents anastasis (Tang et al. 2012).

DNA damage occurs as a consequence of apoptosis but can also cause apoptosis by activating p53, which stimulates expression of pro-apoptotic factors such as Bax (Bouvard et al. 2000). Apoptotic nucleases are also activated in apoptosis. To reverse apoptosis, dying cells need to suppress p53 and apoptotic nucleases, and also repair DNA repair systems. To reverse apoptosis, MDM2 may repress p53 by recruiting ubiquitinated p53 to the proteasome to facilitate its degradation (Momand et al. 1992; Marine and Lozano 2010). Interestingly, transcription of MDM2 increases in cells following removal of the apoptotic stimulus, and inhibiting MDM2 suppresses anastasis (Tang et al. 2012). Expression of Hsp27 can suppress the activity of AIF (Ravagnan et al. 2001). In addition, the expression of ICAD and PARP return to pretreatment levels after anastasis (Tang et al. 2012), so that ICAD can lock CAD and PARP can repair DNA.

Since caspases cause massive destruction by cleaving functional and structural components in dying cells (Taylor et al. 2008; Luthi and Martin 2007), a wide range of new proteins need to be synthesized during reversal of apoptosis, so it is not surprising that new transcription is necessary. Transient suppression of transcription after removal of the apoptotic stimulus suppresses anastasis and results in cell death (Tang et al. 2012), providing empirical evidence that transcription is critical for anastasis.

Perspectives

The discovery of anastasis leads to several key questions. Can anastasis occur in animals, and if so, how can it be detected? What are the physiological consequences of anastasis, and the molecular mechanisms that regulate it? Can targeting anastasis be used to prevent or treat diseases? Currently, one major challenge in studying anastasis is to detect and track it in vivo in order to answer questions such as whether cycles of chemo- or radiation therapy induce anastasis, which in turn contributes to cancer recurrence. Does anastasis cause cancer to develop after repeated non-lethal apoptotic inductions? Can germ cells reverse apoptosis, and acquire mutations from anastasis when animals experience environmental or physiological stress?

Without a specific marker to label cells after reversal of apoptosis, there is no direct way to track anastasis in vivo long term. Time-lapse imaging in live animals is possible in a few tissues such as the retina, but it remains technically challenging to track the same group of cells in the same live animals for days. Ultimately, developing in vivo biosensors to track anastasis long term and identifying molecular hallmarks of anastasis will be critical steps for studying the mechanisms and consequences of anastasis, and harnessing this knowledge in order to improve human health.

References

Abraham MC, Lu Y, Shaham S (2007) A morphologically conserved nonapoptotic program promotes linker cell death in Caenorhabditis elegans. Dev Cell 12:73–86

Abrams JM, White K, Fessler LI, Steller H (1993) Programmed cell death during Drosophila embryogenesis. Development 117:29–43

Aitken RJ, Findlay JK, Hutt KJ, Kerr JB (2011) Apoptosis in the germ line. Reproduction 141:139–150

Arama E, Agapite J, Steller H (2003) Caspase activity and a specific cytochrome C are required for sperm differentiation in Drosophila. Dev Cell 4:687–697

Baehrecke EH (2002) How death shapes life during development. Nat Rev Mol Cell Biol 3:779–787

Bergmann A, Steller H (2010) Apoptosis, stem cells, and tissue regeneration. Sci Signal 3:re8

Betti CJ, Villalobos MJ, Diaz MO, Vaughan AT (2001) Apoptotic triggers initiate translocations within the MLL gene involving the nonhomologous end joining repair system. Cancer Res 61:4550–4555

Blum ES, Abraham MC, Yoshimura S, Lu Y, Shaham S (2012) Control of nonapoptotic developmental cell death in Caenorhabditis elegans by a polyglutamine-repeat protein. Science 335:970–973
Bouvard V, Zaitchouk T, Vacher M, Duthu A, Canivet M, Choisy-Rossi C, Nieruchalski M, May E (2000) Tissue and cell-specific expression of the p53-target genes: bax, fas, mdm2 and waf1/p21, before and following ionising irradiation in mice. Oncogene 19:649–660
Brown JM, Attardi LD (2005) The role of apoptosis in cancer development and treatment response. Nat Rev Cancer 5:231–237
Bruey JM, Ducasse C, Bonniaud P, Ravagnan L, Susin SA, Diaz-Latoud C, Gurbuxani S, Arrigo AP, Kroemer G, Solary E et al (2000) Hsp27 negatively regulates cell death by interacting with cytochrome c. Nat Cell Biol 2:645–652
Buttner S, Eisenberg T, Herker E, Carmona-Gutierrez D, Kroemer G, Madeo F (2006) Why yeast cells can undergo apoptosis: death in times of peace, love, and war. J Cell Biol 175:521–525
Byrne HM (2010) Dissecting cancer through mathematics: from the cell to the animal model. Nat Rev Cancer 10:221–230
Carmona-Gutierrez D, Eisenberg T, Buttner S, Meisinger C, Kroemer G, Madeo F (2010) Apoptosis in yeast: triggers, pathways, subroutines. Cell Death Differ 17:763–773
Chai J, Du C, Wu JW, Kyin S, Wang X, Shi Y (2000) Structural and biochemical basis of apoptotic activation by Smac/DIABLO. Nature 406:855–862
Chen Y, Klionsky DJ (2011) The regulation of autophagy—unanswered questions. J Cell Sci 124:161–170
Chipuk JE, Moldoveanu T, Llambi F, Parsons MJ, Green DR (2010) The BCL-2 family reunion. Mol Cell 37:299–310
Codogno P, Meijer AJ (2005) Autophagy and signaling: their role in cell survival and cell death. Cell Death Differ 12(Suppl 2):1509–1518
Dandel M, Weng Y, Siniawski H, Stepanenko A, Krabatsch T, Potapov E, Lehmkuhl HB, Knosalla C, Hetzer R (2011) Heart failure reversal by ventricular unloading in patients with chronic cardiomyopathy: criteria for weaning from ventricular assist devices. Eur Heart J 32:1148–1160
Datta SR, Dudek H, Tao X, Masters S, Fu H, Gotoh Y, Greenberg ME (1997) Akt phosphorylation of BAD couples survival signals to the cell-intrinsic death machinery. Cell 91:231–241
de Murcia JM, Niedergang C, Trucco C, Ricoul M, Dutrillaux B, Mark M, Oliver FJ, Masson M, Dierich A, LeMeur M et al (1997) Requirement of poly(ADP-ribose) polymerase in recovery from DNA damage in mice and in cells. Proc Natl Acad Sci USA 94:7303–7307
Demirozu ZT, Frazier OH (2012) Remission of chronic, advanced heart failure after left ventricular unloading with an implantable left ventricular assist device. Tex Heart Inst J 39:268–270
Deveraux QL, Takahashi R, Salvesen GS, Reed JC (1997) X-linked IAP is a direct inhibitor of cell-death proteases. Nature 388:300–304
Deveraux QL, Roy N, Stennicke HR, Van Arsdale T, Zhou Q, Srinivasula SM, Alnemri ES, Salvesen GS, Reed JC (1998) IAPs block apoptotic events induced by caspase-8 and cytochrome c by direct inhibition of distinct caspases. EMBO J 17:2215–2223
Ditzel M, Broemer M, Tenev T, Bolduc C, Lee TV, Rigbolt KT, Elliott R, Zvelebil M, Blagoev B, Bergmann A et al (2008) Inactivation of effector caspases through nondegradative polyubiquitylation. Mol Cell 32:540–553
Dlugosz PJ, Billen LP, Annis MG, Zhu W, Zhang Z, Lin J, Leber B, Andrews DW (2006) Bcl-2 changes conformation to inhibit Bax oligomerization. EMBO J 25:2287–2296
Drummond-Barbosa D, Spradling AC (2001) Stem cells and their progeny respond to nutritional changes during Drosophila oogenesis. Dev Biol 231:265–278
Du C, Fang M, Li Y, Li L, Wang X (2000) Smac, a mitochondrial protein that promotes cytochrome c-dependent caspase activation by eliminating IAP inhibition. Cell 102:33–42
Eguchi-Ishimae M, Eguchi M, Ishii E, Miyazaki S, Ueda K, Kamada N, Mizutani S (2001) Breakage and fusion of the TEL (ETV6) gene in immature B lymphocytes induced by apoptogenic signals. Blood 97:737–743

Enari M, Sakahira H, Yokoyama H, Okawa K, Iwamatsu A, Nagata S (1998) A caspase-activated DNase that degrades DNA during apoptosis, and its inhibitor ICAD. Nature 391:43–50

Flanagan L, Sebastia J, Tuffy LP, Spring A, Lichawska A, Devocelle M, Prehn JH, Rehm M (2010) XIAP impairs Smac release from the mitochondria during apoptosis. Cell Death Dis 1:e49

Friedlander RM (2003) Apoptosis and caspases in neurodegenerative diseases. N Engl J Med 348:1365–1375

Fu D, Calvo JA, Samson LD (2012) Balancing repair and tolerance of DNA damage caused by alkylating agents. Nat Rev Cancer 12:104–120

Fuchs Y, Steller H (2011) Programmed cell death in animal development and disease. Cell 147:742–758

Fullwood MJ, Lee J, Lin L, Li G, Huss M, Ng P, Sung WK, Shenolikar S (2011) Next-generation sequencing of apoptotic DNA breakpoints reveals association with actively transcribed genes and gene translocations. PLoS One 6:e26054

Gabai VL, Mabuchi K, Mosser DD, Sherman MY (2002) Hsp72 and stress kinase c-jun N-terminal kinase regulate the bid-dependent pathway in tumor necrosis factor-induced apoptosis. Mol Cell Biol 22:3415–3424

Galluzzi L, Vitale I, Abrams JM, Alnemri ES, Baehrecke EH, Blagosklonny MV, Dawson TM, Dawson VL, El-Deiry WS, Fulda S et al (2012) Molecular definitions of cell death subroutines: recommendations of the Nomenclature Committee on Cell Death 2012. Cell Death Differ 19:107–120

Geske FJ, Lieberman R, Strange R, Gerschenson LE (2001) Early stages of p53-induced apoptosis are reversible. Cell Death Differ 8:182–191

Glücksmann A (1951) Cell deaths in normal vertebrate ontogeny. Biol Rev Camb Philos Soc 26:59–86

Goldstein JC, Waterhouse NJ, Juin P, Evan GI, Green DR (2000) The coordinate release of cytochrome c during apoptosis is rapid, complete and kinetically invariant. Nat Cell Biol 2: 156–162

Goldstein JC, Munoz-Pinedo C, Ricci JE, Adams SR, Kelekar A, Schuler M, Tsien RY, Green DR (2005) Cytochrome c is released in a single step during apoptosis. Cell Death Differ 12:453–462

Golstein P, Kroemer G (2007) Cell death by necrosis: towards a molecular definition. Trends Biochem Sci 32:37–43

Gordon WC, Casey DM, Lukiw WJ, Bazan NG (2002) DNA damage and repair in light-induced photoreceptor degeneration. Invest Ophthalmol Vis Sci 43:3511–3521

Gourlay CW, Du W, Ayscough KR (2006) Apoptosis in yeast—mechanisms and benefits to a unicellular organism. Mol Microbiol 62:1515–1521

Green DR, Kroemer G (2004) The pathophysiology of mitochondrial cell death. Science 305:626–629

Haider N, Narula N, Narula J (2002) Apoptosis in heart failure represents programmed cell survival, not death, of cardiomyocytes and likelihood of reverse remodeling. J Card Fail 8:S512–S517

Hammill AK, Uhr JW, Scheuermann RH (1999) Annexin V staining due to loss of membrane asymmetry can be reversible and precede commitment to apoptotic death. Exp Cell Res 251:16–21

Harlin H, Reffey SB, Duckett CS, Lindsten T, Thompson CB (2001) Characterization of XIAP-deficient mice. Mol Cell Biol 21:3604–3608

Horvitz HR (1999) Genetic control of programmed cell death in the nematode Caenorhabditis elegans. Cancer Res 59:1701s–1706s

Jacobson MD, Weil M, Raff MC (1997) Programmed cell death in animal development. Cell 88:347–354

Johnstone RW, Ruefli AA, Lowe SW (2002) Apoptosis: a link between cancer genetics and chemotherapy. Cell 108:153–164

Juo P, Kuo CJ, Yuan J, Blenis J (1998) Essential requirement for caspase-8/FLICE in the initiation of the Fas-induced apoptotic cascade. Curr Biol 8:1001–1008

Kenis H, Zandbergen HR, Hofstra L, Petrov AD, Dumont EA, Blankenberg FD, Haider N, Bitsch N, Gijbels M, Verjans JW et al (2010) Annexin A5 uptake in ischemic myocardium: demonstration of reversible phosphatidylserine externalization and feasibility of radionuclide imaging. J Nucl Med 51:259–267

Kerr JF, Wyllie AH, Currie AR (1972) Apoptosis: a basic biological phenomenon with wide-ranging implications in tissue kinetics. Br J Cancer 26:239–257

Kroemer G, Martin SJ (2005) Caspase-independent cell death. Nat Med 11:725–730

Kuida K, Zheng TS, Na S, Kuan C, Yang D, Karasuyama H, Rakic P, Flavell RA (1996) Decreased apoptosis in the brain and premature lethality in CPP32-deficient mice. Nature 384:368–372

Kuwana T, Mackey MR, Perkins G, Ellisman MH, Latterich M, Schneiter R, Green DR, Newmeyer DD (2002) Bid, Bax, and lipids cooperate to form supramolecular openings in the outer mitochondrial membrane. Cell 111:331–342

Lam E (2004) Controlled cell death, plant survival and development. Nat Rev Mol Cell Biol 5:305–315

Lazebnik YA, Kaufmann SH, Desnoyers S, Poirier GG, Earnshaw WC (1994) Cleavage of poly(ADP-ribose) polymerase by a proteinase with properties like ICE. Nature 371:346–347

Lettre G, Hengartner MO (2006) Developmental apoptosis in C. elegans: a complex CEDnario. Nat Rev Mol Cell Biol 7:97–108

Li H, Zhu H, Xu CJ, Yuan J (1998) Cleavage of BID by caspase 8 mediates the mitochondrial damage in the Fas pathway of apoptosis. Cell 94:491–501

Li LY, Luo X, Wang X (2001) Endonuclease G is an apoptotic DNase when released from mitochondria. Nature 412:95–99

Liu X, Zou H, Slaughter C, Wang X (1997) DFF, a heterodimeric protein that functions downstream of caspase-3 to trigger DNA fragmentation during apoptosis. Cell 89:175–184

Liu Z, Sun C, Olejniczak ET, Meadows RP, Betz SF, Oost T, Herrmann J, Wu JC, Fesik SW (2000) Structural basis for binding of Smac/DIABLO to the XIAP BIR3 domain. Nature 408:1004–1008

Lockshin RA, Williams CM (1964) Programmed cell death. II. Endocrine potentiation of the breakdown of the intersegmental muscles of silkmoths. J Insect Physiol 10:643–649

Logue SE, Elgendy M, Martin SJ (2009) Expression, purification and use of recombinant annexin V for the detection of apoptotic cells. Nat Protoc 4:1383–1395

Luo X, Budihardjo I, Zou H, Slaughter C, Wang X (1998) Bid, a Bcl2 interacting protein, mediates cytochrome c release from mitochondria in response to activation of cell surface death receptors. Cell 94:481–490

Luthi AU, Martin SJ (2007) The CASBAH: a searchable database of caspase substrates. Cell Death Differ 14:641–650

MacFarlane M, Merrison W, Bratton SB, Cohen GM (2002) Proteasome-mediated degradation of Smac during apoptosis: XIAP promotes Smac ubiquitination in vitro. J Biol Chem 277: 36611–36616

Marine JC, Lozano G (2010) Mdm2-mediated ubiquitylation: p53 and beyond. Cell Death Differ 17:93–102

Mattson MP (2000) Apoptosis in neurodegenerative disorders. Nat Rev Mol Cell Biol 1:120–129

Maxmen A (2012) Malaria surge feared. Nature 485:293

McKechnie NM, Foulds WS (1980) Recovery of the rabbit retina after light damage (preliminary observations). Albrecht Von Graefes Arch Klin Exp Ophthalmol 212:271–283

Milligan SC, Alb JG Jr, Elagina RB, Bankaitis VA, Hyde DR (1997) The phosphatidylinositol transfer protein domain of Drosophila retinal degeneration B protein is essential for photoreceptor cell survival and recovery from light stimulation. J Cell Biol 139:351–363

Milting H, Bartling B, Schumann H, El-Banayosy A, Wlost S, Ruter F, Darmer D, Holtz J, Korfer R, Zerkowski HR (1999) Altered levels of mRNA of apoptosis-mediating genes after mid-term mechanical ventricular support in dilative cardiomyopathy—first results of the Halle Assist Induced Recovery Study (HAIR). Thorac Cardiovasc Surg 47:48–50

Momand J, Zambetti GP, Olson DC, George D, Levine AJ (1992) The mdm-2 oncogene product forms a complex with the p53 protein and inhibits p53-mediated transactivation. Cell 69: 1237–1245

Mrozek A, Petrowsky H, Sturm I, Kraus J, Hermann S, Hauptmann S, Lorenz M, Dorken B, Daniel PT (2003) Combined p53/Bax mutation results in extremely poor prognosis in gastric carcinoma with low microsatellite instability. Cell Death Differ 10:461–467

Murray-Zmijewski F, Slee EA, Lu X (2008) A complex barcode underlies the heterogeneous response of p53 to stress. Nat Rev Mol Cell Biol 9:702–712

Narula J, Pandey P, Arbustini E, Haider N, Narula N, Kolodgie FD, Dal Bello B, Semigran MJ, Bielsa-Masdeu A, Dec GW et al (1999) Apoptosis in heart failure: release of cytochrome c from mitochondria and activation of caspase-3 in human cardiomyopathy. Proc Natl Acad Sci USA 96:8144–8149

Narula J, Haider N, Arbustini E, Chandrashekhar Y (2006) Mechanisms of disease: apoptosis in heart failure—seeing hope in death. Nat Clin Pract Cardiovasc Med 3:681–688

Pandey P, Saleh A, Nakazawa A, Kumar S, Srinivasula SM, Kumar V, Weichselbaum R, Nalin C, Alnemri ES, Kufe D et al (2000) Negative regulation of cytochrome c-mediated oligomerization of Apaf-1 and activation of procaspase-9 by heat shock protein 90. EMBO J 19:4310–4322

Paul C, Manero F, Gonin S, Kretz-Remy C, Virot S, Arrigo AP (2002) Hsp27 as a negative regulator of cytochrome C release. Mol Cell Biol 22:816–834

Peng J, Tan C, Roberts GJ, Nikolaeva O, Zhang Z, Lapolla SM, Primorac S, Andrews DW, Lin J (2006) tBid elicits a conformational alteration in membrane-bound Bcl-2 such that it inhibits Bax pore formation. J Biol Chem 281:35802–35811

Perez-Crespo M, Pintado B, Gutierrez-Adan A (2008) Scrotal heat stress effects on sperm viability, sperm DNA integrity, and the offspring sex ratio in mice. Mol Reprod Dev 75:40–47

Ravagnan L, Gurbuxani S, Susin SA, Maisse C, Daugas E, Zamzami N, Mak T, Jaattela M, Penninger JM, Garrido C et al (2001) Heat-shock protein 70 antagonizes apoptosis-inducing factor. Nat Cell Biol 3:839–843

Reed JC, Paternostro G (1999) Postmitochondrial regulation of apoptosis during heart failure. Proc Natl Acad Sci USA 96:7614–7616

Riedl SJ, Shi Y (2004) Molecular mechanisms of caspase regulation during apoptosis. Nat Rev Mol Cell Biol 5:897–907

Rosenberg SM (2001) Evolving responsively: adaptive mutation. Nat Rev Genet 2:504–515

Ross GM (1999) Induction of cell death by radiotherapy. Endocr Relat Cancer 6:41–44

Salinas LS, Maldonado E, Navarro RE (2006) Stress-induced germ cell apoptosis by a p53 independent pathway in Caenorhabditis elegans. Cell Death Differ 13:2129–2139

Schug ZT, Gonzalvez F, Houtkooper RH, Vaz FM, Gottlieb E (2011) BID is cleaved by caspase-8 within a native complex on the mitochondrial membrane. Cell Death Differ 18:538–548

Scott FL, Denault JB, Riedl SJ, Shin H, Renatus M, Salvesen GS (2005) XIAP inhibits caspase-3 and -7 using two binding sites: evolutionarily conserved mechanism of IAPs. EMBO J 24:645–655

Sheridan C, Martin SJ (2008) Commitment in apoptosis: slightly dead but mostly alive. Trends Cell Biol 18:353–357

Stanulla M, Wang J, Chervinsky DS, Thandla S, Aplan PD (1997) DNA cleavage within the MLL breakpoint cluster region is a specific event which occurs as part of higher-order chromatin fragmentation during the initial stages of apoptosis. Mol Cell Biol 17:4070–4079

Stennicke HR, Jurgensmeier JM, Shin H, Deveraux Q, Wolf BB, Yang X, Zhou Q, Ellerby HM, Ellerby LM, Bredesen D et al (1998) Pro-caspase-3 is a major physiologic target of caspase-8. J Biol Chem 273:27084–27090

Stratton MR, Campbell PJ, Futreal PA (2009) The cancer genome. Nature 458:719–724

Susin SA, Lorenzo HK, Zamzami N, Marzo I, Snow BE, Brothers GM, Mangion J, Jacotot E, Costantini P, Loeffler M et al (1999) Molecular characterization of mitochondrial apoptosis-inducing factor. Nature 397:441–446

Suzuki Y, Nakabayashi Y, Takahashi R (2001) Ubiquitin-protein ligase activity of X-linked inhibitor of apoptosis protein promotes proteasomal degradation of caspase-3 and enhances its anti-apoptotic effect in Fas-induced cell death. Proc Natl Acad Sci USA 98:8662–8667

Tait SW, Green DR (2010) Mitochondria and cell death: outer membrane permeabilization and beyond. Nat Rev Mol Cell Biol 11:621–632

Takemoto K, Nagai T, Miyawaki A, Miura M (2003) Spatio-temporal activation of caspase revealed by indicator that is insensitive to environmental effects. J Cell Biol 160:235–243
Tang HL, Yuen KL, Tang HM, Fung MC (2009) Reversibility of apoptosis in cancer cells. Br J Cancer 100:118–122
Tang HL, Tang HM, Mak KH, Hu S, Wang SS, Wong KM, Wong CS, Wu HY, Law HT, Liu K et al (2012) Cell survival, DNA damage, and oncogenic transformation after a transient and reversible apoptotic response. Mol Biol Cell 23:2240–2252
Taylor RC, Cullen SP, Martin SJ (2008) Apoptosis: controlled demolition at the cellular level. Nat Rev Mol Cell Biol 9:231–241
Tubio JM, Estivill X (2011) Cancer: when catastrophe strikes a cell. Nature 470:476–477
Tyas L, Brophy VA, Pope A, Rivett AJ, Tavare JM (2000) Rapid caspase-3 activation during apoptosis revealed using fluorescence-resonance energy transfer. EMBO Rep 1:266–270
Vandenabeele P, Galluzzi L, Vanden Berghe T, Kroemer G (2010) Molecular mechanisms of necroptosis: an ordered cellular explosion. Nat Rev Mol Cell Biol 11:700–714
Vaughan AT, Betti CJ, Villalobos MJ (2002) Surviving apoptosis. Apoptosis 7:173–177
Vazquez A, Bond EE, Levine AJ, Bond GL (2008) The genetics of the p53 pathway, apoptosis and cancer therapy. Nat Rev Drug Discov 7:979–987
Verhagen AM, Ekert PG, Pakusch M, Silke J, Connolly LM, Reid GE, Moritz RL, Simpson RJ, Vaux DL (2000) Identification of DIABLO, a mammalian protein that promotes apoptosis by binding to and antagonizing IAP proteins. Cell 102:43–53
Wang X (2001) The expanding role of mitochondria in apoptosis. Genes Dev 15:2922–2933
Wang ZQ, Stingl L, Morrison C, Jantsch M, Los M, Schulze-Osthoff K, Wagner EF (1997) PARP is important for genomic stability but dispensable in apoptosis. Genes Dev 11:2347–2358
Wang L, Du F, Wang X (2008) TNF-alpha induces two distinct caspase-8 activation pathways. Cell 133:693–703
Wei MC, Lindsten T, Mootha VK, Weiler S, Gross A, Ashiya M, Thompson CB, Korsmeyer SJ (2000) tBID, a membrane-targeted death ligand, oligomerizes BAK to release cytochrome c. Genes Dev 14:2060–2071
Wu G, Chai J, Suber TL, Wu JW, Du C, Wang X, Shi Y (2000) Structural basis of IAP recognition by Smac/DIABLO. Nature 408:1008–1012
Yang J, Liu X, Bhalla K, Kim CN, Ibrado AM, Cai J, Peng TI, Jones DP, Wang X (1997) Prevention of apoptosis by Bcl-2: release of cytochrome c from mitochondria blocked. Science 275:1129–1132
Yin XM, Oltvai ZN, Korsmeyer SJ (1994) BH1 and BH2 domains of Bcl-2 are required for inhibition of apoptosis and heterodimerization with Bax. Nature 369:321–323
Zelent A, Greaves M, Enver T (2004) Role of the TEL-AML1 fusion gene in the molecular pathogenesis of childhood acute lymphoblastic leukaemia. Oncogene 23:4275–4283
Zhu Y, Yang GY, Ahlemeyer B, Pang L, Che XM, Culmsee C, Klumpp S, Krieglstein J (2002) Transforming growth factor-beta 1 increases bad phosphorylation and protects neurons against damage. J Neurosci 22:3898–3909

Chapter 12
The Transgenerational Effects of Parental Exposure to Mutagens in Mammals

Yuri E. Dubrova

Abstract The review describes the transgenerational effects of parental exposure to ionizing radiation, chemical mutagens and anticancer drugs. The results of some recent animal studies showing elevated mutation rates in the non-exposed offspring of irradiated parents are presented and discussed. The possible mechanisms and implications of transgenerational instability are also discussed.

Introduction

A number of recent studies have clearly demonstrated the existence of non-targeted effects of ionizing radiation (Wright 2010). These effects include mutation induction occurring at tandem repeat DNA loci not directly targeted by ionizing radiation (Sadamoto et al. 1994; Dubrova et al. 1996, 1998), radiation-induced genomic instability among the non-exposed progeny of irradiated cells/organisms (Morgan 2003a, b), and the bystander effect, whereby an elevated mutation rate is detected in the non-exposed cells, which receive signals from the irradiated neighbors (Morgan 2003a, b; Lorimore et al. 2003). Taken together, these observations imply that the non-targeted effects of ionizing radiation can manifest over a period of time much longer than previously thought, and may therefore be regarded as a component of the genetic risk of ionizing radiation.

It should be stressed that the phenomenon of non-targeted effects cannot be explained by the target theory, independently developed in 1947 by Timofeeff-Ressovsky and Zimmer (1947), and Lea (1947). According to the modern interpretation of the target theory, the effects of ionizing radiation on the exposed cell are attributed to the initial DNA damage at the sites (targets) that are either directly hit

Y.E. Dubrova (✉)
Department of Genetics, University of Leicester, University Road, Leicester LE1 7RH, UK
e-mail: yed2@le.ac.uk

D. Mittelman (ed.), *Stress-Induced Mutagenesis*, DOI 10.1007/978-1-4614-6280-4_12,

by photons, or affected by free radicals arising as a result of the localized ionization of water (Lehnert 2007). The main prediction of the target theory is that mutation induction almost exclusively occurs in the directly exposed cells at damaged sites that are unrepaired or mis-repaired. In other words, everything, including the recognition of radiation-induced damage, its repair, and mutation induction, happens in the exposed cells. There is another important prediction of the target theory, postulating that the magnitude of mutation induction is directly related to the amount of radiation-induced DNA damage and the ability of exposed cell to repair it, meaning that the yield of radiation-induced mutations strongly depends on the dose- and dose-rate of exposure (UNSCEAR 2001). The target theory also lays the foundation of a linear-no-threshold risk model, whereby the risk of human exposure to ionizing radiation is linearly related to the dose of exposure without a threshold (BEIR 2006). Given that the above-mentioned non-targeted effects of ionizing radiation manifest either in the progeny of irradiated cells, or in neighboring cells that received signals from the exposed cells, these effects cannot therefore be attributed to direct targeting, and do not arise at the sites initially damaged by ionizing radiation.

In this review, the non-targeted, transgenerational effects of parental exposure to ionizing radiation and chemical mutagens, including genomic instability manifesting in the offspring of exposed parents, are presented and discussed. The phenomenon of transgenerational genomic instability can be defined as an increased rate of mutation observed in the non-exposed offspring of irradiated parents (Dubrova 2003).

Transgenerational Effects in the Offspring of Irradiated Parents

The studies on the transgenerational effects of parental exposure to ionizing radiation are designed to test the hypothesis that non-targeted effects induced in the germ line of directly exposed parents could manifest in the offspring, affecting their mutation rates, cancer predisposition, and other characteristics. As far as the design of studies analyzing transgenerational genomic instability is concerned, it substantially differs from that used to investigate mutation induction in the germ line of exposed parents. Given that these studies aim to establish whether the mutation rate in the offspring of parents exposed to ionizing radiation or chemical mutagens is elevated, the frequency of mutations occurring in the offspring after fertilization should be evaluated. This can be achieved either by sampling the offspring tissues, which would allow establishing mutation frequencies in their germ line and somatic tissues, or, alternatively, by profiling the second-generation offspring, thus scoring mutations occurred in the F_1 germ line and passed to their offspring (Fig. 12.1).

Apart from the design of transgenerational studies, there is another essential issue dealing with the choice of techniques for mutation detection in the offspring. It is clear that these techniques should equally detect mutations occurring in the germ line and somatic tissues, which is not the case for the majority of current approaches for monitoring mutation induction in eukaryotes. For example, the specific locus test developed in the late 1950s by William Russell was designed to

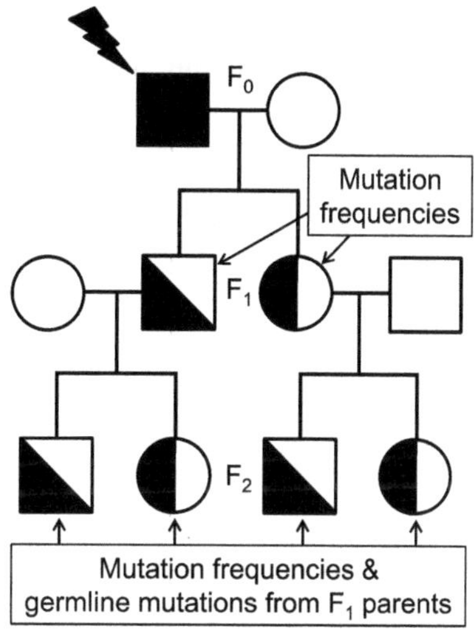

Fig. 12.1 The design of transgenerational study

evaluate mutation induction in the germ line of exposed parents (Searle 1974). Although using this approach, the genetic effects of high-dose exposure to ionizing radiation were comprehensively established (UNSCEAR 2001; Searle 1974), its sensitivity is quite low, which effectively makes it useless for the analysis of relatively moderate transgenerational changes occurring in the offspring germ line. That is why the first experimental evidence for unusually elevated mutation rates in the first-generation offspring of exposed parents was obtained using other techniques. In these studies, a substantially elevated incidence of mosaicism was detected in the F_1 offspring of male drosophila exposed to a number of alkylating agents (Mathew 1964; Lee et al. 1970). Another piece of evidence for transgenerational effects was obtained by Luning and coworkers, who studied the frequency of dominant lethal mutations in the germ line of non-exposed offspring of irradiated male mice (Luning et al. 1976). It should be noted that although the above-mentioned studies provided the first indication of transgeneration destabilization of the F_1 genome, for a long time, these studies remained inconclusive and, unfortunately, forgotten.

The foundation of transgenerational studies was laid in the 1990s, when we and others proposed that expanded simple tandem repeat (ESTR) loci could provide a sensitive approach for monitoring radiation-induced mutation in the mouse germ line (Sadamoto et al. 1994; Dubrova et al. 1993). ESTRs consist of relatively short repeats, 5–10 bp long, and belong to the most unstable loci in the mouse genome with mutation rate as high as 0.05 per cell division (Bois et al. 1998; Hardwick et al. 2009). Spontaneous and induced mutation at these loci is attributed to gain or loss of a certain number of repeats. Initially, we and others used a pedigree-based approach similar to that of the specific locus test, which requires the profiling of all parents and their offspring, and therefore detects ESTR mutations occurring in the parental germ line (Sadamoto et al. 1994; Dubrova et al. 1993, 1998, 2000a; Barber et al. 2000, 2004; Vilarino-Guell et al. 2003; Dubrova 2005). Later on, we developed a single-molecule PCR technique capable of detecting ESTR mutants in DNA samples extracted from any mouse tissue, including the male germ line (Yauk et al. 2002). This technique involves diluting bulk genomic DNA and amplifying multiple samples of DNA, each containing approximately one amplifiable ESTR molecule. ESTR mutants are detected by Southern blot hybridization, following agarose-gel electrophoresis. The results of our studies show that the sensitivity of the single-molecule PCR technique for ESTR mutation detection in the germ line of male mice exposed to ionizing radiation, chemical mutagens, and anticancer drugs is very close to that of a pedigree-based approach (Yauk et al. 2002; Glen et al. 2008). Most importantly, the dose–response for ESTR mutation induction in male mice exposed to ionizing radiation and chemical mutagens is remarkably similar to that obtained by the specific-locus test (Dubrova et al. 1998, 2000a; Vilarino-Guell et al. 2003; Glen et al. 2008). Given that high frequencies of spontaneous and induced mutations at these loci allow mutation induction to be evaluated in very small population samples, we therefore used these loci for the analysis of transgenerational instability.

The first evidence for elevated ESTR mutation rates in the F_1 germ line was obtained by analyzing the offspring of male mice exposed to fission neutrons (Dubrova et al. 2000b). According to the results of this study, ESTR mutation rates show highly significant elevation in the germ line of the male and female offspring of irradiated male mice. To further validate these unexpected results, we later designed a three-generation study specifically aimed at analyzing whether radiation-induced genomic instability may be restricted to a certain inbred strain of mice (Barber et al. 2002). In this study, male mice from three different inbred strains—CBA/H, C57BL/6J, and BALB/c—were exposed either to fission neutrons or acute X-rays and ESTR mutation rates were evaluated in the germ line of their first- and second-generation offspring. Our results show that exposure to fission neutrons or acute X-rays can equally destabilize the F_1/F_2 genomes, thus implying that the transgenerational effects of paternal exposure to high- and low-linear energy transfer sources is likely to be similar. Most importantly, our data demonstrate that the phenomenon of radiation-induced genomic instability is not restricted to a certain inbred strain of mice, as ESTR mutation rates were more or less equally elevated in the first- and second-generation offspring of all three strains. The results of these two studies therefore provided the first experimental evidence for the effects of paternal irradiation on transgenerational genome destabilization.

The results of our early studies raised a number of important questions regarding the effects of parental exposure on the genome stability of their offspring. First of all, it remained to be established whether paternal exposure to other mutagens could destabilize the F_1 genomes. Given that exposure to ionizing radiation produces an extremely wide spectrum of DNA lesions, ranging from base damage to double-strand breaks (Frankenberg-Schwager 1990), it was suggested that the delayed instability detected in the progeny of irradiated cells may be initiated by a specific subset of DNA lesions, namely, complex double-strand DNA breaks (Limoli et al. 1997). If correct, than exposure to a vast majority of chemical mutagens may not result in transgenerational effects detected in the offspring. To this end, we designed two studies aimed to establish the transgenerational effects of parental exposure to chemical mutagens.

In the first study, the transgenerational effects of paternal treatment by the alkylating agent ethylnitrosourea were evaluated in two strains of mice (Dubrova et al. 2008). According to our results, ESTR mutation rates at two expanded simple tandem repeat loci are significantly elevated in the F_1 germ line of both strains. Given that in contrast to irradiation, exposure to ENU mainly causes alkylation of DNA at the N- and O-positions, resulting predominantly in base substitution mutations (Shibuya and Morimoto 1993), our data therefore suggest that transgenerational instability is not attributed to a specific sub-set of DNA lesions, such as double strand breaks, and is most probably triggered by a generalized DNA damage. In the second study we evaluated the transgenerational effects of paternal treatment by three commonly used chemotherapeutic agents—cyclophosphamide, mitomycin C and procarbazine, exposure which results in a diverse spectrum of DNA lesions (Glen and Dubrova 2012). Thus cyclophosphamide forms DNA adducts—mostly N^7-substituted guanine derivatives and phosphotriester adducts (Anderson et al. 1995); exposure to mitomycin C mainly causes sequence-specific DNA adducts, including alkylated monoadducts and cross-linked adducts (Tomasz 1995); and procarbazine inhibits DNA polymerase and reacts directly with DNA (Goerne et al. 2008). The results of our study show that, despite a spectacular diversity of DNA lesions caused by these anticancer drugs, paternal exposure to all of them can significantly destabilize F_1 genomes. Our results are in line with the results of previous studies showing that exposure to some chemical carcinogens and mutagens can result in a delayed increase in mutation rate in somatic cells (Limoli et al. 1997; Bardelli et al. 2001; Li et al. 2001; Gowans et al. 2005), or can affect the fitness of the offspring of exposed male rats (Hales et al. 1992). Taken together, these findings show that relatively high-dose acute exposure to a variety of germ-line mutagens can result in transgenerational instability manifesting in the offspring.

To establish whether transgenerational effects are tissue-specific, the single-molecule PCR approach was used to establish ESTR mutation frequencies in DNA samples extracted from a number of tissues taken from the first-generation offspring of irradiated male mice (Barber et al. 2006, 2009; Hatch et al. 2007; Abouzeid Ali et al. 2012). According to the results of these studies, the frequency of ESTR mutation is equally elevated in the F_1 germ line (sperm) and somatic tissues (bone marrow, spleen, and brain). The same is also true for the offspring of male mice exposed to anticancer drugs, where ESTR mutation frequency is significantly elevated in DNA samples extracted from sperm and bone marrow (Glen and Dubrova 2012). Our data are in line with the results of previous studies on the transgenerational effects of paternal irradiation in mice rats. Thus, significant transgenerational increases in somatic mutation rates were detected by studying the frequency of chromosome aberrations (Vorobtsova 2000; Slovinska et al. 2004), micronuclei (Fomenko et al. 2001), and *lac*I mutations (Luke et al. 1997). The analysis of somatic mutation at protein-coding genes also revealed that their frequency is significantly elevated in the F_1 offspring of irradiated male mice (Barber et al. 2006; Shiraishi et al. 2002). Taken together, the aforementioned results imply that that the transgenerational effects are attributable to genome-wide destabilization, which manifests in many, and possibly all, tissues. Further analysis of transgenerational effects manifesting in the germ line and somatic tissues has provided additional important evidence for the genome-wide nature of this phenomenon. In these studies, the transgenerational effects were evaluated in the offspring conceived by the exposed males and non-irradiated females (Glen and Dubrova 2012; Barber et al. 2002, 2006; Shiraishi et al. 2002; Niwa and Kominami 2001). According to their results in the offspring of irradiated male mice, the mutation rate is similarly elevated at the alleles derived from the exposed and non-exposed parents.

We have also analyzed whether the destabilization of the F_1 genome may be affected by the stage of paternal irradiation. In these studies, male mice were mated over a considerable period of time following exposure to acute X-rays during adulthood, ranging from just few days up to 12 weeks (Dubrova et al. 2000b; Barber et al. 2002, 2006; Hatch et al. 2007; Abouzeid Ali et al. 2012). We also established the effects of in utero irradiation on the manifestation of transgenerational effects (Barber et al. 2009). In this study, the in utero-irradiated male mice (12 days of gestation) were mated to non-exposed females, and the frequency of ESTR mutation was evaluated in their first-generation offspring. According to the results of the aforementioned studies, the magnitude of the increases in ESTR mutation frequency in the germ line and somatic tissues of F_1 offspring of male mice, irradiated during adulthood or prenatally, does not significantly differ. These data suggest that some as yet unknown marks of paternal irradiation can survive even massive epigenetic reprogramming during the early stages of development and, being passed to the offspring, destabilize their genomes.

We have recently compared the transgenerational effects among the offspring of male and female mice irradiated either in utero (Barber et al. 2009), or during adulthood (Abouzeid Ali et al. 2012). Surprisingly enough, the results of both studies provide no evidence for elevated ESTR mutation frequencies in the germ line and somatic tissues of the first-generation offspring of irradiated females. In sharp contrast to this, the mutation frequencies in the offspring of males from the same inbred strain of mice, exposed to the same dose of acute X-rays, are significantly elevated. This data implies that maternal irradiation may not affect the genomic stability of offspring. It should be noted that, as there is no plausible explanation for such dramatic differences between the transgenerational effects of paternal and maternal irradiation, future studies are clearly need to investigate this important issue in detail.

Possible Mechanisms Underlying Transgenerational Instability

It should be noted that, to date, the mechanisms underlying radiation-induced genomic instability and transgenerational effects remains unknown. However, in a number of publications is has been suggested that these phenomena cannot be ascribed to conventional mechanisms such as a mutator phenotype, and are most likely a result of epigenetic events (Morgan 2003a, b; Dubrova 2003). This conclusion is based on the two sets of experimental data, showing that (1) in vitro and in vivo radiation-induced genomic instability persists over a long period of time after the initial exposure and (2) the number of cells/organisms manifesting radiation-induced genomic instability is too high to be explained by the direct targeting of any group of genes.

As far as the transgenerational effects of paternal exposure are concerned, simple estimates clearly show that there is no room for conventional Mendelian patterns of inheritance. Given that exposure to any mutagen randomly induces mutation in the paternal genome, one may therefore assume that the offspring can inherit a number of mutations in genes involved in DNA repair, DNA damage checkpoints, and apoptosis. The presence of such mutations can compromise the offspring genomes, resulting in transgenerational instability. Given that no more than 1,000 genes in the mammalian genome are involved in the genome stability maintenance, this hypothesis appears to be highly implausible. As the frequency of radiation-induced mutation in the germ line of male mice exposed to 1 Gy of acute X-rays (the dose used in our studies) is approximately 3×10^{-6} per locus (Searle 1974), just a tiny percentage of the offspring ($1{,}000 \times 3 \times 10^{-6} = 0.003$) can inherit mutations affecting this set of genes. According to our data, up to 100% of the offspring of exposed male mice manifest genomic instability (Dubrova et al. 2000b, 2008; Barber et al. 2002, 2006, 2009; Glen and Dubrova 2012; Hatch et al. 2007; Abouzeid Ali et al. 2012). In other words, the transgenerational effects are attributed to epigenetic changes occurring in the germ line of exposed males, and their manifestation in the offspring.

As already mentioned, the phenomenon of transgenerational instability is attributed to a genome-wide destabilization, and affects a number of endpoints, including tandem repeat DNA loci, protein coding genes, and chromosomal aberrations. If so, then the F_1 genomes should contain a variety of types of DNA damage. To establish whether this is the case, we evaluated the amount of endogenous single- and double-strand DNA breaks, measured by the alkaline Comet and γ-H2AX assays in the F_1 bone marrow and spleen cells (Barber et al. 2006). According to our data, the amount of endogenous DNA damage in both tissues shows highly significant elevation in the offspring of irradiated male mice from two different inbred strains. As the two tissues contain a substantial proportion of cells with a high mitotic index, the abnormally high level of DNA damage most probably reflects a continuous DNA replication-related process. The main argument for this is that the lifespan of cells containing deleterious lesions, such as single- and double-strand breaks, is restricted, as these types of DNA damage are not compatible with replication (Friedberg et al. 2006). It therefore appears that transgenerational instability is an ongoing process occurring in multiple adult tissues.

Our data on the abnormally high level of DNA lesions observed in the offspring of irradiated males provides a plausible explanation for a genome-wide destabilization of the F_1 genome. Indeed, the increased amount of endogenous single- and double-strand DNA breaks can affect the rate of chromosomal aberrations and gene mutations in any replicating tissue. The presence of a persistent subset of mutations in the F_1 may either be attributed to oxidative stress, or reflect their compromised repair of a certain type of endogenous damage. The involvement of oxidative stress/inflammatory-type response in the delayed increases in mutation rates in the progeny of irradiated cells has long been suspected (Morgan 2003a; Lorimore et al. 2003). It is well known that reactive oxygen species represent one of the major sources of endogenous DNA damage, including single- and double-strand breaks, abasic sites, and a variety of nucleotide modifications (Jackson and Loeb 2001). Using the FPG Comet assay, we measured the level of oxidatively damaged nucleotides in the offspring of irradiated male mice, and compared it with that in controls (Barber et al. 2006). Surprisingly enough, the amount of oxidative nucleotide damage in the offspring of irradiated males was indistinguishable from that in control.

Given that epigenetic changes such as promoter hypermethylation of some DNA-repair genes can substantially compromise the efficiency of DNA repair in cancer cells and destabilize microsatellite loci (Esteller 2007), it therefore appears plausible that this may also be the case for transgenerational effects. Irradiation may somehow affect the epigenetic landscape of a number of DNA-repair genes in the germ line of exposed males. If such epigenetic changes can be transmitted to the fertilized egg and survive epigenetic reprogramming during early development, then the efficiency of DNA repair in the offspring would be compromised. To verify whether this is be the case, we measured the efficiency of repair of single-strand DNA breaks and other alkali labile sites in the F_1 ex-vivo irradiated bone marrow samples. Our data clearly showed that the efficiency of DNA repair in the offspring of irradiated parents and controls is similar (Barber et al. 2006). Together with the results of an early study comparing the spectra of directly induced and delayed mutations at the *hprt* locus (Little et al. 1996), these data imply that radiation-induced genomic instability may result from the enhancement of the process of spontaneous mutation, and therefore cannot be attributed to compromised DNA repair.

Another possible explanation for transgenerational effects is that this phenomenon may be related either to replication stress or to compromised apoptosis. As far as the former explanation is concerned, there has been some experimental evidence for the effects of replication stress on the accumulation of DNA damage in human precancerous cells (Bartkova et al. 2005; Gorgoulis et al. 2005). According to the results of these studies, deregulated DNA replication (replication stress) leads to the multiplicity of DNA alterations, which, in turn, may affect mutation rates in these cells. On the other hand, if, for some reason, in the offspring of exposed males the efficiency of apoptosis were compromised, this would relax the selection against the cells containing endogenous DNA damage, and thus destabilize their genomes. A detailed analysis of the expression profiles in the F_1 tissues should elucidate the still unknown mechanisms underlying the phenomenon of radiation-induced genomic instability.

Impact

According to the exiting paradigm in radiation protection, the cornerstone of which is the aforementioned target theory, mutation induction in directly exposed cells is regarded as the main component of the genetic risk of ionizing radiation for humans (UNSCEAR 2001). In essence, there are two components: mutation induction in the germ line of exposed parents, and the manifestation of radiation-induced mutations in the offspring, i.e., whether these mutations are dominant or recessive. The data presented here raise the important issue of the delayed transgenerational effects in the offspring of exposed parents, and suggest that these effects may represent a significant component of the genetic risk of human exposure to environmental mutagens. Although a recent UNSCEAR report has concluded that the existing experimental data on non-targeted effects, including those on transgenerational instability "are insufficient to justify modifications of current risk estimates" (UNSCEAR 2008), there are a number of lessons to be learned from the animal data presented here.

As already discussed, destabilization of the offspring's genomes results in the accumulation of mutations at protein-coding genes, as well as chromosomal aberrations. These observations imply that this destabilization may affect a number of health-related traits in the offspring. Indeed, the results of two studies show that the rate of dominant lethal mutations is significantly elevated in the germ line of the F_1 offspring of male mice and rats, exposed either to ionizing radiation (Luning et al. 1976), or to the anticancer drug cyclophosphamide (Hales et al. 1992). These data are in line with the results of other studies showing decreased proliferation of early embryonic cells, and the increased frequency of malformations in the F_2 offspring of irradiated parents is also consistent with these observations (Wiley et al. 1997; Pils et al. 1999).

There have also been a number of publications on transgenerational carcinogenesis, where the incidence of cancer in the offspring of male mice and rats exposed to ionizing radiation and chemical mutagens was analyzed (Nomura 1983; Vorobtsova et al. 1993; Lord et al. 1998a, b; Hoyes et al. 2001). The results of these studies show that the incidence of cancer among the carcinogen-challenged offspring of irradiated males is significantly elevated. Given that tumor progression is attributed to accumulation of oncogenic mutations (Loeb et al. 2003), the aforementioned data on the elevated frequencies of mutations at protein-coding genes and chromosome aberrations provides a plausible explanation for these effects.

In conclusion, it should be stressed that despite the results of the aforementioned animal studies, the experimental evidence for transgenerational instability in humans still remains highly controversial. For example, a recent publication showed an elevated frequency of chromosomal aberrations among the children of fathers exposed to post-Chernobyl radioactive contamination (Aghanjanyan and Suskov 2009; Aghajanyan et al. 2011). On the other hand, using the same technique, Tawn and coworkers (2005) failed to detect any significant changes among the children of cancer-radiotherapy survivors. For many reasons, the comparison of human and animal data appears to be problematic, mainly because the doses of paternal exposure analyzed in these studies dramatically differed. It therefore remains to establish to what extent parental exposure to mutagens, including ionizing radiation can result in transgenerational destabilization of the genomes of their children.

Acknowledgment This work was supported by grants from the European Commission (NOTE, Contract Number 036465), Cancer Research UK (Contract Number C23612/A9483 and C23912/A12262), the Wellcome Trust (Contract Number 091106/Z/10/Z), and the EMF Biological Research Trust (Contract Number BRT 11/42).

References

Abouzeid Ali HE, Barber RC, Dubrova YE (2012) The effects of maternal irradiation during adulthood on mutation induction and transgenerational instability in mice. Mutat Res 732:21–25

Aghajanyan A, Kuzmina N, Sipyagyna A, Baleva L, Suskov I (2011) Analysis of genomic instability in the offspring of fathers exposed to low doses of ionizing radiation. Environ Mol Mutagen 52:538–546

Aghanjanyan A, Suskov I (2009) Transgenerational genomic instability in children of irradiated parents as a result of the Chernobyl accident. Mutat Res 671:52–57

Anderson D, Bishop JB, Garner RC, Ostrosky-Wegman P, Selby PB (1995) Cyclophosphamide: review of its mutagenicity for an assessment of potential germ cell risks. Mutat Res 330:115–181

Barber R, Plumb MA, Smith AG, Cesar CE, Boulton E, Jeffreys AJ, Dubrova YE (2000) No correlation between germline mutation at repeat DNA and meiotic crossover in male mice exposed to X-rays or cisplatin. Mutat Res 457:79–91

Barber R, Plumb MA, Boulton E, Roux I, Dubrova YE (2002) Elevated mutation rates in the germline of first- and second-generation offspring of irradiated male mice. Proc Natl Acad Sci U S A 99:6877–6882

Barber RC, Miccoli L, van Buul PPW, Burr KL-A, van Duyn-Goedhart A, Angulo JF, Dubrova YE (2004) Germline mutation rates at tandem repeat loci in DNA-repair deficient mice. Mutat Res 554:287–295

Barber RC, Hickenbotham P, Hatch T, Kelly D, Topchiy N, Almeida G, Jones GGD, Johnson GE, Parry JM, Rothkamm K, Dubrova YE (2006) Radiation-induced transgenerational alterations in genome stability and DNA damage. Oncogene 25:7336–7342

Barber RC, Hardwick RJ, Shanks ME, Glen CD, Mughal SK, Voutounou M, Dubrova YE (2009) The effects of *in utero* irradiation on mutation induction and transgenerational instability in mice. Mutat Res 664:6–12

Bardelli A, Cahill DP, Lederer G, Speicher MR, Kinzler KW, Vogelstein B, Lengauer C (2001) Carcinogen-specific induction of genetic instability. Proc Natl Acad Sci USA 98:5770–5775

Bartkova J, Horejsi Z, Koed K, Kramer A, Tort F, Zieger K, Guldberg P, Sehested M, Nesland JM et al (2005) DNA damage response as a candidate anti-cancer barrier in early human tumorigenesis. Nature 434:864–870

BEIR (2006) Health risks from exposure to low levels of ionizing radiation: BEIR VII, phase 2. National Academy Press, Washington, DC

Bois P, Williamson J, Brown J, Dubrova YE, Jeffreys AJ (1998) A novel unstable mouse VNTR family expanded from SINE B1 element. Genomics 49:122–128

Dubrova YE (2003) Radiation-induced transgenerational instability. Oncogene 22:7087–7093

Dubrova YE (2005) Radiation-induced mutation at tandem repeat DNA loci in the mouse germline: spectra and doubling doses. Radiat Res 163:200–207

Dubrova YE, Jeffreys AJ, Malashenko AM (1993) Mouse minisatellite mutations induced by ionizing radiation. Nat Genet 5:92–94

Dubrova YE, Nesterov VN, Krouchinsky NG, Ostapenko VA, Neumann R, Neil DL, Jeffreys AJ (1996) Human minisatellite mutation rate after the Chernobyl accident. Nature 380:683–686

Dubrova YE, Plumb M, Brown J, Fennelly J, Bois P, Goodhead D, Jeffreys AJ (1998) Stage specificity, dose response, and doubling dose for mouse minisatellite germ-line mutation induced by acute radiation. Proc Natl Acad Sci U S A 95:6251–6255

Dubrova YE, Plumb M, Brown J, Boulton E, Goodhead D, Jeffreys AJ (2000a) Induction of minisatellite mutations in the mouse germline by low-dose chronic exposure to γ-radiation and fission neutrons. Mutat Res 453:17–24

Dubrova YE, Plumb M, Gutierrez B, Boulton E, Jeffreys AJ (2000b) Transgenerational mutation by radiation. Nature 405:37

Dubrova YE, Hickenbotham P, Glen CD, Monger K, Wong H-P, Barber RC (2008) Paternal exposure to ethylnitrosourea results in transgenerational genomic instability in mice. Environ Mol Mutagen 49:308–311

Esteller M (2007) Cancer epigenomics: DNA methylomes and histone-modification maps. Nat Rev Genet 8:286–297

Fomenko LA, Vasil'eva GV, Bezlepkin VG (2001) Micronucleus frequency is increased in bone marrow erythrocytes from offspring of male mice exposed to chronic low-dose gamma irradiation. Biol Bull 28:419–423

Frankenberg-Schwager M (1990) Induction, repair and biological relevance of radiation-induced DNA lesions in eukaryotic cells. Radiat Environ Biophys 29:273–292

Friedberg EC, Walker GC, Siede WR, Wood D, Schultz RA, Ellenberger T (2006) DNA repair and mutagenesis, 2nd edn. ASM, Washington, DC

Glen CD, Dubrova YE (2012) Exposure to anticancer drugs can result in transgenerational genomic instability in mice. Proc Natl Acad Sci USA 109:2984–2988

Glen CD, Smith AG, Dubrova YE (2008) Single-molecule PCR analysis of germ line mutation induction by anticancer drugs in mice. Cancer Res 68:3630–3636

Goerne R, Bogdahn U, Hau P (2008) Procarbazine—a traditional drug in the treatment of malignant gliomas. Curr Med Chem 15:1376–1387

Gorgoulis VG, Vassiliou LV, Karakaidos P, Zacharatos P, Kotsinas A, Liloglou T, Venere M, Ditullio RA Jr, Kastrinakis NG, Levy B et al (2005) Activation of the DNA damage checkpoint and genomic instability in human precancerous lesions. Nature 434:907–913

Gowans ID, Lorimore SA, McIlrath JM, Wright EG (2005) Genotype-dependent induction of transmissible chromosomal instability by gamma-radiation and the benzene metabolite hydroquinone. Cancer Res 65:3527–3530

Hales BF, Crosman K, Robaire B (1992) Increased postimplantation loss and malformations among the F_2 progeny of male rats chronically treated with cyclophosphamide. Teratology 45:671–678

Hardwick RJ, Tretyakov MV, Dubrova YE (2009) Age-related accumulation of mutations supports a replication-dependent mechanism of spontaneous mutation at tandem repeat DNA loci in mice. Mol Biol Evol 26:2647–2654

Hatch T, Derijck AAHA, Black PD, van der Heijden GW, De Boer P, Dubrova YE (2007) Maternal effects of the *scid* mutation on radiation induced transgenerational instability in mice. Oncogene 26:4720–4724

Hoyes KP, Lord BI, McCann C, Hendry JH, Morris ID (2001) Transgenerational effects of preconception paternal contamination with ^{55}Fe. Radiat Res 156:488–494

Jackson AL, Loeb LA (2001) The contribution of endogenous sources of DNA damage to the multiple mutations in cancer. Mutat Res 477:7–21

Lea DE (1947) Actions of radiations on living cells. Cambridge University Press, New York

Lee WR, Sega GA, Bishop JB (1970) Chemically induced mutations observed as mosaics in Drosophila melanogaster. Mutat Res 9:323–336

Lehnert S (2007) Biomolecular action of ionizing radiation. Taylor & Frances, New York

Li C-Y, Little JB, Hu K, Zhang L, Dewhirst MW, Huang Q (2001) Persistent genetic instability in cancer cells induced by non-DNA-damaging stress exposures. Cancer Res 61:428–432

Limoli CL, Kaplan MI, Phillips JW, Adair GM, Morgan WF (1997) Differential induction of chromosomal instability by DNA strand-breaking agents. Cancer Res 57:4048–4056

Little JB, Li C, Nagasawa H, Pfenning T, Vetrovs H (1996) Genomic instability and radiation mutagenesis. J Chim Phys 93:157–164

Loeb LA, Loeb KR, Anderson JP (2003) Multiple mutations and cancer. Proc Natl Acad Sci USA 100:776–781

Lord BI, Woolford LB, Wang L, Stones VA, McDonald D, Lorimore SA, Papworth D, Wright EG, Scott D (1998a) Tumour induction by methyl-nitroso-urea following preconceptional paternal contamination with plutonium-239. Br J Cancer 78:301–311

Lord BI, Woolford LB, Wang L, McDonald D, Lorimore SA, Stones VA, Wright EG, Scott D (1998b) Induction of lympho-haemopoietic malignancy: impact of preconception paternal irradiation. Int J Radiat Biol 74:721–728
Lorimore SA, Coates PJ, Wright EG (2003) Radiation-induced genomic instability and bystander effects: inter-related nontargeted effects of exposure to ionizing radiation. Oncogene 22:7058–7069
Luke GA, Riches AC, Bryant PE (1997) Genomic instability in haematopoietic cells of F_1 generation mice of irradiated male parents. Mutagenesis 12:147–152
Luning KG, Frolen H, Nilsson A (1976) Genetic effects of ^{239}Pu salt injections in male mice. Mutat Res 34:539–542
Mathew C (1964) The nature of delayed mutations after treatment with chloroethyl methanesuphonate and other alkylating agents. Mutat Res 1:163–172
Morgan WF (2003a) Non-targeted and delayed effects of exposure to ionizing radiation: I. Radiation-induced genomic instability and bystander effects in vitro. Radiat Res 159:567–580
Morgan WF (2003b) Non-targeted and delayed effects of exposure to ionizing radiation: II. Radiation-induced genomic instability and bystander effects in vivo, clastogenic factors and transgenerational effects. Radiat Res 159:581–596
Niwa O, Kominami R (2001) Untargeted mutation of the maternally derived mouse hypervariable minisatellite allele in F_1 mice born to irradiated spermatozoa. Proc Natl Acad Sci USA 98:1705–1710
Nomura T (1983) X-ray-induced germ-line mutation leading to tumors. Its manifestation in mice given urethane post-natally. Mutat Res 121:59–65
Pils S, Muller W-U, Streffer C (1999) Lethal and teratogenic effects in two successive generations of the HLG mouse strain after radiation exposure of zygotes—association with genomic instability? Mutat Res 429:85–92
Sadamoto S, Suzuki S, Kamiya K, Kominami R, Dohi K, Niwa O (1994) Radiation induction of germline mutation at a hypervariable mouse minisatellite locus. Int J Radiat Biol 65:549–557
Searle AG (1974) Mutation induction in mice. Adv Radiat Biol 4:131–207
Shibuya T, Morimoto K (1993) A review of the genotoxicity of 1-ethyl-1-nitrosourea. Mutat Res 297:3–38
Shiraishi K, Shimura T, Taga M, Uematsu N, Gondo Y, Ohtaki M, Kominami R, Niwa O (2002) Persistent induction of somatic reversions of the pink-eyed unstable mutation in F-1 mice born to fathers irradiated at the spermatozoa stage. Radiat Res 157:661–667
Slovinska L, Elbertova A, Misurova E (2004) Transmission of genome damage from irradiated male rats to their progeny. Mutat Res 559:29–37
Tawn EJ, Whitehouse CA, Winther JF, Curwen GB, Rees GS, Stovall M, Olsen JH, Guldberg P, Rechnitzer C, Schrøder H, Boice JD Jr (2005) Chromosomal analysis in childhood cancer survivors and their offspring—no evidence for radiotherapy-induced persistent genomic instability. Mutat Res 583:198–206
Timofeeff-Ressovsky NW, Zimmer KG (1947) Das Trefferprinzip in der Biologie. S. Herzel, Leipzig
Tomasz M (1995) Mitomycin C, small fast and deadly (but very selective). Chem Biol 2:575–579
UNSCEAR (2001) Hereditary effects of radiation. United Nations, New York
UNSCEAR (2008) Effects of ionizing radiation. Annex C. Non-targeted and delayed effects of exposure to ionizing radiation. United Nations, New York
Vilarino-Guell C, Smith AG, Dubrova YE (2003) Germline mutation induction at mouse repeat DNA loci by chemical mutagens. Mutat Res 526:63–73
Vorobtsova IE (2000) Irradiation of male rats increases the chromosomal sensitivity of progeny to genotoxic agents. Mutagenesis 15:33–38
Vorobtsova IE, Aliyakparova LM, Anisimov VN (1993) Promotion of skin tumors by 12-O-tetradecanoylphorbol-13-acetate in two generations of descendants of male mice exposed to X-ray irradiation. Mutat Res 287:207–216

Wiley LM, Baulch JE, Raabe OG, Straume T (1997) Impaired cell proliferation in mice that persists across at least two generations after paternal irradiation. Radiat Res 148:145–151

Wright EG (2010) Manifestations and mechanisms of non-targeted effects of ionizing radiation. Mutat Res 687:28–33

Yauk CL, Dubrova YE, Grant GR, Jeffreys AJ (2002) A novel single molecule analysis of spontaneous and radiation-induced mutation at a mouse tandem repeat locus. Mutat Res 500:147–156

Chapter 13
Revisiting Mutagenesis in the Age of High-Throughput Sequencing

Subhajyoti De and R. Matthew Ward

Abstract Mutagenesis is a fundamental biological process by which changes in DNA sequences occur in vivo. On one hand, mutagenesis facilitates adaptation and drives evolution, and on the other hand it leads to diseases such as cancer. Recent advances in high-throughput DNA-sequencing technologies, coupled with efficient bioinformatics tools, have allowed analyses of genomic data at an unprecedented scale. These emerging sequencing technologies have not only helped speed up genome analyses; they have also provided novel insights into mutagenesis and its implications for evolution, diseases, and biotechnology. The versatility of these technologies has initiated a new paradigm of biological research in diverse areas of research such as microbiology, evolutionary biology, human genetics and cancer.

"Today we are learning the language in which God created life," said former US president Bill Clinton when announcing the initial publication of the reference human genome (Lander et al. 2001) in 2000. It was widely anticipated that in the next 20–30 years, this new knowledge would transform our understanding of the biology of life, our perception about what causes diseases, and how we treat patients. Today, after just 10 years, that prediction appears to fail, but in a positive sense! We are already beginning to observe an extraordinary transformation of biology- and health-related research, partly due to emerging high-throughput sequencing and other genomic technologies.

DNA is the molecule that carries genetic information for most, if not all, known forms of life. Mutations or changes in DNA sequence alter that information content, which can result in phenotypic consequences. Deleterious effects include disease such as cancer, while beneficial effects lead to adaptation driving evolution.

S. De (✉)
Department of Medicine, University of Colorado School of Medicine
and University of Colorado Cancer Center, Aurora, CO 80045, USA
e-mail: subhajyoti.de@ucdenver.edu

R.M. Ward
Virginia Bioinformatics Institute, Virginia Tech, Blacksburg, VA 24061, USA

D. Mittelman (ed.), *Stress-Induced Mutagenesis*, DOI 10.1007/978-1-4614-6280-4_13,

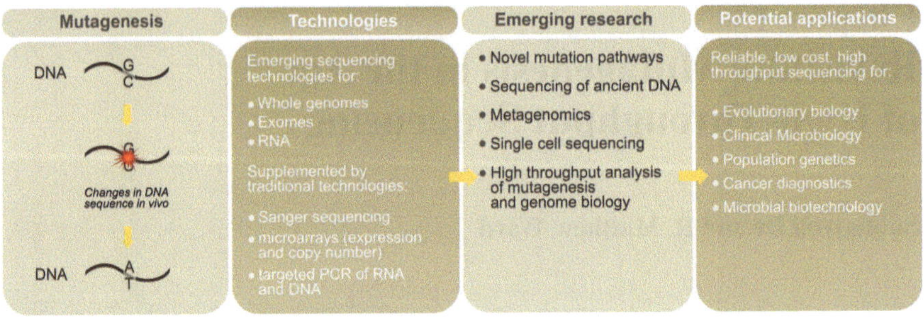

Fig. 13.1 Emerging high-throughput sequencing technologies complement traditional genomic technologies, and facilitate new directions in basic and translational research

Mutagenesis—i.e., the mechanism of in vivo mutations in DNA—has been studied for a very long time, but only recently have high-throughput sequencing technologies allowed us to study mutations on a genome-wide scale. Here, we provide a brief snapshot of recent developments in high-throughput sequencing technologies, and their implications to mutation research in the areas of microbiology, human genetics, and cancer, as well as potential applications in biotechnology, drug development, and personalized medicine (Fig. 13.1).

High-Throughput Sequencing and Other Genomic Technologies

DNA sequencing and genomic analysis began in the 1970s. Walter Fiers and his coworkers were the first to sequence the complete genes and genome of Bacteriophage MS2 (Min Jou et al. 1972; Fiers et al. 1976). In 1973, Gilbert and Maxam reported the sequence of 24 base-pairs using a method known as wandering-spot analysis (Gilbert and Maxam 1973). Soon after, the techniques developed by Maxam and Gilbert (1977), and Fred Sanger (Gilbert and Maxam 1973) became the standard choice for DNA sequencing in the scientific community. The throughput of these early sequencing techniques was modest. The MPSS, invented at Lynx Therapeutics in the 1990s, was the first "next-generation" sequencing technology, but it met with limited commercial success. During the 1990s, and through the mid-2000s, DNA microarray technology provided an intermediate-throughput approach for genomic analyses such as genotyping and copy number alteration analyses. Polony sequencing, developed by George Church and coworkers, was among the first of the next-generation sequencing technologies; polony sequencing was used to sequence a full bacterial genome with high throughput, and at a much lower cost than the traditional Sanger sequencing (Shendure et al. 2005). Since then, a new generation of companies, such as 454 Life Sciences (pyrosequencing), Illumina (Solexa sequencing), Applied Biosystems (SOLiD technology), and, more recently, Life Technologies (Ion semiconductor sequencing) and Oxford Nanopore (Nanopore sequencing), have laid out new, high-throughput, and inexpensive sequencing strategies. Due to these innovations, DNA

Table 13.1 A list of the commonly used software for analyses of high-throughput sequencing data, as reviewed by Treangen and Salzberg (2011)

Objective	Software	Website
SV or CNV detection	BreakDancer	http://sourceforge.net/projects/breakdancer
	CNVnator	http://sv.gersteinlab.org/cnvnator/
	He et al. (2011)	Reference (He et al. 2011)
	PEMer	http://sv.gersteinlab.org/pemer
	VariationHunter	http://compbio.cs.sfu.ca/strvar.htm
SNP detection	GATK	http://www.broadinstitute.org/gsa/wiki/index.php/Downloading_the_GATK
	SAMtools	http://samtools.sourceforge.net
	SOAPsnp	http://soap.genomics.org.cn/soapsnp.html
	Sniper	http://kim.bio.upenn.edu.ezp-prod1.hul.harvard.edu/software/sniper.shtml
	VarScan	http://varscan.sourceforge.net
Short-read alignment	Bowtie	http://bowtie-bio.sourceforge.net
	BFAST	http://bfast.sourceforge.net
	Burrows–Wheeler Aligner (BWA)	http://bio-bwa.sourceforge.net
	mrFAST	http://mrfast.sourceforge.net
	SOAPAligner	http://soap.genomics.org.cn/soapaligner.html
De novo assembly	Allpaths-LG	http://www.broadinstitute.org/software/allpaths-lg/
	CABOG	http://wgs-assembler.sf.net
	SGA	http://github.com/jts/sga
	SOAPdenovo	http://soap.genomics.org.cn/soapdenovo.html
	Velvet	http://www.ebi.ac.uk/~zerbino/velvet

sequencing efficiency has increased by approximately 100,000-fold versus the initial sequencing of the human genome (Lander et al. 2001).

These new technologies have also fueled the development of new bioinformatics algorithms for sequence analysis: e.g., alignment of sequences to reference genomes, de novo assembly of genomes, and detection of single nucleotide variants and structural variants. A list of common bioinformatics software for sequence analysis is provided in the Table 13.1. These algorithms differ in their speed, accuracy, and memory requirements, and their performance may depend on the genomic sequence context itself. Many of the "next-generation" sequencing strategies produce short sequence reads (50–150 bp). One common problem associated with mapping short sequence reads is that a substantial proportion of the human genome is covered by repetitive sequences, and thus a read may map to multiple locations in the genome. The proportion of short sequence reads (>25 bp) that are uniquely mapable to the human genome is typically reported to be 70–80%, although this number depends on the read length, the availability of paired-end reads, and the choice of software for alignment (Treangen and Salzberg 2011). The choice of alignment and estimation of sequencing errors can potentially affect single nucleotide variant calls. One may argue that assigning reads to the site of their best alignment is perhaps the simplest way to solve this problem, but it may not always be correct. Along a similar line, short sequence reads limit our ability to detect structural variations such as insertions, deletions, and rearrangements in repetitive regions. Recently, He and colleagues have described a new method to detect CNVs in repetitive regions using information from read pairs and depth of coverage (He et al. 2011). For a detailed discussion on read mapping, alignment and variant calls see (Treangen and Salzberg 2011).

Some other recent works have focused on analyses of intra-sample heterogeneity using bulk or single-cell sequence data. Shah and colleagues developed one of the earliest software tools, SNVmix for identifying single nucleotide variants using sequence reads derived from the whole sample (Goya et al. 2010). In an alternative approach, Navin and colleagues (2011) have described a high-throughput sequencing-based approach for identifying copy number alterations from single cells, and used that approach to uncover intra-tumor genetic heterogeneity. Fan and colleagues (2010) outlined a microfluidic device-based approach for determining haplotype from single cells. Taken together, the emerging genomics technologies are beginning to initiate a new paradigm of research in diverse areas such as microbial genetics, human genetics and cancer.

Microbial Genomics

Microbes, such as bacteria, are one of the simplest forms of life. There are approximately five nonillion (5×10^{30}) bacteria on the earth (Whitman et al. 1998), forming a biomass that exceeds that of all plants and animals. Even though a handful of microbes have been used as model organisms for a very long time, most microbes are not yet cultivatable, and thus little is known about their biology. Even for model microbes, until recently, our knowledge was primarily driven by gene-centric approaches. Recent advances in genomic technologies have not only allowed us to gather new knowledge about the biology of known microbes, but have also enabled the first attempts to capture the complexity of the microbiomes of the air, soil, water, and our bodies.

Escherichia coli has long been used as a model system. Taking advantage of high-throughput sequencing, Lee and colleagues showed that in short-term adaptive laboratory evolution (up to 40–50 days), *E. coli*, under growth rate selection pressure, undergoes approximately $10^{11.2}$ total cumulative cell divisions in the population to produce a new stable growth phenotype; but continuous exposure to a low level of a mutagen accelerates this timescale and leads to a superior growth rate phenotype with a much larger number of mutations (Lee et al. 2011). In another evolutionary experiment, Lin and colleagues (2011) investigated patterns of genomic instability in *E. coli* under stress. In starving *E. coli*, gene amplification is induced by stress, and regulated by its general stress response mechanism. Lin and colleagues observed that, under stress, chromosomal structural changes occur in bursts of instability that have the potential to drive accelerated evolution. These studies are some of the emerging examples that rely on the power of high-throughput sequencing to identify novel biological principles that would otherwise be difficult, if not impossible, to identify with older genomics approaches.

The current genomic technologies have made it possible to rapidly sequence the genomes of pathogenic microbes that have implications for public health. For instance, using large-scale sequencing approaches, 209 complete genomes of the human influenza A virus have recently been catalogued, providing a comprehensive picture of the evolution of influenza viruses, and of their pattern of transmission through human and animal populations (Ghedin et al. 2005). In another major initiative, all known human rhinovirus genomes have recently been sequenced, revealing their viral structure and evolutionary properties (Palmenberg et al. 2009). More recently, a draft genome of *Yersinia pestis* was built from the victims of the Black Death from London between 1347 and 1351, which provided valuable insights into the evolution of the bacterium, the likely sources of its increased virulence at the time of the Black Death, and the risk of a tragic epidemic in today's society (Bos et al. 2011). Rapid evolution of microbial genomes can confer selective drug resistance relatively quickly. A recent high-throughput study has demonstrated how genomic plasticity within lineages of recombinogenic bacteria such as *Streptococcus pneumoniae* permits adaptation to clinical interventions over remarkably short time scales (Croucher et al. 2011). Taken together, these works highlight how clinical microbiology can benefit from innovations in genomic technologies.

As mentioned, most of the microbes present in the environment are not culturable under laboratory conditions. High-throughput sequencing approaches have made it possible to capture the diversity of such microbial fauna. Recent works have revealed the patterns of genomic and functional adaptation present in the microbiome of the ocean's surface (Yooseph et al. 2010). The authors report the existence of two microbial groups: one composed of a small number of taxa that are frequently abundant in picoplanktonic communities, and the other consisting of many microbial taxa that are rarely found. Analysis of the sequence content of the second group implied that these organisms are slow growing, survive in energy-limited environments, and are capable of rapid growth in energy-rich environments.

In a human body, there is almost an order of magnitude more bacterial cells than human cells, with large numbers of bacteria on the skin, in the gut, and inside the genitalia. A large initiative, named the Human Microbiome Project (HMP), is in progress; an improved understanding of the human microbiome and its aggregate genetic information is expected to shed new light on human health and disease (Cho and Blaser 2012). The emerging genomic technologies have opened up new facets of microbiology research with important implications to public health, diseases, and biotechnology.

Personal Genomes, Population Genetics, and Evolutionary Genomics

Mutations that arise in germ cells are propagated to offspring, leading to genetic differences between the parents and child, variation in a population, and, eventually, genetic divergence between species. Even though germ-line mutations have been studied for a long time, tremendous progress has been made in this area since the initial sequencing and analysis of the human genome and other eukaryote genomes, primarily due to a series of innovations in genomics technologies.

Germ-line mutation analyses are studied typically within the framework of personal, family-based, population-based, and species-level genomic studies. The initial sequencing of the first human genome (Lander et al. 2001) was a decade-long international initiative with an overall cost of approximately $3 billion. Since then, emerging genomics technologies have made enormous progress, and it is now possible to sequence individual genomes within a single laboratory in a few hours. The first personal genome to be sequenced was that of Craig Venter (HuRef) (Levy et al. 2007), followed by that of James Watson (Wheeler et al. 2008). Since then, several personal genomes have been sequenced. The X Prize Foundation has announced that the Archon X Prize for Genomics, with a monetary value of $10 million, would be awarded to "the first team that can build a device and use it to sequence 100 human genomes within 10 days or less, with an accuracy of no more than one error in every 100,000 bases sequenced, with sequences accurately covering at least 98% of the genome, and at a recurring cost of no more than $10,000 (US) per genome." The goal, which seemed impossible even a few years back, appears well within reach today, primarily due to remarkable advancements in sequencing technologies.

In the last few years, a vast majority of the common SNPs were interrogated in genome-wide studies for statistical associations with a large panel of complex traits, including several common diseases, and it became apparent that only a limited amount of the heritable component of any complex trait could be explained by the common variants (Frazer et al. 2009). A whole array of genomics, statistical, and computational strategies is being developed that enable systematic annotation of both coding and non-coding variants (Cooper and Shendure 2011; Bansal et al. 2010). Population-based studies, spearheaded by the 1000 Genomes Project (2010), are among the beneficiaries of emerging genomics technologies. The pilot phase of the 1000 Genomes Project has already been complete, and the consortium presented a map of human genome variation from population-scale sequencing with an unprecedented resolution, providing a catalogue of millions of single nucleotide variants, indels, and structural variants (1000GPC 2010). Interestingly, it was found that, on average, a human individual carries approximately 250–300 loss-of-function variants in annotated genes, and 50–100 variants previously implicated in inherited disorders. The current emphasis of other population-based projects such as case-control and family-based genome-wide association studies is to identify rare variants associated with complex traits using deep sequencing (Frazer et al. 2009). For example, recently whole exome sequencing has identified loss-of-function mutations in ABCC9 in Cantu syndrome (van Bon et al. 2012), in CEP57 in mosaic variegated aneuploidy syndrome (Snape et al. 2011), and in POP1 in novel skeletal dysplasia (Glazov et al. 2011). Advances in genomics technologies and efficient data mining techniques are expected to bring valuable insights into the genetics of complex traits, including many common diseases (Cooper and Shendure 2012).

The classic parent–offspring trio and twin-studies have started using high-throughput genomics approaches to identify germ-line mutations not only in candidate genes, but also in the whole genome (Ott et al. 2012; Silva et al. 2011). Often, but interestingly, not always, these studies have lead to identification of functional disease-causing mutations. A peculiar example was a recent study that performed systematic genome, epigenome, and RNA sequencing of monozygotic twins discordant for multiple sclerosis, and reported a lack of genetic, epigenetic, or transcriptome differences that could explain disease discordance (Baranzini et al. 2010), indicating that the origin of the disease might be more complex than previously thought.

Species-level studies have also benefitted from the recent advance in sequencing technologies. The number of representative genomes of different species that have been sequenced is increasing steadily. Systematic analysis of these genomes will provide insights into highly conserved and rapidly evolving regions of the genome, genetic exchange, population bottlenecks, and genetic changes that have functional consequences critical for adaptation and speciation. For instance, recent sequencing of the gorilla genome provided interesting insights into evidence for parallel acceleration in both human and chimpanzee lineages, particularly for the genes involved in hearing (Scally et al. 2012). A recent initiative, the Genome 10K project, aims to collect the genome sequences of 10,000 vertebrate species, approximately one for every vertebrate genus (G10KCoS 2009). This project is expected to capture the enormous genetic diversity of vertebrate species, and create a remarkable resource for the scientific community, and for worldwide conservation efforts.

Somatic Mosaicism

The advancement of genomic technologies has also rekindled enthusiasm for systematic assessments of somatic mutations, with a coverage and precision difficult to imagine even a few years earlier (De 2011). A multicellular organism has many cells in the body whose origin trace back to a single fertilized cell. From the fertilization of an egg until the death of the individual, somatic cells in the body are constantly exposed to various mutagenic factors that can introduce point mutations, as well as large amplifications, deletions, and ploidy change, such that cells from different tissues, or even within the same tissue, differ genetically. The presence of sequence-level variation between cells in the same individual is referred to as "somatic mosaicism", a term first coined in 1956 by C.W. Cotterman (1956). Early works reported characterization of somatic mosaicism in both animals (Carlson and Southin 1963) and plants (Sastry et al. 1965). Many endogenous factors such as double-stranded breaks, inefficient DNA repair, unbalanced chromosomal segregation, and some exogenous factors, such as nicotine and UV exposure, can contribute to the generation of somatic mutations, thereby leading to somatic mosaicism. Advances in the high-throughput genomics technologies have allowed us to revisit somatic mosaicism with a greater coverage and precision.

Different mutagenic processes lead to alterations of different sizes. Early studies, based on normal and malignant lymphocyte cells, estimated mutation rates to be approximately 10^{-7} to 10^{-8} per base pair per generation (Seshadri et al. 1987). Ongoing works are expected to demonstrate whether the single nucleotide mutation rate is invariant across tissue-types and developmental stages, or whether various factors need to be taken into account before extending these estimates to other studies. Recently it has been reported that de novo, locus-specific rates for genomic alterations are between 10^{-6} and 10^{-4} per locus per generation (Itsara et al. 2010; van Ommen 2005; Lupski 2007), much higher than the point mutation rate. In addition to point mutations and copy number alterations, chromosome ploidy changes also occur during development (Vanneste et al. 2009). Somatic mutations, especially large events such as ploidy changes, are probably selected against, but recent findings based on high-throughput technologies highlight that they are more common than previously thought (De 2011).

Technology development has aided the characterization of somatic mutations. For instance, the copy number and single nucleotide variant calling algorithms routinely used in cancer genomics can also be used to detect somatic mosaicism. It has recently been found that mobile elements such as L1 and Alu elements are active during neuronal differentiation, probably under regulation by epigenetic mechanisms (Dewannieux et al. 2003; Muotri et al. 2010; Singer et al. 2010). Recently, a new sequencing-based technique for detecting young retrotransposon insertions has been described (Iskow et al. 2010), and such approaches can be powerful tools for detecting their activity in somatic tissues.

The mitochondrial genome also accumulates mutations throughout the lifetime of an individual, and mtDNA typically has a higher substitution rate than does genomic DNA (Cavelier et al. 2000). Many mtDNA alterations are associated with diseases. Recent developments in quantifying mitochondrial mutations using high-throughput sequencing technology (He et al. 2010; Li et al. 2010) are particularly encouraging because similar strategies can be used to detect somatic mosaicism in the genomic DNA of normal tissue.

It has recently been proposed that cell competition occurs when cells of different genotypes share the same developing compartment, with one genotype displacing the other as a result of a proliferative advantage (Baker 2011). Studies of cell competition in Drosophila have identified an active process of cell assassination and corpse engulfment, and roles for the Myc and Warts/Hippo tumor suppressor pathways (Baker 2011; Menendez et al. 2010). It remains to be seen whether cell competition is pervasive in individuals with somatic mosaicism. In any case, somatic mutations, together with epigenetic variations and environmental perturbations, can give rise to phenotypic heterogeneity among cells within a tissue, and might have important implications for normal development, and for diseases such as cancer. Emerging genomic technologies, including high-throughput sequencing, can be powerful tools for research along this line.

Cancer

Cancer is a disease condition marked by uncontrolled growth of abnormal cells in the body, representing a special case of somatic mosaicism (Weinberg 2006), and can be thought of as a "disease of the genome" (Mardis 2012a). Even though the molecular biology of cancer started in the mid-twentieth century, early research was primarily geared towards oncogenes and tumor-suppressor genes. The advent of high-throughput sequencing has enabled mapping of the mutational landscapes of cancer genomes in an unbiased manner, allowing identification of novel cancer-related genes, mutagenic patterns, and molecular mechanisms.

Microarray- and sequencing-based studies have recently catalogued somatic amplifications and deletions in a large number of cancer samples from different cancer-types. Large-scale analyses of these somatic copy number variations have allowed us to identify genomic regions that are frequently altered in cancer-types. For example, 158 regions of variation were found in 3,131 cancer specimens, including genes related to apoptosis regulation (which were found to be essential to some cancer cells) and the NF-κB pathway, as well as regions common to multiple cancer types (Beroukhim et al. 2010). Many of these regions hold known or likely cancer-related genes, opening possibilities that many more novel cancer-related genes can be indentified in an unbiased manner.

A cytogenetically normal acute myeloid leukemia genome was one of the first cancer genomes to be completely sequenced (Ley et al. 2008). Using Illumina/Solexa technology, Ley and colleagues sequenced the tumor and matched normal tissue obtained from the same patient's skin, and catalogued genes with somatically acquired mutations. Two of them were known mutations with implications in tumor progression, and eight were novel mutations of unknown function that were present in virtually all tumor cells at both presentation and relapse. This study established whole-genome sequencing as an unbiased method for identifying cancer-initiating mutations at single-base pair resolution. Since then, hundreds of cancer genomes have been sequenced, mostly in the premiere genome centers in the USA, England, and China, including the sequencing of other cancer types such as pancreatic cancer (Mardis 2012b). The new sequencing technologies have made it possible to identify mutations that are present in pre-malignant lesions, primary tumors, and metastases. It is reported that several of the mutations that were present in metastasis during relapse had previously been observed in a minor sub-populations of cells in the primary tumor during presentation, indicating an early origin of the metastatic clones (Ding et al. 2010; Shah et al. 2009). More recent studies have uncovered intra-tumor heterogeneity between different locations in a tumor mass. As mentioned above, Navin and colleagues used analysis of copy number via tools such as flow cytometry, whole genome amplification, and next-generation sequencing to identify three clonal subpopulations in one polygenomic tumor and, in another, a single clonal expansion responsible for tumor formation and metastasis (Navin et al. 2011; Gerlinger et al. 2012). Another study by Gerlinger et al. (2012) used exome sequencing, chromosome aberration analysis, and ploidy profiling on spatially separated samples to identify divergence in 26 of 30 samples, and examined the consequences of heterogeneity. Interestingly, this study also identified phenotypically convergent loss-of-function mutations in tumor-suppressor genes. The resolution and magnitude of intra-tumor heterogeneity uncovered using emerging genomic technologies has been truly unprecedented, and implies that intra-tumor heterogeneity might lead to underestimation of the mutational landscape of tumor genomes obtained from single tumor-biopsy samples, and present major challenges to personalized treatment strategies (Gerlinger et al. 2012).

High-throughput sequencing strategies have also led to characterization of novel mechanisms behind tumor initiation. Classical models of tumor initiation suggest sequential accumulation of somatic mutations (Weinberg 2006). In contrast, using next-generation sequencing, it was found that, in some cases, tens to hundreds of genomic rearrangements occur in a one-off cellular crisis, termed chromothripsis, involving complex rearrangements within or between one or several chromosomes (Stephens et al. 2011). The signature of chromothripsis can be seen in more than 2% of all cancers, across many subtypes; is present in ~25% of bone cancers; and can result in multiple lesions that cause cancer via mechanisms such as the loss of tumor suppressor genes (Stephens et al. 2011).

Many individuals carrying tumor masses are expected to have some tumor cells and cell-free tumor DNA in circulation. Recently, it has been possible to identify and characterize circulating tumor cells and cell-free tumor DNA using sequencing technologies (Pantel et al. 2008; Schwarzenbach et al. 2011). Characterization of circulating tumor cells and cell-free DNA will have important implications for clinical diagnostics, since the circulation of such "tumor stem cells" in the bloodstream and bone marrow leads to the metastasis, growth, and, ultimately, the spread of tumor cells to other organs and tissues, a process that is a common cause of death in cancer patients. Characterization of circulating tumor cells should be particularly important for monitoring advanced-stage patients who have had tumors removed (Pantel et al. 2008). Similarly, changes in the circulation of cell-free DNA and other nucleic acids is associated with tumor progression, suggesting their use as biomarkers in cancer screening, prognosis, and monitoring of therapy (Schwarzenbach et al. 2011).

Taken together, not only have high-throughput approaches made it possible to characterize mutational landscapes of a large number cancer genomes, exomes, and transcriptomes—in terms of variation such as nucleotide substitutions, insertions and deletions, copy number variations, chromosomal rearrangements, and the incorporation of foreign DNA—they have also allowed us to estimate intra-tumor genetic heterogeneity, identify novel cancer genes, and identify mechanisms associated with tumorigenesis, and may eventually contribute to diagnosis and treatment (Meyerson et al. 2010).

Personalized Medicine and Biotechnology

With the falling cost, simplicity, and versatility of usage of high-throughput sequencing and related technologies, we are just beginning to see their applications in different fields of biotechnology and personalized medicine. For instance, sequencing of genomes of cancer patients can identify novel, cryptic variants in cancer susceptibility genes, in addition to providing unbiased information on the spectrum of mutations in a cancer genome, leading to customized treatment (Pasche and Absher 2011). A landmark example is the case of a 39-year-old female patient with acute myeloid leukemia, whose cancer genome was sequenced at the Washington University Genome Center. Massively parallel paired-end sequencing allowed identification of novel mutations not identifiable using cytogenetic techniques. These identified mutations consequently changed the treatment plan for the patient and prolonged her survival (Welch et al. 2011). Usage of high-throughput sequencing in personalized diagnostics is expected to become more commonplace in the near future.

An emerging application of high-throughput sequencing is the characterization of circulating tumor cells and circulating tumor DNA in the plasma of cancer patients (Pantel et al. 2008; Schwarzenbach et al. 2011). Similar approaches have also been used to identify ploidy changes or disease causing mutations in fetal DNA in circulation in maternal plasma during pregnancy (Barrett et al. 2012). Even though these innovations are still maturing, high-throughput sequencing of circulating DNA has potential to be a powerful noninvasive tool for the diagnosis of cancer and neonatal diseases.

Previously, gene-centric approaches took years to identify targets of small molecules, critical residues in the target genes, and the scope of resistance. Recently, high-throughput sequencing has shown promise to expedite the process enormously. One example was presented by Wacker et al. (2012), who performed high-throughput sequencing of the human colon cancer cell line, HCT-116, which is deficient in mismatch repair and has low expression of multidrug-resistant (MDR) pumps, to screen for the target of drugs in human cells by examining resistance mechanisms. Their strategy involves inducing evolution of a cell-population in the presence of a moderate dosage of a chosen drug, and isolating multiple-drug-resistant clones from genetically heterogeneous human cells, followed by sequencing to identify clusters of mutations that mark the drug target. Unlike many other target identification approaches, this method does not rely on chemical modifications of the drug of interest, and offers a wider range of applications.

Outlook

Recent advances in DNA-sequencing technology, coupled with efficient bioinformatics analysis tools, have made it possible to analyze genomic data at an unprecedented scale. We are just beginning to see the beneficial effects of the new genomic technologies in diverse areas of research, ranging from microbiology and evolutionary biology to human genetics and cancer. These new technologies have not only helped speed up genome analyses, they have also provided novel insights into mutagenic processes and their implications in evolution, disease, and biotechnology.

References

1000GPC (2010) A map of human genome variation from population-scale sequencing. Nature 467:1061–1073

Baker NE (2011) Cell competition. Curr Biol 21:R11–R15

Bansal V, Libiger O, Torkamani A, Schork NJ (2010) Statistical analysis strategies for association studies involving rare variants. Nat Rev Genet 11:773–785

Baranzini SE, Mudge J, van Velkinburgh JC, Khankhanian P, Khrebtukova I, Miller NA, Zhang L, Farmer AD, Bell CJ, Kim RW et al (2010) Genome, epigenome and RNA sequences of monozygotic twins discordant for multiple sclerosis. Nature 464:1351–1356

Barrett AN, McDonnell TC, Chan KC, Chitty LS (2012) Digital PCR analysis of maternal plasma for noninvasive detection of sickle cell anemia. Clin Chem 58(6):1026–1032

Beroukhim R, Mermel CH, Porter D, Wei G, Raychaudhuri S, Donovan J, Barretina J, Boehm JS, Dobson J, Urashima M et al (2010) The landscape of somatic copy-number alteration across human cancers. Nature 463:899–905

Bos KI, Schuenemann VJ, Golding GB, Burbano HA, Waglechner N, Coombes BK, McPhee JB, DeWitte SN, Meyer M, Schmedes S et al (2011) A draft genome of Yersinia pestis from victims of the Black Death. Nature 478:506–510

Carlson EA, Southin JL (1963) Chemically induced somatic and gonadal mosaicism in Drosophila. I. Sex-linked lethals. Genetics 48:663–675

Cavelier L, Jazin E, Jalonen P, Gyllensten U (2000) MtDNA substitution rate and segregation of heteroplasmy in coding and noncoding regions. Hum Genet 107:45–50

Cho I, Blaser MJ (2012) The human microbiome: at the interface of health and disease. Nat Rev Genet 13:260–270

Cooper GM, Shendure J (2011) Needles in stacks of needles: finding disease-causal variants in a wealth of genomic data. Nat Rev Genet 12:628–640

Cooper GM, Shendure J (2012) Needles in stacks of needles: finding disease-causal variants in a wealth of genomic data. Nat Rev Genet 12:628–640

Cotterman CW (1956) Somatic mosaicism for antigen A2. Acta Genet Stat Med 6:520–521

Croucher NJ, Harris SR, Fraser C, Quail MA, Burton J, van der Linden M, McGee L, von Gottberg A, Song JH, Ko KS et al (2011) Rapid pneumococcal evolution in response to clinical interventions. Science 331:430–434

De S (2011) Somatic mosaicism in healthy human tissues. Trends Genet 27:217–223

Dewannieux M, Esnault C, Heidmann T (2003) LINE-mediated retrotransposition of marked Alu sequences. Nat Genet 35:41–48

Ding L, Ellis MJ, Li S, Larson DE, Chen K, Wallis JW, Harris CC, McLellan MD, Fulton RS, Fulton LL et al (2010) Genome remodelling in a basal-like breast cancer metastasis and xenograft. Nature 464:999–1005

Fan HC, Wang J, Potanina A, Quake SR (2010) Whole-genome molecular haplotyping of single cells. Nat Biotechnol 29:51–57

Fiers W, Contreras R, Duerinck F, Haegeman G, Iserentant D, Merregaert J, Min Jou W, Molemans F, Raeymaekers A, Van den Berghe A et al (1976) Complete nucleotide sequence of bacteriophage MS2 RNA: primary and secondary structure of the replicase gene. Nature 260:500–507

Frazer KA, Murray SS, Schork NJ, Topol EJ (2009) Human genetic variation and its contribution to complex traits. Nat Rev Genet 10:241–251

G10KCoS (2009) Genome 10K: a proposal to obtain whole-genome sequence for 10,000 vertebrate species. J Hered 100:659–674

Gerlinger M, Rowan AJ, Horswell S, Larkin J, Endesfelder D, Gronroos E, Martinez P, Matthews N, Stewart A, Tarpey P et al (2012) Intratumor heterogeneity and branched evolution revealed by multiregion sequencing. N Engl J Med 366:883–892

Ghedin E, Sengamalay NA, Shumway M, Zaborsky J, Feldblyum T, Subbu V, Spiro DJ, Sitz J, Koo H, Bolotov P et al (2005) Large-scale sequencing of human influenza reveals the dynamic nature of viral genome evolution. Nature 437:1162–1166

Gilbert W, Maxam A (1973) The nucleotide sequence of the lac operator. Proc Natl Acad Sci USA 70:3581–3584

Glazov EA, Zankl A, Donskoi M, Kenna TJ, Thomas GP, Clark GR, Duncan EL, Brown MA (2011) Whole-exome re-sequencing in a family quartet identifies POP1 mutations as the cause of a novel skeletal dysplasia. PLoS Genet 7:e1002027

Goya R, Sun MG, Morin RD, Leung G, Ha G, Wiegand KC, Senz J, Crisan A, Marra MA, Hirst M et al (2010) SNVMix: predicting single nucleotide variants from next-generation sequencing of tumors. Bioinformatics 26:730–736

He Y, Wu J, Dressman DC, Iacobuzio-Donahue C, Markowitz SD, Velculescu VE, Diaz LA Jr, Kinzler KW, Vogelstein B, Papadopoulos N (2010) Heteroplasmic mitochondrial DNA mutations in normal and tumour cells. Nature 464:610–614

He D, Hormozdiari F, Furlotte N, Eskin E (2011) Efficient algorithms for tandem copy number variation reconstruction in repeat-rich regions. Bioinformatics 27:1513–1520

Iskow RC, McCabe MT, Mills RE, Torene S, Pittard WS, Neuwald AF, Van Meir EG, Vertino PM, Devine SE (2010) Natural mutagenesis of human genomes by endogenous retrotransposons. Cell 141:1253–1261

Itsara A, Wu H, Smith JD, Nickerson DA, Romieu I, London SJ, Eichler EE (2010) De novo rates and selection of large copy number variation. Genome Res 20:1469–1481

Lander ES, Linton LM, Birren B, Nusbaum C, Zody MC, Baldwin J, Devon K, Dewar K, Doyle M, FitzHugh W et al (2001) Initial sequencing and analysis of the human genome. Nature 409:860–921

Lee DH, Feist AM, Barrett CL, Palsson BO (2011) Cumulative number of cell divisions as a meaningful timescale for adaptive laboratory evolution of Escherichia coli. PLoS One 6:e26172

Levy S, Sutton G, Ng PC, Feuk L, Halpern AL, Walenz BP, Axelrod N, Huang J, Kirkness EF, Denisov G et al (2007) The diploid genome sequence of an individual human. PLoS Biol 5:e254

Ley TJ, Mardis ER, Ding L, Fulton B, McLellan MD, Chen K, Dooling D, Dunford-Shore BH, McGrath S, Hickenbotham M et al (2008) DNA sequencing of a cytogenetically normal acute myeloid leukaemia genome. Nature 456:66–72

Li M, Schonberg A, Schaefer M, Schroeder R, Nasidze I, Stoneking M (2010) Detecting heteroplasmy from high-throughput sequencing of complete human mitochondrial DNA genomes. Am J Hum Genet 87:237–249

Lin D, Gibson IB, Moore JM, Thornton PC, Leal SM, Hastings PJ (2011) Global chromosomal structural instability in a subpopulation of starving Escherichia coli cells. PLoS Genet 7:e1002223

Lupski JR (2007) Genomic rearrangements and sporadic disease. Nat Genet 39:S43–S47

Mardis ER (2012a) Genome sequencing and cancer. Curr Opin Genet Dev 22:245–250

Mardis ER (2012b) Applying next-generation sequencing to pancreatic cancer treatment. Nat Rev Gastroenterol Hepatol 9(8):477–486

Maxam AM, Gilbert W (1977) A new method for sequencing DNA. Proc Natl Acad Sci USA 74:560–564

Menendez J, Perez-Garijo A, Calleja M, Morata G (2010) A tumor-suppressing mechanism in Drosophila involving cell competition and the Hippo pathway. Proc Natl Acad Sci USA 107:14651–14656

Meyerson M, Gabriel S, Getz G (2010) Advances in understanding cancer genomes through second-generation sequencing. Nat Rev Genet 11:685–696

Min Jou W, Haegeman G, Ysebaert M, Fiers W (1972) Nucleotide sequence of the gene coding for the bacteriophage MS2 coat protein. Nature 237:82–88

Muotri AR, Marchetto MC, Coufal NG, Oefner R, Yeo G, Nakashima K, Gage FH (2010) L1 retrotransposition in neurons is modulated by MeCP2. Nature 468:443–446

Navin N, Kendall J, Troge J, Andrews P, Rodgers L, McIndoo J, Cook K, Stepansky A, Levy D, Esposito D et al (2011) Tumour evolution inferred by single-cell sequencing. Nature 472:90–94

Ott J, Kamatani Y, Lathrop M (2012) Family-based designs for genome-wide association studies. Nat Rev Genet 12:465–474

Palmenberg AC, Spiro D, Kuzmickas R, Wang S, Djikeng A, Rathe JA, Fraser-Liggett CM, Liggett SB (2009) Sequencing and analyses of all known human rhinovirus genomes reveal structure and evolution. Science 324:55–59

Pantel K, Brakenhoff RH, Brandt B (2008) Detection, clinical relevance and specific biological properties of disseminating tumour cells. Nat Rev Cancer 8:329–340

Pasche B, Absher D (2011) Whole-genome sequencing: a step closer to personalized medicine. JAMA 305:1596–1597

Sastry GR, Cooper HB Jr, Brink RA (1965) Paramutation and somatic mosaicism in maize. Genetics 52:407–424

Scally A, Dutheil JY, Hillier LW, Jordan GE, Goodhead I, Herrero J, Hobolth A, Lappalainen T, Mailund T, Marques-Bonet T et al (2012) Insights into hominid evolution from the gorilla genome sequence. Nature 483:169–175

Schwarzenbach H, Hoon DS, Pantel K (2011) Cell-free nucleic acids as biomarkers in cancer patients. Nat Rev Cancer 11:426–437

Seshadri R, Kutlaca RJ, Trainor K, Matthews C, Morley AA (1987) Mutation rate of normal and malignant human lymphocytes. Cancer Res 47:407–409

Shah SP, Morin RD, Khattra J, Prentice L, Pugh T, Burleigh A, Delaney A, Gelmon K, Guliany R, Senz J et al (2009) Mutational evolution in a lobular breast tumour profiled at single nucleotide resolution. Nature 461:809–813

Shendure J, Porreca GJ, Reppas NB, Lin X, McCutcheon JP, Rosenbaum AM, Wang MD, Zhang K, Mitra RD, Church GM (2005) Accurate multiplex polony sequencing of an evolved bacterial genome. Science 309:1728–1732

Silva S, Martins Y, Matias A, Blickstein I (2011) Why are monozygotic twins different? J Perinat Med 39:195–202

Singer T, McConnell MJ, Marchetto MC, Coufal NG, Gage FH (2010) LINE-1 retrotransposons: mediators of somatic variation in neuronal genomes? Trends Neurosci 33:345–354

Snape K, Hanks S, Ruark E, Barros-Nunez P, Elliott A, Murray A, Lane AH, Shannon N, Callier P, Chitayat D et al (2011) Mutations in CEP57 cause mosaic variegated aneuploidy syndrome. Nat Genet 43:527–529

Stephens PJ, Greenman CD, Fu B, Yang F, Bignell GR, Mudie LJ, Pleasance ED, Lau KW, Beare D, Stebbings LA et al (2011) Massive genomic rearrangement acquired in a single catastrophic event during cancer development. Cell 144:27–40

Treangen TJ, Salzberg SL (2011) Repetitive DNA and next-generation sequencing: computational challenges and solutions. Nat Rev Genet 13:36–46

van Bon BW, Gilissen C, Grange DK, Hennekam RC, Kayserili H, Engels H, Reutter H, Ostergaard JR, Morava E, Tsiakas K et al (2012) Cantu syndrome is caused by mutations in ABCC9. Am J Hum Genet 90:1094–1101

van Ommen GJ (2005) Frequency of new copy number variation in humans. Nat Genet 37:333–334

Vanneste E, Voet T, Le Caignec C, Ampe M, Konings P, Melotte C, Debrock S, Amyere M, Vikkula M, Schuit F et al (2009) Chromosome instability is common in human cleavage-stage embryos. Nat Med 15:577–583

Wacker SA, Houghtaling BR, Elemento O, Kapoor TM (2012) Using transcriptome sequencing to identify mechanisms of drug action and resistance. Nat Chem Biol 8:235–237

Weinberg RA (2006) The biology of cancer. Garland Science, New York

Welch JS, Westervelt P, Ding L, Larson DE, Klco JM, Kulkarni S, Wallis J, Chen K, Payton JE, Fulton RS et al (2011) Use of whole-genome sequencing to diagnose a cryptic fusion oncogene. JAMA 305:1577–1584

Wheeler DA, Srinivasan M, Egholm M, Shen Y, Chen L, McGuire A, He W, Chen YJ, Makhijani V, Roth GT et al (2008) The complete genome of an individual by massively parallel DNA sequencing. Nature 452:872–876

Whitman WB, Coleman DC, Wiebe WJ (1998) Prokaryotes: the unseen majority. Proc Natl Acad Sci USA 95:6578–6583

Yooseph S, Nealson KH, Rusch DB, McCrow JP, Dupont CL, Kim M, Johnson J, Montgomery R, Ferriera S, Beeson K et al (2010) Genomic and functional adaptation in surface ocean planktonic prokaryotes. Nature 468:60–66

Index

D. Mittelman (ed.), *Stress-Induced Mutagenesis*, DOI 10.1007/978-1-4614-6280-4,

MIX
Papier aus verantwortungsvollen Quellen
Paper from responsible sources
FSC® C105338

If you have any concerns about our products,
you can contact us on
ProductSafety@springernature.com

In case Publisher is established outside the EU,
the EU authorized representative is:
Springer Nature Customer Service Center GmbH
Europaplatz 3, 69115 Heidelberg, Germany

Printed by Libri Plureos GmbH
in Hamburg, Germany